The
Intimate
Lives
of the
FOUNDING
FATHERS

THOMAS
FLEMING

Author of
THE PERILS OF PEACE

"Thomas Fleming is one of our most interesting
scholars of the revolutionary period, and in his insightful
latest work he does not disappoint." — Jay Winik,
author of ***The Great Upheaval*** and ***April 1865***

Dear One –
Merry Christmas 2010
I hope you enjoy this –
Love always,
Judy

Additional Praise for
The Intimate Lives of the Founding Fathers

"With his ample gifts as a novelist and his brilliant historical reach, the esteemed Thomas Fleming never disappoints. . . . A remarkable achievement—and hard to put down."

—Brenda Wineapple, author of
White Heat: The Friendship of Emily Dickinson and Thomas Wentworth Higginson

"Tom Fleming is a rare combination—a fine historian and a fine writer. His assessment of George Washington's relationships with Sally Fairfax and Martha Custis is right on target."

—Peter R. Henriques, author of
Realistic Visionary: A Portrait of George Washington

"In this solid, sometimes titillating account, novelist and historian Fleming draws parallels to today's media obsession with our leaders' sex lives. . . . Showing the more human and sometimes unlikable sides of our founders, the author writes good history." —*Publishers Weekly*

"Combines new scholarship and fluent prose to examine the women in the Founders' lives—and their historical effects."

—*American History* magazine

"Analyzing history in the context of our Founders' personal relationships makes for fascinating reading." —*American Spirit*

"A fascinating, truly behind-close-doors look at the birth of our nation."

—Book of the Month Club

"This is better than any history book you've ever read."

—Curledup.com

"[A] relentlessly fascinating book. . . . Addictively readable, informative, fascinating, engaging, revelatory, and provocative. Which, you might say, is just another way of saying—it's a Thomas Fleming book."

—*FrontPage* magazine

Also by Thomas Fleming

—

The Intimate Lives

❧ *of the* ❧

FOUNDING FATHERS

Thomas Fleming

HARPER

NEW YORK · LONDON · TORONTO · SYDNEY

 Smithsonian Books

HARPER

A hardcover edition of this book was published in 2009 by HarperCollins Publishers.

THE INTIMATE LIVES OF THE FOUNDING FATHERS. Copyright © 2009 by Thomas Fleming. All rights reserved. Printed in the United States of America. No part of this book may be used or reproduced in any manner whatsoever without written permission except in the case of brief quotations embodied in critical articles and reviews. For information address HarperCollins Publishers, 10 East 53rd Street, New York, NY 10022.

HarperCollins books may be purchased for educational, business, or sales promotional use. For information please write: Special Markets Department, HarperCollins Publishers, 10 East 53rd Street, New York, NY 10022.

FIRST HARPER PAPERBACK PUBLISHED 2010.

Designed by Mary Austin Speaker

Library of Congress Cataloging-in-Publication Data has been applied for.

ISBN 978-0-06-113913-0

10 11 12 13 14   OV/RRD   10 9 8 7 6 5 4 3 2 1

The heart has its reasons,
of which the mind knows nothing.

— BLAISE PASCAL

Contents

———

Acknowledgments

*A*book of this dimension leaves the writer indebted to a wide range of people. I want to express my gratitude, first, to the numerous librarians who have assisted me, beginning with my early work on the American Revolution at Yale University. More recently, Lewis Daniels, the librarian at the Westbrook Public Library in the Connecticut town of the same name (where I spend my summers), has been especially helpful in tracking down books in the state's research libraries for me. Equal gratitude goes to Mark Bartlett and his staff at the New York Society Library. Another thank you in this regard is warmly extended to W. Gregory Gallagher of the Century Association library—and to the librarians of the New York Historical Society.

Special thanks must go to Mary Thompson, the research historian at Mount Vernon, who gave me several hours of her time during a visit. Also helpful was Washington scholar Peter Henriques, who generously shared with me some of his research. Several geneticists advised me on exploring the world of chromosomes and haplotypes, notably Kenneth Kidd, MD, of Yale University; Brian Ference, MD, formerly of Yale, now in private practice in Michigan; and Dr. Edwin Knights, who has written widely on the subject and practices in New Hampshire. Also helpful have been science writer Steven Corneliussen and his associates, biostatisticians William Blackwelder and David Douglas, who helped me probe the arcane world of probability.

Steven Bernstein, who is writing a history book of his own, found time to explore several collections in the Library of Congress and other libraries in Washington, DC. He also played a part in tracking down hitherto unknown facts about Samuel F. Wetmore, ghostwriter extraordinary. My son, Richard Fleming, with his computer skills and his access to the Columbia University Library, gave me significant aid on a variety of topics, ranging from Thomas Jefferson and Benjamin Franklin to Joseph Pulitzer. Another son, attorney Thomas J. Fleming, took time from his busy practice to advise me on the logic, legal and otherwise, of certain arguments in the text. Most helpful of all, with her mastery of computer research as well as her editorial experience, was the most important woman in my life, Alice Mulcahey Fleming. Also invaluable was the advice and counsel of my editor, Elisabeth Kallick Dyssegaard, as the manuscript evolved over the past two years.

— Thomas Fleming

Introduction

With leaks and wandering emails, talk shows and tell-all aides, the private lives of today's politicians seem to have become public property. Whether this may eventually unravel the republic is frequently debated in the media. Not a few columnists and late-night gurus maintain that the best and brightest are now loath to enter politics.

Still, the number of politicians has not noticeably declined. Nor are we the first generation to take a more than passing interest in the personal lives of our elected leaders. Convinced that historical perspective might be the best answer to the Götterdämmerung tone that the discussion sometimes takes, I decided to explore the roles of women in the lives of the first group of American politicians to win fame—George Washington, Benjamin Franklin, Alexander Hamilton, John Adams, Thomas Jefferson, and James Madison. Collectively, most historians agree, these are the founding fathers, the men who made the greatest contribution to the birth of the nation.

I was soon watching a young George Washington riven with desire for the wife of his close friend. I stood with Thomas Jefferson at the bedside of his dying wife, Martha Wayles, as he sobbed a fateful promise that he would never marry again. I saw a youthful Alexander Hamilton imbibe a toxic mix of fear and anger in his psyche when his headstrong mother banished his hapless father from her bed.

As one Jefferson biographer has remarked, every man carries on a life-long dialogue with his mother, sometimes in his conscious mind, more often in his unconscious. Mothers have an especially strong influence in the shadowy realm of emotions. George Washington and Thomas Jefferson seem to have inherited their mothers' temperaments. Some historians think John Adams's mother, Susanna Boylston Adams, was a manic depressive, who passed on the illness to her favorite son.

Although the women in these famous lives spoke 150 years before feminism entered the American vocabulary, their independent voices will surprise many people. The men and women of 1776 were far more candid and realistic about sexual desire and marital relationships than Americans of the twenty-first century realize. They gave serious thought to the ancient conflict between the sexes and talked and wrote about it in ways that still have relevance today.

This was evident from the novels they read and the stories that were printed in the newspapers. Samuel Richardson's *Pamela* was the most popular novel of the era. This story of a servant girl's rise to wealth and power proved that virtue was rewarded and simultaneously delivered titillating descriptions of a young woman agonizing over sexual desire. When fifteen-year-old Betsy Hanford of Virginia married wealthy fifty-one-year-old John Cam, the local newspaper reported, "She is to have a chariot and there is to be no padlock put upon her mind."

The women of 1776 had high expectations from marriage. They wanted not only affection but respect as persons. For a lucky few, these essentials could blend into near adoration. One Virginian began his letters to his wife, with "My dearest life" and declared that she "blessed the earth" with her presence. At the same time, essays and letters about unhappy marriages frequently appeared in the newspapers. One correspondent in the *Virginia Gazette* blamed these misfortunes on women who spent too much of a man's money on luxury, and on men who for the sake of beauty or wealth married "a fury" or an "ideot [*sic*]."

We will see how strongly the founders, especially those primary political rivals Alexander Hamilton and Thomas Jefferson, stressed the importance of a happy marriage in a man's life. Thanks to his five years in France as America's ambassador, Jefferson was able to compare American and European marriage customs and found America's far superior. Fidelity was virtually unknown among the French upper classes. Jefferson advised young Americans to abandon dreams of a grand tour, lest they acquire the Old World's attitude toward women.

Recent decades of scholarship in herstory have made us aware of a dark side of women's lives in the eighteenth century. Other than from private tutors, they had almost no educational opportunities. Divorce was seldom granted by the courts, and a woman's property was legally controlled by her husband. On the eve of America's independence, we will see protofeminist Abigail Adams protesting these inequalities in a famous letter to her husband, John—and his less well-known, extremely unsatisfying reply.

A woman also had little control over her reproductive life. Pregnancy and childbirth were dangerous. Equally troubling was the awful infant mortality rate. The primitive medicine of the era made childhood almost as perilous. By late middle age, Martha Washington had lost all four of her children to death. Martha Wayles Jefferson lost four out of six children in ten years. Benjamin Franklin's marriage was poisoned by his wife's bitterness over the death of their four-year-old son, Frankie, while his hated illegitimate half-brother, William, thrived.

As Franklin's story makes clear, the founders' marriages were not without controversy. Did George Washington recklessly pursue other women after he married Martha, as the British and later his American political enemies claimed? Is there convincing proof that Thomas Jefferson had a long-term sexual relationship with his mulatto slave, Sally Hemings? Why did the curvaceous well-off widow, Dolley Payne Todd, marry pint-sized, sickly James Madison? As first lady, Dolley had to deal with vicious rumors about her sex life. She met them with a shrewdness that wives of contemporary politicians might well emulate

All the women in the founders' lives had to confront and cope with fame. This little-understood phenomenon transformed many aspects of their private lives into public dramas. The fame that the founders sought and won was not the same as our modern version of it, mere celebrity. For them, fame was an enormously serious matter, involving a man's place in history. It was reserved for founders or rescuers of nations or givers of laws and required the approval of men of judgment and intelligence.

All the founders were aware that the Revolutionary upheaval and the task of creating a nation gave them unique opportunities to win this ultimate accolade. "You and I," John Adams wrote to a Virginia friend in 1777, "have been sent into life at a time when the greatest lawgivers of antiquity would have wished to live." In 1778, Alexander Hamilton wrote a pamphlet attacking certain congressmen who were using their positions to get rich. Hamilton could not understand how a man could succumb to such a "mean pursuit" when he had a chance to be THE FOUNDER OF AN EMPIRE [Hamilton's capitals]. "A man of virtue and ability, dignified with so precious a trust, would rejoice that fortune had given him birth at [such] a time."

Washington, debating whether to risk his fame by endorsing the dubious idea of a constitutional convention in 1787, told a friend, "To see this country happy is so much the wish of my soul, nothing on this side of Elysium can be placed in competition with it."

For some women this kind of fame could be a stimulant. But it could also be a beast in the jungle, almost an evil spirit. This was especially true of Abigail Adams, who had to endure years of agonizing loneliness while her husband pursued diplomatic fame in Europe. Even worse was Elizabeth Hamilton's ordeal when her husband defended his fame as the creator of the new nation's financial system by making a public confession of his infidelity with a Philadelphia temptress.

When we explore the wives' influence on these famous lives, we discover evidence so strong, it can easily be asserted in some cases—that of John Adams, for instance—that the great man would never have earned his place on fame's ladder without the woman at his side. The same conclusion is even more true for James Madison. While Martha Washington destroyed all but a handful of the personal letters she exchanged with George, there is evidence that she was by no means a mere fellow-traveler on Washington's journey to fame.

Knowing and understanding the women in their lives adds pathos and depth to the public dimensions of the founding fathers' political journeys. We do them no dishonor when we explore how often public greatness emerged in spite of personal pain and secret disappointment. Far from diminishing these men and women, an examination of their intimate lives will enlarge them for our time. In their loves and losses, their hopes and fears, they are more like us than we have dared to imagine.

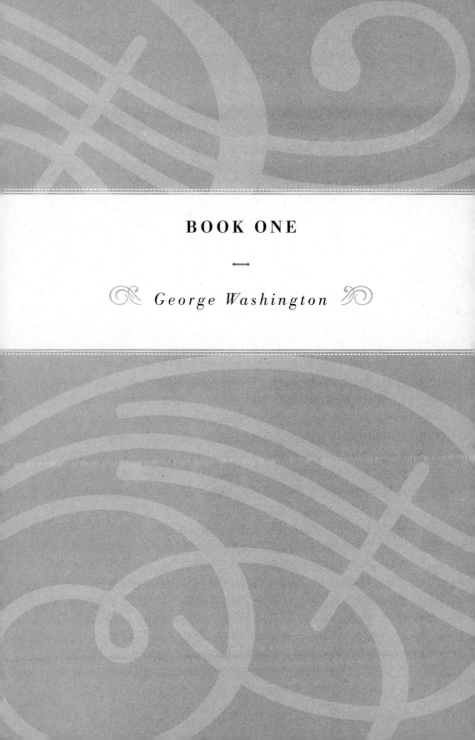

BOOK ONE

George Washington

THE AGONIES OF HONOR

On March 30, 1877, the *New York Herald,* one of the largest newspapers in America, printed a lengthy love letter that had been written on September 12, 1758. Not exactly hot news, you might say. Had the editors lost their collective marbles? The *Herald*'s editors did not think so. Nothing they printed that day created more of a sensation among their readers. The letter was from George Washington. Here is the heart of its text, exactly as it was printed:

Tis true I profess myself a votary of love. I acknowledge that a lady is in the case and further I confess that this lady is known to you. Yes, Madame, as well as she is to one who is too sensible to her charms to deny the power whose influence he feels and must ever submit to. I feel the force of her amiable beauties and the recollection of a thousand tender passages that I could wish to obliterate till I am bid to revive them. But experience, alas, sadly reminds me how impossible this is, and evinces an opinion which I have long entertained, that there is a destiny which has control of our actions, not to be resisted by the strongest efforts of human nature. You have drawn me, dear Madam, or rather have I drawn myself into an honest confession of a simple fact. Misconstrue not my meaning; doubt it not nor expose it. The world has no business to know the object of my love declared in this manner to you when I want to conceal it. One thing above all things in this world I wish to

*know, and only one person of your acquaintance can solve me that, or guess
my meaning. But adieu to this till happier times, if I shall ever see them.*

In this welter of indirection and hinted meanings was George Washington crying out, "I love you! Do you love me?" The *Herald*'s headline was: "A Washington Romance." Beneath it was a subtitle: "A Letter from General Washington Acknowledging The Power of Love." Then came an introduction to the text:

> *In a collection of rare and autograph letters which will be sold by Bangs
> & Co. this afternoon we find the accompanying letter written by General
> Washington at the age of twenty six and never before made public. The pres-
> ent owner purchased it in England some years ago for the sum of L15. The
> letter is addressed to Mrs. Sarah Fairfax at Belvoir. This lady was a Miss
> Cary, to whom George Washington once offered his hand but was refused for
> his friend and comrade, George William Fairfax. Irving asserts that it was a
> sister of Mrs. Fairfax, Miss Mary Cary, after Mrs. Edward Ambler.*

"Irving" refers to Washington Irving, author of an acclaimed five-volume biography of Washington. But the *Herald* reporter dismisses Irving's assertion by citing an article that was published in *Scribner's* magazine in June of 1876, in which a Fairfax descendant insisted it was Sally Cary, Mrs. Fairfax, for whom Washington "had a tenderness." He quotes from the article: "It is fair to say that papers which have never been given to the public set this question beyond a doubt. Mrs. George William Fairfax, the object of George Washington's early and passionate love, lived to an advanced age in Bath, England . . . Upon her death at the age of eighty-one, letters, still in possession of the Fairfax family, were found among her effects, showing that Washington had never forgotten the influence of his youthful disappointment."

Next came a gaffe that underscores why newspapers are often called history's first draft. The reporter noted that in the sentences preceding the confession that he was "a votary of love," Washington rebuked Sally Fairfax for suggesting in a letter to him that he was preoccupied with "the animating prospect of possessing Mrs. Custis." The reporter blithely dismissed this reference to Washington's future wife: "It is hardly probable that Washington means to express his love for Mrs. Custis, for her husband was then living—in fact did not die until twenty years after the date of this letter." Readers who had access to Washington Irving's biography swiftly discovered that Daniel Parke Custis had been dead more than a year when Washington wrote this September 12, 1758, letter to Sally Fairfax. Worse, on or about June 5, 1758, he had become engaged to marry Martha Dandridge Custis.[1]

For Americans who regarded George Washington as a virtual incarnation of divinity—and they were numerous in 1877—the letter created consternation. It was only three months after the fervent yearlong celebration of the one hundredth anniversary of independence, in which Washington had been portrayed as the ultimate hero. Today, the Bangs auction house would have been rubbing its hands with unconcealed glee and kiting the price of the manuscript into the stratosphere. In 1877, no bidding took place. Bangs merely announced that the letter had been sold for $13. Even in 1877, when the dollar was worth perhaps thirty of our depreciated dollars, the price is much too low to be believable. Rumor long maintained that the purchaser was J. P. Morgan, but no evidence has been discovered to support that assertion. Whoever he was, the buyer evidently felt he was performing a patriotic act by removing the letter from sight.

For the next eighty years, the original letter remained unexamined by scholars, which spurred violent arguments about its authenticity and meaning. Some people were eager to dismiss it as a forgery. But the first two collectors of Washington's papers reluctantly decided to include the newspaper text because the style was so unmistakably authentic and no one could produce an adequate reason for forging it. Not until the late 1950s did a determined Washington biographer find the original in the files of Harvard's Houghton Library. That discovery has not prevented people from continuing to disagree over its meaning.[2]

Some historians have argued it is a good-humored joke, the sort of risqué banter that men and women often exchanged in the eighteenth century. John C. Fitzpatrick, who spent several decades on his monumental edition of Washington's papers, maintained that the letter was a paean of praise for Martha Custis. He took ferocious issue with those who said that Washington was professing his passion for Mrs. Fairfax in spite of being engaged to Martha. If they were correct, Fitzpatrick wrote, every decent person would be forced to conclude that George Washington was "a worthless scoundrel."[3]

II

Sally Cary Fairfax was the daughter of one of the wealthiest planters in Virginia, Wilson Cary, possessor of a splendid estate at Ceelys, on the James River overlooking Hampton Roads, not far from Newport News. The family also enjoyed comfortable town houses in Williamsburg and Hampton. Wilson Cary's father had been rector of the College of William and Mary; the younger man had studied there and at Trinity College in England. His houses were stocked with the latest English books and magazines, and he took pleasure in teaching Sally and her three younger sisters how to read and write French. He was one of the leaders of the colony's legislature, the House of Burgesses, which meant that each year the Carys enjoyed the brilliant social season in Williamsburg while the burgesses were in session. It was a world of fancy balls, lavish dinners, and witty conversation that made the pursuit of happiness a fact of life long before it became a phrase in a political declaration.[4]

An anecdote passed down in the Cary family gives a glimpse of Sally that suggests she was the center of male attention at an early age. She was returning to the family's Williamsburg house while one of the colonial wars with France was raging and the town was patrolled by sentries. One of these soldiers demanded to hear the password of the night from Sally's coachman. The man was flummoxed into silence. Sally stamped her foot and cried, "But I am Miss Sally Cary!" The sentry gulped and said, "Pass." It seems that the officer of the watch was an admirer and had made "Sally Cary" the password as a compliment to the young lady.[5]

At eighteen Sally married George William Fairfax, son of William Fairfax, the proprietor of Belvoir, on the Potomac River not far from Mount Vernon. The bride enjoyed the dizzying expectation that one day she might become not merely the mistress of this fine mansion but the wife of a bona fide nobleman. George William stood a better than even chance of becoming the next Lord Fairfax. This would not only entitle him to sit in the House of Lords in Parliament and preside over a vast English estate; it would make him owner of five million acres in northern Virginia that King Charles II had given to a maternal ancestor of Lord Fairfax in 1673.

This potentially glorious future was probably the best explanation for the match. William Fairfax had sent his son to England at the age of six to be educated by his family, describing him as a "poor West India boy."

George William was the product of a marriage that Fairfax had made in the Bahamas with the obscure widow of a British artillery major. Someone launched the rumor that the woman had Negro blood. For fifteen years, George William endured the unlovely experience of his wealthy English relatives eyeing his skin color and debating whether he was a mulatto.[6]

The result was a slight, rather timid young man with a dour, down-turned mouth surmounted by a strong hooked nose and shrewd close-set eyes. His letter to an influential Fairfax kinsman in England reporting his engagement to Sally did not exactly seethe with passion. He made it sound like it was something he had worked into his schedule while attending the House of Burgesses. He described Sally as an "amiable person" and reported that he had obtained her and her father's consent to an early marriage. He closed by noting that "Col. Cary wears the same coat-of-arms as Lord Hunsden." (The first Lord Hunsden was a cousin of Queen Elizabeth, Henry Carey.) This apparently was as important to George William as Sally's charms.[7]

A portrait of Sally painted by a not very talented artist around this time reveals a slim, dark-haired young woman most people would call handsome rather than beautiful. But the narrow face is striking nonetheless: the deep-set dark eyes emanate a subtly mocking intelligence; the nose is strong and the mouth, firm and confident. Her waist is narrow and her bosom ample. It is not hard to imagine her as the leader of some lively revels.

Sally came to Belvoir as a bride in 1748 and soon met sixteen-year-old George Washington. He was a frequent guest at nearby Mount Vernon, where his older half brother, Lawrence, was happily married to Anne Fairfax, George William's older sister. Lawrence was doing his utmost to rescue George, already six feet tall, from the clutches of his headstrong widowed mother, Mary Ball Washington, who was trying to convert her oldest son into a surrogate husband and father figure for her four younger children.

III

George's father, Augustine Washington, had died in 1743, when George was eleven; later George mournfully remarked that he had only a blurred recollection of this huge, muscular man, a sort of rural Hercules famous for his feats of strength. For much of the time, Augustine had been an absentee father, traveling between his scattered farms and an iron works that required a great deal of his attention. Enterprise was in Augustine Washington's blood. Since the arrival of the first Washington in Virginia in 1657, a refugee from the English civil war, the males had made a habit of marrying well and acquiring land. Augustine Washington had continued this tradition, expanding his holdings from 1,740 acres at the time of his first marriage to almost 11,000 acres at his death.[8]

Compared with the Carys, the Byrds, the Lees, the Randolphs, and the other first families of Virginia, the Washingtons remained "middling gentry." Their house on 250-acre Ferry Farm, across the Rappahannock River from Fredericksburg, where George grew up, was an eight-room frame structure, not even faintly comparable to the stately brick mansions such as Robert Carter's Nomini Hall or William Byrd's Westover. Augustine Washington owned forty-nine slaves. Robert "King" Carter had over 600 toiling on his 60,000 acres. It is easy to imagine young George Washington's awe when he visited Belvoir. Elegant English-made couches, chairs, and tables filled the parlor. At Ferry Farm, the parlor contained three beds.[9]

George's mother, Mary Ball, was Augustine Washington's second wife. Her mother was illiterate, and Mary received no education worth mentioning. She was a physically imposing woman, large and vigorous, with an explosive temper. One man described her as "majestic." One of George's boyhood playmates said he was "ten times more afraid of her" than he was of his own parents. One day, Mary stood up in her carriage on Fredericksburg's main street and cursed and lashed a slave because he had mishandled the horse. As her oldest son, George was exposed at an early age to his mother's tantrums. Worse, he inherited her violent temper.[10]

There are more than a few intimations that George Washington's home life with this turbulent woman was unhappy before as well as after his father's death. Second marriages, especially ones in which children of the first wife must be dealt with, are often uneasy. Augustine Washington's decision to send his two sons by his first marriage, Augustine and Lawrence, to school in England at considerable expense may have been motivated in part by Mary's sharp tongue and short temper.

In one of his boyhood notebooks, in which George laboriously copied such things as *Rules of Civility and Decent Behavior in Company,* he included in his strong, firm script a poem, "True Happiness." It was a portrait of domestic tranquility. The "truly bless'd" enjoy a "good estate" with productive soil, a warm fire in the hearth, a simple diet, "constant friends," a healthy mind and body—and "a quiet wife, a quiet soul." Almost certainly, this portrait was the opposite of what George encountered on Ferry Farm. Were it not for the intervention of Lawrence Washington, it is dismaying to think what George Washington might have become.[11]

IV

Fourteen years older than George, Lawrence had inherited Mount Vernon from his father. His marriage to Anne Fairfax, William Fairfax's daughter, had catapulted him from middling gentry into the heady stratosphere of Virginia's aristocracy. Fairfax was the cousin and land agent for the sixth Lord Fairfax, who owned those five million acres between the Potomac and Rappahannock rivers. Virginians had done everything in their power to invalidate Charles II's generosity, but the courts had upheld the royal prerogative, thereby making Fairfax the most influential name in Virginia. In swift succession after his marriage, Lawrence became adjutant of the Virginia militia and a member of the House of Burgesses. Add the polish of his English education and his love of martial glory and it is easy to see how he became a formidable figure on young George's horizon.[12]

Thanks to Lawrence, George became a close friend of George William Fairfax. Seven years older than George, Fairfax accepted the elongated teenager as a companion at fox hunts at Belvoir and on trips into the Shenandoah Valley, where more Fairfax acres were being surveyed for sale. Under his genteel influence, Washington was soon spending the money he earned as a surveyor on stylish clothes and feeling at ease in the elegant atmosphere of Belvoir. To Mary Ball's mounting exasperation, George spent more and more of his time on the banks of the Potomac.

Then came tragedy. Lawrence Washington was stricken with tuberculosis and slowly died before George's grief-stricken eyes. In a desperate attempt to regain his health, Lawrence journeyed to the island of Barbados with George, hoping a winter spent in warm sunshine might restore him. The experiment proved fruitless for Lawrence—and doubly painful for George. He caught a bad case of smallpox, which he barely survived. Lawrence was as generous to his younger brother in death as he had been in life. He named George the heir of the Mount Vernon estate, if Anne

Fairfax Washington predeceased him. In the meantime, George leased the house and lands from her for a modest sum. When Anne died in 1761, he became Mount Vernon's owner.[13]

V

Mount Vernon was young George's refuge from Mary—and nearby Belvoir was a place where he met some of the most sophisticated young women in Virginia. Sally Cary Fairfax's sisters and numerous friends were frequent visitors. George soon proved himself more than vulnerable to their charms. One belle, who remains nameless, inspired some of the worst poetry ever committed by an adolescent:

> *O, ye Gods, why should my poor resistless heart*
> *Stand to oppose thy might and power*
> *At last to surrender to Cupid's feather'd dart*
> *And now lays bleeding every hour*
> *For her that's pitiless of my grief and woes*
> *And will not on me pity take.*

This atrocity may have been committed on behalf of "a lowland beauty" who particularly tormented George. He told his friend Robin during a sojourn in the Shenandoah Valley at Lord Fairfax's hunting lodge that there was "an agreeable young lady" living in the house, but every time he looked at her, he thought of the "lowland beauty," which was only "adding fuel to the fire." There seems to be little doubt that George was powerfully attracted to the opposite sex—hardly surprising for a healthy, vigorous teenager.[14]

George's romantic emotions slowly acquired a darker tinge. Sally Fairfax seems to have been a coquette who tantalized, teased, and dominated the men around her. She soon realized that one of her conquests was George Washington—a discovery that did not displease her. The contrast between the tall, muscular Washington and her short, precise courtier husband, whose greatest talent was assiduous flattery of his superiors, could not have been more complete. As they performed together in amateur theatrics, danced minuets in Belvoir's ballroom, and exchanged gossip about the amorous doings of their contemporaries, George Washington fell violently in love with his close friend's wife.

One of their favorite plays was *Cato,* written by the celebrated essayist and poet Joseph Addison in 1713. It was the most popular drama of the century; more to the point, it had two parts made to order for lovers and would-be lovers. Marcia was Cato's devoted daughter; Juba was a North African warrior who rallied to Cato's side when he resisted the rise of Julius Caesar. Marcia confessed her love for Juba, but Cato refused his approval because he was a mere colonial. Juba nevertheless remained devoted to the untouchable beauty.[15]

VI

By the time George realized what was happening to him emotionally, he was on his way to becoming Virginia's best-known soldier. Grown to his full six feet two and one half inches, he stood, in the words of one eyewitness, "as straight as an Indian." Thanks to his Fairfax connections, he was appointed a major of the militia at age twenty. The following year he won a skirmish against a French patrol that became the opening shots of the world's first global conflict, the Seven Years' War. Next, in spite of strenuous objections from his mother, George become a favorite aide of British general Edward Braddock and miraculously survived the rout of his army of regulars when they marched into western Pennsylvania to oust the French from Fort Duquesne, where Pittsburgh now stands. Ignoring four bullets through his coat and two horses killed under him, Washington was among the few who distinguished himself on that chaotic battlefield.

He staggered back to Mount Vernon a very tired man. A letter from William Fairfax reveals how closely the residents of Belvoir followed Washington's military career: "Your safe return gives an uncommon joy to us and will no doubt be sympathized by all lovers of heroick [*sic*] virtue," Fairfax wrote. He hoped a Saturday night's rest would refresh the weary warrior enough to enable him to come to Belvoir in the morning.

Sally added a saucy note, cosigned by two visiting women friends, accusing the hero of "great unkindness in not visiting us this night. I assure you that nothing but being satisfied that our company would be *disagreeable* would prevent us from trying if our legs would not carry us to Mount Vernon this night; but if you will not come to us, tomorrow morning, very early, we shall be at Mount Vernon." This is a letter from a woman who knows she has a certain gentleman virtually at her beck and call.[16]

VII

Washington soon became the colonel of a regiment of Virginia regulars that struggled to defend the 700-mile-long frontier against French and Indian incursions. George William Fairfax wrote him admiring letters, vowing that he would be honored to serve under his leadership. But Fairfax never got around to volunteering, even when his younger brother Bryan joined the seemingly endless and extremely dangerous wilderness war and another brother became an officer in the British regular army.

Washington's relationship to Sally Fairfax during these years resembled a roller-coaster ride. He wrote her letters from the frontier, hoping she would honor him with a reply. But when his messages became too emotional, she abruptly ordered him to stop writing to her. At another point, she apparently banished him from Belvoir. He accepted this treatment with remarkable patience.[17]

Late in 1757, Washington suffered a physical collapse and staggered home to Mount Vernon, seriously ill with dysentery, a nameless fever—probably malaria—and a cough that reminded him alarmingly of Lawrence's fatal malady. He took to his bed in Mount Vernon and asked Sally to obtain various medicines a local doctor had recommended. George William Fairfax had sailed for England to deal with his difficult relatives. His father, William, had recently died, giving his British cousins a chance to bring up the ruinous suspicion that George William was a mulatto. Washington's younger brother, John Augustine ("Jack"), and his wife, who had been staying at Mount Vernon as caretakers, were away. Sally brought Washington his medicines—special wines, jellies, and other delicacies beyond the ability of Mount Vernon's kitchen. Was it during these months that some or all of "the thousand tender passages" occurred that Washington struggled to forget in the letter he wrote a year later?[18]

We simply do not know. The other letters Washington and Sally exchanged have all been destroyed, except for one or two fragments and a puzzling note that he wrote to her when he first arrived at Mount Vernon in 1757. It is as laconic and impersonal as one can imagine, asking for help with his medicines. In mid-February Washington received a letter from George William, reporting he had survived the perils of the wintry Atlantic. He forwarded it to Sally, adding: "When you are at leisure to favor us with a visit, we shall endeavor to partake as much as possible of the joy you receive on this occasion." This does not breathe deep passion, to say the least; it also suggests that Sally's visits had been infrequent.[19]

During these months, Washington was a very sick man. He wrote a plaintive letter to the doctor who had treated him on the frontier, James Craik, reporting that he was not getting better. Dr. Craik replied that he was not surprised; his malady had "corrupted the whole mass of blood." The physician ordered the patient to stay in bed and avoid any and all exertion, saying, "The fate of your friends and country [he meant Virginia] are in a manner dependent on your recovery." This was flattering stuff, but not the sort of message that inspired a depressed, anxious man to become an impassioned lothario.[20]

VIII

Something else has to be factored into the situation at this point: Washington's relationship to the Fairfaxes. William Fairfax had been almost as much a substitute father as Lawrence. After Lawrence's death, William had regarded George as a member of the family and used his considerable power to push his military career whenever possible. When William Fairfax died, Washington had left his regiment and journeyed over the mountains to his funeral, ignoring the dysentery that was already making his life difficult. In a letter to his brother Jack, he remarked, "To that family I am under many obligations, particularly to the old gentleman." To some extent these obligations extended to George William Fairfax. He had befriended George, the teenage country bumpkin, as Lawrence's brother, and their relationship had remained close for the previous decade. There is not a hint in any of Washington's letters of a change in opinion or attitude, even when he emerged as Virginia's most notable military leader.

In fact, it can be argued with some force that this role of military hero only made the possibility of George realizing his desire for Sally Cary Fairfax more remote. In a sense George had become the man Lawrence might have been—and that only intensified his sense of obligation to the Fairfaxes. Honor was the brightest word in Lawrence Washington's vocabulary—a beacon that both guarded and guided his conduct. The thought of doing something that Lawrence would have judged grossly dishonorable was a more than believable reason why George chained his desire deep within himself. It was another lesson in the harsh school of self-control in which destiny seemed to be matriculating him.

This does not mean that George Washington was inhibited by puritanical views of sexual conduct. Puritanism was almost as foreign to eighteenth-century Virginia as Mohammedanism. Life on the frontier was by no means devoid of women. Every eighteenth-century army had "camp women" who were married or pretended to be married to soldiers and followed them into the war zone. One of his officers wrote to Washington while he was home on leave, wondering if he was "plunged in delight . . . & enchanted by charms even stranger to the Ciprian Dame." A Ciprian Dame was an available woman, sometimes a prostitute. Another officer wrote him from South Carolina, telling him that the local women lacked "the enticing heaving throbbing alluring . . . plump breasts common with our northern belles." Such letters have enabled some writers to imagine a blazing covert affair between Sally and George that lasted months or even years.[21]

Far stronger is evidence that suggests Washington struggled to put Sally out of his mind and future. George pursued several other women, notably strong-willed Mary Philipse, heiress to a swath of the Hudson River Valley. But his efforts were halfhearted—proof, it might seem, either of his longing for Sally or of Mary's temperamental resemblance to Mary Ball Washington. This was the situation in March 1758 when the ailing bachelor, still convinced that he was in his final days, mounted his horse and rode slowly to Williamsburg to see Dr. John Anson, the best physician in Virginia, hoping against hope that this medico might have a cure but fearing that he would deliver a death sentence. Before he departed, George told his British superior on the frontier, Colonel John Stanwix, that he had "ruined [my] constitution" and was thinking of "quitting my command." He was convinced that he had tuberculosis and foresaw little but "approaching decay."[22]

To Washington's amazement and delight, Dr. Anson assured him that he was recovering nicely and had prospects of living to a vigorous old age. The reincarnated patient strode into Williamsburg's spring sunshine and began thinking about what to do with the rest of his life. One of his first thoughts was marriage. The sequence inclines this writer to wonder if during the long winter of his illness, he and Sally had not told each other—or at least hinted—that they realized their love had no future.

Another scenario, perhaps more likely, has Washington reaching this glum but unavoidable conclusion during the long, lonely night hours in his sickbed.

Realistically, in 1758 Virginia, there was no way that Colonel George Washington could marry Sally Cary Fairfax. It would have triggered an immense scandal that would have made them both social outcasts. A clandestine affair could easily have led to the same result. Either way, Washington would have exposed himself to a ruinous lawsuit from her outraged husband. Lurking in the background of both their minds was the memory of an earlier sex scandal: Lawrence Washington had sued a neighbor, accusing him of raping Anne Fairfax before her marriage. The lawsuit had been reported in salacious detail in newspapers throughout Virginia, Maryland, and Pennsylvania. Before it was over, everyone wished Lawrence had never mentioned the incident to anyone, no matter how much the vile deed may have haunted his wife.[23]

IX

With marriage on his mind, Colonel Washington rode from Williamsburg to the nearby estate known as The White House, on the Pamunkey River, to visit Martha Dandridge Custis. They undoubtedly knew each other already. The elite society of eastern Virginia was fairly small, and Martha and her husband, Daniel Parke Custis, had participated in its lively social world with enthusiasm. George and Martha had probably met and may have danced at more than one fancy-dress ball in Williamsburg.

A recent widow, Martha was receiving a veritable stream of suitors. In Virginia during these years, money was frankly accepted as a significant item in a marriage. Newspapers regularly stated the amount of a bride's net worth. Elizabeth Stith, for instance, was described as "a very amiable lady with a fortune of a thousand pounds sterling." Cash was often the baldly stated reason for mingling youth and age. Such women aspired to—and often expected—a certain amount of respect and independence. In Martha Dandridge Custis's case, she could expect a great deal of respect and virtually unlimited freedom of choice. She was the richest widow in Virginia.

George was more than pleased with Martha's warm, affable manner and was even more charmed when she invited him to stay overnight. He played cheerfully with her two children, John, four, and Patsy, two. As he departed, he tipped her servants extravagantly, a sure sign that he wanted their comments about him to be favorable.[24]

A week later, he returned for another visit and something seems to have been arranged. On May 4, the colonel ordered a ring from Philadelphia and a suit of "superfine" broadcloth from London to fit a "tall man." By this time he was back on the frontier, once more in command of his regiment. They were soon part of another British army committed to ousting the French from Fort Duquesne.[25]

X

Here the aftershock of the turmoil stirred by the *New York Herald*'s publication of the letter to Sally Fairfax intrudes on our story. On July 20, 1758, George Washington supposedly wrote in a letter to Martha Custis:

We have begun our march for the Ohio. A courier is starting for Williamsburg and I embrace the opportunity to send a few words to one whose life is now inseparable from mine. Since that happy hour when we made our pledge to each other, my thoughts have been continually going to you as to another Self. That an all-powerful Providence may keep us both in safety is the prayer of your ever faithful and affectionate friend.[26]

A number of reputable historians have concluded this letter is a forgery. It was included in the first two editions of Washington's papers, but John C. Fitzpatrick noted tersely, "The location of the original is not known." Perhaps most important, the statement about beginning a march to the Ohio is wrong. On July 20, much to Washington's exasperation, the British army was still sitting on the edge of the wilderness, debating which route to take. He was not even sure he and his troops would be included in the expedition. Furthermore, the word "courier" was never used by Washington during these years; he preferred "express." Other words in the letter strike similar false notes. Perhaps most convincing, the style is extravagantly emotional from a man who has spent comparatively little time with Mrs. Custis.

Where did the letter come from? Was it concocted in an attempt to counter the 1877 revelation of the letter to Sally Fairfax? We know this much:

it first appeared in 1886 in a sentimental biography of Martha Washington titled *Mary and Martha, Mother and Wife of George Washington*. The author, Benjamin Lossing, claimed he had seen the original, but it was never found by anyone else. It seems likely that the letter was forged by someone who was trying to make the letter to Sally seem like a fake. It might have been written by a Washington family descendant, who imposed it on Lossing, or by someone else with patriotic motives such as the mystery purchaser of the letter to Sally Fairfax in 1877.[27]

XI

These explanations enable us to see Colonel Washington grumbling and cursing in Fort Cumberland, Maryland, while the British ignored his advice on the best way to march on Fort Duquesne. Meanwhile, the evidence of George's plans for the future were unfolding at Mount Vernon, which Washington had decided to expand and rebuild with all possible speed. George William Fairfax, back from Europe, was asked to help with advice and supervision. This inevitably led to Washington telling him about his engagement to Martha Custis. Fairfax naturally told his wife about this interesting change in the fortunes of their mutual friend. Into an envelope with a letter from George William about the Mount Vernon renovation, Sally slipped a letter of her own.

With that mocking style she preferred, Sally apparently teased Washington about his complaints that the campaign was moving too slowly. Was he impatient because he had become a "votary of love"? She was of course referring to his engagement to Martha Custis. But the lonely warrior, facing an Indian-rife wilderness in which there was a strong possibility of a bullet with his name on it, read a very different meaning into the inquiry. What came back to Sally was nothing less than an explosion—a jumbled cry of anguish from a man who could bury his feelings no longer. As usual, Sally was discreet. Her answer was apparently indirect; some historians think she pretended Washington was joking. The letter is lost. We have only Washington's answer:

> *Do we still misunderstand the true meaning of each other's letters? I think it must appear so tho I would fain hope the contrary as I cannot speak plainer without—But I'll say no more and leave you to guess the rest.*

He gloomily added that he was almost certain the expedition to the Ohio would end in disaster. Then he added words that had deep meaning for both of them: "I should think my time more agreeable [*sic*] spent, believe me, in playing a part in Cato with the company you mention and myself doubly happy in being the Juba to such a Marcia as you must make."

He closed with some offhand speculation on the marital plans of several friends but made no mention of his own. Then came a last burst of emotion:

> *One thing more and I have done. You ask if I am not tired at the length of your letter? No, Madam, I am not, nor never can be while the lines are an inch asunder to bring you in haste to the end of the paper. You may be tired of mine by this. Adieu, dear Madam, you possibly will hear something of me or from me before we shall meet. Believe me that I am most unalterably, your most obedient and obliged. . . .*[28]

In his surviving letters to Sally, Washington never before wrote "most unalterably." Once more he was telling her the secret that they would share for the rest of their lives. They were lovers that destiny had tragically separated, as history had forever parted Marcia and Juba.[29]

XII

Four months after he revealed this passionate longing, George Washington married Martha Custis. If romance was not uppermost in his mind, there is evidence that Martha felt a few tremors. Her first husband, Daniel Parke Custis, had been fifteen years older than she—and he was by most accounts a rather pathetic (though extremely handsome) man, browbeaten all his life by a miserly father. Towering Colonel Washington was not only Virginia's foremost soldier, but he must have been a breathtaking sight in the suit of superfine blue cotton broadcloth that he had ordered from England for his wedding. To the end of her life, Martha saved a piece of her wedding dress—deep yellow brocaded satin threaded with silver—and the white gloves her new husband had worn to the ceremony.[30]

PARTNER IN LOVE AND LIFE

We know—or at least suspect—that Martha Dandridge Custis was more than a little pleased by George Washington's proposal. But this does not tell us much about her. There were undoubtedly a great many women in Virginia who would have felt a few shivers at the thought of being embraced by the famous Colonel Washington. For some people, Martha's wealth has complicated and even distorted her and the marriage. Even when Washington was in his second term as president, a man inflamed by the politics of the day shouted at him, "What would you have been if you had not married the widow Custis?"

This crass view of the match as a marriage of ultimate convenience has led others to annotate with various degrees of wryness what Washington obtained when he said "I do." Daniel Parke Custis had died without a will, which left Martha in charge of his 17,438-acre estate, all prime land within a forty-mile radius of Williamsburg. An early estimate of its value was 23,622 pounds—well over three million dollars in today's money. A third of that sum was Martha's and the rest would go to her two children, so the entire enterprise was her responsibility for decades to come. The moment George became her husband, Martha's one-third was his property and he became the administrator of the children's portions. From a cash-short landowner with a few thousand acres, Washington, like his brother Lawrence before him, ascended into Virginia's upper class.

The cynics who note this miraculous transformation often forget that it was a marriage of convenience for Martha, too. She knew little or nothing about managing a huge estate. It involved dealing with overseers, worries over runaway or recalcitrant slaves, and problems with livestock and tenant farmers and with shipping tobacco to London when insurance rates were skyrocketing because of the ongoing war with France. Moreover, the Custis estate had a worrisome lawsuit looming on the legal horizon, stemming from the will of Daniel Parke Custis's rakehell grandfather. He had named as one of his heirs an illegitimate daughter he had fathered in Antigua, and her descendants were threatening to sue in Virginia for a share of his estate in that colony. If the final verdict went the wrong way, Martha could lose almost all the property she had inherited.

This was another reason why she wanted and needed a man of cool judgment in her life. Her choice of Colonel Washington, who had managed large affairs during the war and was intimate with the ruling politicians of Virginia, indicated that she had not a little common sense in her makeup, which the tremors of possible romance did not by any means obscure.[1]

II

Historians and historical novelists have long disagreed about Martha's looks. Some report her as ugly, or at best plain. One 1784 visitor to Mount Vernon even found fault with her "squeaky" voice. Unfortunately for Martha, she did not become a subject for first-rate portrait painters—or a personage to be studied by random visitors—until she was long past her youth. Most of her likenesses were painted when she was in her sixties and the wife of the first president. One writer sourly wondered why she persisted in wearing those silly mobcaps that made her look so old. Martha was, of course, merely dressing her age.[2]

Luckily, a traveling painter named John Wollaston has left us a portrait of Martha in 1757, when she was still Mrs. Custis. She was unquestionably an attractive young woman, with large hazel eyes and curly brown hair. Her forehead was wide and fine; her strong nose and firmly rounded chin added an air of self-confidence, which was equally visible in her small, firm mouth. She was barely five feet tall, but her figure was full and even eye-catching. One recent biographer called her "a pocket Venus, a petite cuddlesome armful." By the time she married Washington, Martha had gained enough weight to be considered plump by some people. But the added pounds did not diminish her physical charm.

III

Even more important in appreciating Martha Dandridge Custis's appeal to her husband is her personality. Almost everyone who met her was pleased by her warm, relaxed manner. Very early in life, she revealed a startling capacity to charm the male sex. The Dandridges were middling gentry, like the Washingtons. Everyone was agog when Daniel Parke Custis, one of the richest men in the colony, fell in love with Martha. The Dandridges had very little cash to spare for a dowry.

Daniel's father, Colonel John Custis, already famous for his foul temper, vowed to disinherit his son if he married a penniless Dandridge. For several months, this edict stalemated matters. Daniel had no appetite for arguing with his headstrong parent. John Custis worsened matters by threatening to leave his entire estate to Jack, a mulatto boy he had fathered with Alice, one of his slaves. Daniel—and everyone else—knew he was more than capable of such a bizarre gesture. Soon the imbroglio was the talk of Virginia, with gossips gleefully reporting John Custis's latest outrageous remark. At one point he gave silver engraved with the Custis coat of arms to an innkeeper's wife to make sure it would never be owned by "any Dandridge's daughter." The woman displayed the gift in her Williamsburg tavern.[3]

Who could possibly resolve such an ugly contretemps? Martha Dandridge decided to try. She rode to Williamsburg and confronted the old tyrant in his house on Francis Street. Face to face, the colonel turned into a paper tiger, and then into a pussycat. He was impressed by Martha's courage—and pleased by her calm, even temper and the direct, sensible way she talked to him about herself and his son. Daniel sent one of his friends, a lawyer named Power, to see his father. He discovered that Colonel Custis had changed his mind about the marriage. The Custises were so rich, his son did not need a dowry to marry Martha Dandridge.[4]

The lawyer rushed Daniel the astounding news: "I am empowered by your father to let you know he heartily and willingly consents to your marriage with Miss Dandridge." Power attributed this miraculous transformation to "a prudent speech" that Martha had made to the colonel. "He is

[now] as much enamored with her character as you are with her person,"
Power continued. "Hurry down [here] immediately for fear he should
change the strong inclination he has to your marrying directly."[5]

IV

This episode tells us a lot about Martha Dandridge Custis Washington.
She may have lacked Sally Fairfax's education and interest in art and literature, but she was no plain Jane who was happy to be a humble echo of
her outsized second husband. George had not been the only candidate for
Mrs. Custis's affections. Before Colonel Washington made his first visit,
Martha had been pursued by one of the richest men in Virginia, Charles
Carter, son of fabulously wealthy Robert "King" Carter. Charles's wife
had died about six months before Daniel Custis, and he frankly confessed
that a widower's life made him miserable.

"Mrs. C___s is now the object of my wish," Charles Carter told his
brother. He praised Martha's beauty and—especially significant—her
"uncommon sweetness of temper." Although he was twenty-three years
older than Martha, Carter hoped "to raise a flame in her breast." He was
still a vigorous man who dressed well and was the ultimate social insider.
But Carter had fathered no less than a dozen children, and ten of them
were still living at home. Martha, again exercising her gifts of common
sense as well as frankness, told him she hoped to have more children. She
wondered whether she—or any other woman—was capable of managing
such a huge family, with its inevitable jealousies between half brothers
and /or sisters. It is not hard to see how Colonel George Washington, single,
childless, and almost exactly her own age, had a far greater appeal to this
practical young woman.[6]

V

The newlyweds spent the first few months of their married life at Martha's White House plantation and the Custis house in Williamsburg while
George attended the House of Burgesses. Martha undoubtedly glowed
with pride when he received the unanimous thanks of the legislature for
his five years of service on the frontier. They joined in the parties and
balls that enlivened the little colonial capital. Both loved to dance and

performed all the popular steps of the day, from stately minuets to more intimate allemandes to energetic American jigs and reels that often sent European visitors fleeing to the sidelines, claiming that the "irregular and fantastical" style threatened their "sinews." The Washingtons were young, rich, and with every reason in the world to enjoy themselves.

Not until April 2, 1759, did they take the road to Mount Vernon. Martha and the children traveled in the Custis family coach; Washington rode his horse beside them. Behind them came wagons that carried twelve slave servants from the White House, including a cook, a waiter, a seamstress, and a laundress. In other wagons were no fewer than six beds, several chests of drawers, linens, silverware, two sets of china, and dozens of pieces of kitchenware. In still more wagons rattled 120 bottles of wine, casks of rum and brandy, and numerous hams, plus a large supply of cheeses and sugar. Martha was obviously operating on the assumption that setting up a household in a bachelor establishment such as Mount Vernon was equivalent to a venture into the wilderness. Colonel Washington may have added to this impression with an anxious letter he rushed ahead to his overseer, urging him to "get out the chairs and tables," clean the rooms, start fires in the fireplaces, and make a point of polishing the stairs "to make it look well."[7]

At Mount Vernon, Martha soon began making lists of furniture and other expensive items for Robert Cary, the London merchant who handled such purchases for the Custises and many other wealthy Virginians. One of her most interesting orders was a bedroom set featuring a seven-and-a-half-foot-tall canopied bed with blue (Martha's favorite color) curtains and matching coverlet, wallpaper and window curtains, plus four chair bottoms of the same color "to make the whole furniture of this room uniformly handsome and genteel." Mr. Cary was soon being inundated with similar orders to convert the rest of Mount Vernon from its bachelor bareness to a comfortable, attractive home. Also on the purchase list were amenities such as "a pipe of the best old wine from the best house in Madeira." A pipe, if it survived the high seas without being tapped by thirsty sailors, would deliver 126 gallons of Colonel Washington's favorite wine.[8]

The colonel wrote these orders in his firm, legible hand. He also showed no hesitation in buying Martha virtually unlimited quantities of the best and finest lace, wool, and satin to be made into attractive dresses, riding suits, and cloaks. Satin slippers, black gloves for the winter, and white gloves for the summer arrived in multiple numbers. Although Washington could not carry a tune, he loved to hear Martha and others sing. Not long after their arrival at Mount Vernon, he inscribed *Martha Washington, 1759* in a songbook—perhaps the first time he wrote her married name. Was he thinking with considerable satisfaction that she was no longer Martha Dandridge or Martha Custis?

Toward the end of their yearlong buying spree, Washington wrote a letter to Richard Washington, the English merchant with whom he had previously done business. His conscience was a bit troubled by the way he had deserted him for Mr. Cary, and he included a modest number of purchases to reassure him that he had not been forgotten. "I am now I believe fix'd at this seat [Mount Vernon] with an agreeable consort for life," he wrote, "and hope to find more happiness in retirement than I ever experienced in a wide and bustling world." These are the words of a contented man. To Washington, who had grown up in a household where Mary Ball Washington specialized in being disagreeable, Martha's sunny disposition was something to treasure. He began to realize that marrying her was one of the best decisions of his life. Soon in private conversation he was calling her "Patsy"—the intimate nickname of her girlhood.[9]

VI

It did not take Washington long to see that four-year-old Jack and two-year-old Patsy were central to Martha's happiness, and he did everything in his power to show her that he cared for them. Expensive clothes for the two children, as well as numerous toys, flowed off the ships that docked in nearby Alexandria and were trundled up the road to Mount Vernon. Later he bought one of the finest spinets made in England for Patsy and a good violin for Jack. Fashion dolls dressed in the latest mode also arrived regularly for Patsy as she matured into a pretty brunette. George was pleased when Martha began calling him "Poppa" and encouraged the children to do likewise.

Only one of Martha's letters to George has survived. Fortunately, it tells us a good deal about the progress of their marriage from convenience to deepening love. She wrote it in 1767, while George was in Williamsburg attending a session of the House of Burgesses.

March 30, 1767

> *My Dearest*
>> *It was with very great pleasure that I see in your letter that you got safely down. We are all very well at this time but it still is rainey and wett. I am sorry you will not be at home as soon as I expected you. I had rather my sister did not come up so soon as May would be much plasenter time than April. We wrote you last post as I have nothing new to tell you I must conclude myself*
>> *Your most affectionate*
>
> *Martha Washington*[10]

The only shadow on their happiness was Martha's anxiety about Jack and Patsy. She had lost two children to early deaths, and the thought of losing either of them terrified her. As she slowly realized that she and George were unlikely to have any children, Martha's anxiety intensified. They had expected to have a brood. No one knows why they remained childless, but reasonable speculation suggests two possibilities. Martha may have had difficult deliveries with one or more of her four children that left her unable to conceive again, or Washington's bout with smallpox in the West Indies may have left him sterile.

Gently, with great forbearance and understanding, Washington tried to help Martha deal with her almost uncontrollable maternal anxiety. He wanted her to accompany him to Williamsburg and to visit other planters in Virginia and Maryland who were anxious to entertain the famous colonel and his wife. At one point he suggested they leave Jack home and take Patsy with them on a two-week visit to his brother Jack Washington and his wife. Martha told her sister it was "a trial to see how well I could stay without him."

The experiment was not a success. She was constantly listening for the thud of a horse's hoofs, which she was sure would be a messenger reporting Jack was ill or worse. Even the bark of a dog made her start violently. Her imagination kept conjuring images of Jack lying sick in his bed or writhing in the road after falling from his horse.

VII

As Jack Custis grew older, Washington began worrying about his education. A private tutor had taught him how to read and write and do arithmetic. But Jack was going to be a very wealthy man, and Washington thought he should have a far more extensive intellectual background to play a leading role in Virginia society. All his life, Washington suffered pangs of inferiority over his limited schooling. He wanted to make sure Jack did not became a man with similar regrets. After consulting various friends, Washington enrolled Jack in a boarding school run by the Reverend Jonathan Boucher, a well-regarded Anglican clergyman.

Washington had scarcely persuaded Martha to part with the fourteen-year-old boy when calamity struck. Twelve-year-old Patsy, who seemed to be maturing into a very pretty young girl, suddenly collapsed in what George and Martha at first thought was a fainting spell. But her twitches and gasps and groans soon made them realize it was a convulsion. A few days later, she collapsed again in another "fitt." Washington sent to Alexandria for a doctor, who glumly informed them that Patsy was an epileptic. Over the next several years, the Washingtons consulted eight different doctors. But there was no cure, and the drastic medicines the medical men forced down the poor girl's throat only made her nauseous and morose.

For Martha it was a dismaying blow. It meant Patsy would probably never marry. Several of the doctors warned them that the seizures might grow worse in years to come. This proved to be the case. Soon Patsy was having two seizures in a single day. A mournful Washington knew what this meant for Martha. He told a friend, "The unhappy situation of her daughter has to some degree fixed her eyes upon [Jack] as her only hope."

Meanwhile, Colonel Washington was receiving letters from the Reverend Boucher informing him that Jack, now seventeen, was close to being expelled from his school. "I never did in my life know a youth so exceedingly indolent, or so surprisingly voluptuous; one wd suppose nature had intended him for some Asian prince," the clergyman ranted at one point. Jack was rich and knew it. So did many of his friends and acquaintances. He had far too many invitations to "visits, balls and other scenes of pleasure." Worse, Jack had "a propensity to the sex." It was hardly surprising after these revelations to learn that Jack "does not much like books." More and more, Dr. Boucher had begun to think that only his "fervent prayers" would make Jack "if not very clever, what is much better, a good man."[11]

The clergyman opined that forcing Jack to leave his horses at Mount Vernon might keep him at least in proximity to his books. Martha flew to her son's defense, saying she thought he had done nothing that merited such punishment. She resented the idea of confining him like a criminal. In early 1771, at the close of the Christmas holidays, Washington assured Boucher that Jack was returning to school "with a determination of applying close attention to his studies." But he was forced to add that Jack would be a few days late because he wanted a little more time for "his favorite amusement of hunting."[12]

VIII

In 1773, shortly after Martha and George and Patsy returned from Williamsburg following the spring legislative session, Jack confided to Martha stunning news about his future plans. He had gotten engaged to a Maryland belle, Nelly Calvert, without asking his mother's or his stepfather's permission. Jack could, of course, have committed far worse indiscretions. In his family background lurked the example of his maternal grandfather, Daniel Parke, who was a womanizer of epic proportions, especially after he became governor of the West Indies island of Antigua. He was murdered by a group of outraged citizens, in part because of his pursuit of virtually every female on the island.

Jack was hardly imitating his grandfather in his pursuit of Nelly Calvert. But for a young man worth tens of thousands of pounds—an undoubted millionaire in today's currency—to marry without consulting his parents was serious enough to justify Martha's surprise. As for George, he could scarcely control his anger. Not only did he believe marriage was central to a man's happiness, but there was a great deal of money involved. That large fact stirred worries about sincerity and honesty, especially if there was a disparity in the bride and bridegroom's wealth. Moreover, Jack was still a minor; he could not marry without their permission. Before the distressed parents could do or say anything, Jack arrived at Mount Vernon with his fiancée's father, Benedict Calvert, and the latter's good friend, Sir Robert Eden, the governor of Maryland.

Ostensibly, Governor Eden was there to discuss with Washington ways to make the Potomac River a link to the West by dredging it and building canals. But there was little doubt Eden knew he added some social stature to his friend Calvert by inviting him along. Calvert was an illegitimate son of the colony's proprietor, the fifth Lord Baltimore. Nobody held that against him. He had married the daughter of a former Maryland governor and prospered well enough to preside at Mount Airy, a comfortable plantation near Annapolis, and sit in the colony's legislature.

After the two men departed, Washington wrote Calvert a letter that leaves no doubt that he was still very angry with Jack. The subject was "of no small embarrassment to me," he began. He was aware that Jack had "paid his addresses" to "Miss Nelly" and that her "amiable qualifications stand confess'd by all hands." He would be "wanting in candor" if he did not admit that an "alliance with your family would be pleasing" to Jack's "family." He might as well have written the literal truth behind that statement: Martha and young Patsy were delighted.

Washington then spent a paragraph making it clear that an immediate alliance would not be pleasing to him. Jack's "youth, inexperience and unripened education is & will be insuperable obstacles in my eye." As his guardian, he felt he had an "indispensable duty" to insist on Jack completing his education. He had enrolled him in King's College in New York. At the same time, Washington admitted he had no desire to break up the match; he wanted only to postpone it. He was going to recommend to Jack "with the warmth that becomes a man of honor (notwithstanding he did not vouchsafe to consult either his mother or me) that he consider himself as much engaged to your daughter as if the indissoluble knot was tied."

Washington followed this blend of smoldering rage and soothing assurances with a brief summary of Jack's impressive net worth in land, slaves, and cash. He hoped this information would inspire Calvert to "do something genteel by your daughter" in the matter of a dowry. Then came words of virtual capitulation. He hoped that Calvert and his wife and daughters would "favor us" with a visit to Mount Vernon.[13]

Jack galloped off with this letter to Mount Airy and returned with a warm reply assuring Washington that Calvert agreed "it was too early for Mr. Custis to enter upon the matrimonial state." But he hoped the coming separation would "only delay, not break off the intended match." Unfortunately, he had ten children and feared Nelly's dowry would be "inconsiderable." That bad news delivered, he smoothly assured Washington that the Calverts would be glad to visit Mount Vernon while Jack was studying hard and otherwise maturing in distant New York.[14]

Washington personally escorted Jack to New York, a journey that grew into a two-week series of dinners and receptions with the elite of Maryland, Pennsylvania, and New Jersey. It is hard to decide whether Washington was hoping to impress Jack with the importance that can come from achieving some distinction in life or whether he simply found the attention paid to him impossible to refuse. Fame was already swirling around the edges of his life. The trip ended with the stepfather and son being feted at a dinner given by Washington's friend from frontier warfare days, General Thomas Gage, now commander in chief of the British Army in America.

At King's College, the president, Myles Cooper, greeted them as if they were visiting royalty. Jack was soon telling his "Dear Momma" that he liked the way he was being treated. He was the only student who had dinner with the faculty and joined them in "all their recreations." He had a comfortable three-room suite, with a separate room for his slave body servant, Joe.

IX

Back in Mount Vernon, if Washington had hopes of terminating the match with Nelly Calvert, they vanished before his eyes as Nelly totally charmed Martha. They were similar types of women, warm and cordial by instinct; Nelly added to this pleasing temperament a lustrous brunette beauty. She not only mesmerized Martha, she became Patsy's best friend. The two young women, roughly the same age, became inseparable. Martha buoyed their spirits by inviting numerous other belles from nearby plantations to join them in the evenings for music and dancing. George's younger brother Jack and his wife, Hannah, and two of their children mingled with these visitors. It was a happy family gathering, marred only by the scorchingly hot weather.

On Saturday, June 19, they enjoyed a festive family dinner that kept everyone at the table until about five o'clock. Patsy and Nelly went upstairs, talking about Jack. Patsy ran into her bedroom to get a letter he had written to her. A thud and a strangled cry brought Nelly to the door. Patsy was lying on the floor, writhing in another epileptic seizure. Nelly called downstairs for help. Several people, including her stepfather, helped lift Patsy onto her bed. Almost instantly it became apparent that this was not a mild attack. Patsy's breathing grew labored and suddenly dwindled. In less than two minutes, "without uttering a word, a groan or scarce a sigh," Washington later wrote, she was dead.

Martha wept uncontrollably for hours. Washington's sorrow—and possibly his tears—matched hers. It was visible in a letter he wrote to Martha's brother-in-law, Burwell Bassett, the following day, after Patsy's funeral:

> *It is an easier matter to conceive than to describe the distress of this family, especially that of the unhappy parent of our dear Patcy Custis, when I inform you that yesterday removed the sweet innocent girl into a more happy & peaceful abode than any she has met with in the afflicted path she hitherto has trod. . . . This sudden and unexpected blow has reduced my poor wife to the lowest ebb of misery, which is increased by the absence of her son.*[15]

Washington wished he could persuade Martha's mother, Fanny Dandridge, to move to Mount Vernon permanently, but no argument could entice this lady to join them. He had to depend on frequent visits from the Fairfaxes and his brother Jack and his wife, as well as the continuing presence of Nelly Calvert, to console Martha. He did his best to join them in this almost impossible task.

For the next several months, Washington stayed close to Mount Vernon, canceling a trip to the West with Lord Dunmore, the new governor of Virginia, on which he had hoped to add to the thousands of acres he already owned beyond the Blue Ridge Mountains. He frequently persuaded Martha to join him in a light carriage, supposedly to visit his outlying farms. "Rid with Mrs. Washington to Muddy Hole, Doeg Run and Mill Plantations," he wrote in his diary on one of these days—terse testimony to his attempts to offer Martha his company and sympathy. He became very fond of Nelly Calvert, who stayed at Mount Vernon for most of the summer of 1773. Her tact and skill in comforting Martha soon created a bond that made her a substitute daughter.

X

For a while Jack Custis was a consolation. His letters from New York were cheerful and full of determination to study hard and acquire the education Washington wanted him to have. "I hope the progress I make . . .

will redown not only to my own credit, but to the credit of those who have been instrumental in placing me here," he told Washington. He made a point of thanking his stepfather for "the parental care and attention you have always & upon all occasions manifested toward me."[16]

In September, after only three months of study, King's College gave Jack a vacation—so he claimed. Washington arranged to meet him in Annapolis for the annual horse races and festivities connected with the meeting of the state's assembly. Jack joined him for these revels and had a joyful reunion with Nelly Calvert. He and Washington stayed at Governor Eden's mansion and spent five days enjoying balls, the theater, and the racetrack. They came home in good spirits to Mount Vernon, where Jack was greeted with fervent kisses by his delighted mother.

During the next several weeks, Washington's pleasure at presiding over this joyous reunion turned into angry disappointment. He was a busy man; Patsy's death meant the legal transfer of much valuable property to Jack's estate and to Martha's holdings as well. Meanwhile, Jack was telling his mother how much he loved Nelly and how badly they both wanted to console her for Patsy's loss by giving her grandchildren. Jack persisted in this campaign while he and Martha and Nelly traveled to Williamsburg with Washington to arrange for the transfer of Patsy's property and give Martha a chance to visit her mother and other nearby relatives. On the way home to Mount Vernon, Martha told Washington that Jack had obtained her permission to abandon his education and marry Nelly as soon as possible. All her relatives agreed with her decision.

Washington was infuriated, but what could he do? He had no desire to play the villain and oppose the young lovers, much less curtly inform Martha that she should wait patiently for grandchildren. Jack was so rich, it was difficult if not impossible to argue that a classical education was a vital necessity for him. Back at Mount Vernon, the defeated colonel wrote a letter to Myles Cooper, the president of King's College, informing him of Jack's decision. "I have yielded," he all but growled, "contrary to my judgment & much against my wishes, to his quitting college . . . having his own inclinations—the desires of his mother & the acquiescence of almost all his relatives to encounter, I did not care, as he is the last of the family, to push my opposition too far; & therefore have submitted to a kind of necessity."[17]

Two months later, a resigned Colonel Washington summoned a warm smile as Jack married Nelly at her family home, Mount Airy. Martha, still in mourning for Patsy, did not come with him. The newlyweds made Mount Vernon one of their first destinations. Their brimming happiness undoubtedly gladdened Washington's heart as much as Martha's. He could only console himself with the thought that he had done everything in his power to make Jack a man worthy of his wealth and potential importance. Another consolation was Nelly herself. Everyone, even Jack's grumpy former schoolmaster, The Reverend Boucher, agreed she was an exceptional young woman, as intelligent as she was beautiful.

XI

Another event that stirred deep emotions in the master of Mount Vernon was a decision made by his neighbors, George William Fairfax and his wife, Sally. They were moving to England, perhaps permanently. George William had inherited property from a relative, and the bequest was being challenged in the courts by another member of the family. Behind the lawsuit lurked the accusation that George William had Negro blood and was disqualified from inheriting the dukedom when the now aged Lord Fairfax died. Like most English lawsuits of the era, this wrangle might take years to resolve. Washington could do little but extend his warmest wishes for success and promise to keep a close watch on a darkened, silent Belvoir.

A glum Washington noted in his diary that on July 8, 1773, George William and Sally came to Mount Vernon "to take leave of us." The next day, he and Martha "went to Belvoir to see them take shipping." In Sally's trunks were the two tormented letters he had written to her fifteen years ago. How distant, how strange that yearning soldier must have seemed to Washington now! He was a different man, leading a different life, rich in peace and contentment. There were sorrows such as Patsy's death and frustrations such as Jack Custis's willful ways; disappointments occurred in almost every life. But he no longer lived on the brink of sudden death, clutching at the mere confession of Sally's love as a consolation.[18]

XII

In the closing weeks of 1773, the problem of America's relationship with England abruptly intruded on George Washington and his fellow Virginians. On December 16, a group of Bostonians disguised as Indians dumped 342 chests of British tea into the harbor to protest Parliament's tax on it. Tea was the only item still on the mother country's revenue list; American boycotts and strenuous denunciations by pamphleteers had persuaded the imperial legislature to abandon all the others. But everyone knew the tea tax had been kept to "maintain the right" to extract cash from the defiant Americans. Boston agitators led by Samuel Adams had struck in the night to let the world know they were determined to resist any and all taxation without representation.

Most Virginians, including Washington, denounced the tea party as vandalism. No one but Yankee fanatics worried about the tea tax. Most of the 80,000 pounds of the brew drunk in Virginia was smuggled from the West Indies or England itself, where tax evasion was a national industry. But as the Virginians read their newspapers, they soon realized this tea was a special case, imported under a monopoly set up by Parliament to give the almost bankrupt British East India company some badly needed revenue. The tea would have sold at a price below even that of smuggled tea—probably inducing thousands of people to save a few pennies while affirming Parliament's right to tax Americans.

When the British responded to the destruction of the tea by closing the port of Boston, and making General Thomas Gage the royal governor of Massachusetts, backed by several regiments of regulars, opinion in Virginia underwent a radical change. Especially alarming on this list of what the British called the "Coercive Acts" was a ukase cancelling Massachusetts's right to elect the governor's council. Henceforth its members would be appointed by the king. Another law specified that anyone accused of treason would be tried in England, not in the American colonies. Washington voted wholeheartedly with Virginia's House of Burgesses to protest these encroachments on Massachusetts's rights and proclaim Virginia's solidarity with their fellow Americans. Soon he was one of seven delegates chosen to represent Virginia in a general congress that met in Philadelphia to discuss the crisis.

In the middle of this political turmoil, Washington had a painful personal duty thrust on him. George William and Sally Fairfax asked him to supervise the sale of Belvoir's furnishings. George William's lawsuit looked more and more interminable, and the couple had decided it might be better to stay in England permanently. They thought it would give their argument more weight in court. George bought mahogany chests and tables, mirrors, and bedclothes, no doubt including Sally's own. There was some consolation in bringing these purchases to Mount Vernon, but it was painful to see dozens of strangers buying up his old friends' possessions from rooms where he had enjoyed so many happy hours. In retrospect, there was a fitting finality to the sale. It was a kind of farewell to Washington's youth. But he did not see it that way at the time. He felt only sadness and regret.

History was taking charge of George Washington's life. On August 30, Edmund Pendleton and Patrick Henry, two of the other Virginia delegates to the Congress, arrived at Mount Vernon to join him for the journey to Philadelphia and the first meeting of the Continental Congress. At dinner Martha listened to them discuss the confrontation with England. Pendleton was by nature a cautious man but Patrick Henry was his usual fiery self, determined to assert America's rights no matter what the consequences. The next morning, Martha watched them depart with an uneasy mixture of pride and anxiety.[19]

XIII

In little more than a year, the quarrel with George III and his revenue-hungry Parliament led to bloodshed in Massachusetts. Sam Adams and his cousin John Adams, anxious to win the support of the rest of the country, backed Virginia's Colonel Washington to head the impromptu New England army that rushed to besiege the British inside Boston. Three days after he received his commission from Congress, Washington wrote one of the most difficult—and revealing—letters of his life. It began with words that testified that Martha had become far more than an agreeable consort.

> *My Dearest:*
>
> *I am now set down to write you on a subject which fills me with inexpressible concern—and this concern is greatly aggravated and increased when I reflect on the uneasiness I know it will give you—It has been determined in Congress that the whole army raised for the defence of the American cause shall be put under my care and that it is necessary for me to proceed immediately to Boston to take upon me the command of it. You may believe me, my dear Patcy, when I assure you, in the most solemn manner, that, so far from*

seeking this appointment I have used every endeavour in my power to avoid
it, not only from my unwillingness to part with you and the family, but from
a consciousness of its being a trust too great for my capacity, and that I should
enjoy more real happiness and felicity in one month with you, at home, than
I have my most distant prospect of reaping abroad, if my stay was to be seven
times seven years.[20]

Those words bear witness to the deep and abiding happiness George
Washington had achieved in his sixteen years of marriage to Martha Dandridge Custis. In the next few harried days, he made it clear that Martha's
peace of mind remained one of his foremost concerns. Washington wrote
letters to Burwell Bassett, Jack Custis, and Jack Washington, in which he
admitted "my absence will be a cutting stroke" upon Martha. He begged
them to visit her as often as possible in the months to come. He had no
idea that he was embarking on a venture that would keep him away from
Mount Vernon for eight years.

On June 23, about to depart for Boston, Washington found time for one
more hasty but equally revealing note:

> *My Dearest:*
>
> *As I am within a few minutes of leaving this city [Philadelphia] I could*
> *not think of departing without dropping you a line. . . . I go fully trusting in*
> *that Providence which has been more bountiful to me than I deserve, and in*
> *full confidence of a happy meeting with you sometime in the fall. I have not*
> *time to add more, as I am surrounded with company. . . . I retain an unalter-*
> *able affection for you which neither time or distance can change. . . .*
>
> *With the utmost truth & sincerity Yr entire*
>
> *Geo Washington*[21]

FROM GREAT SOMEBODY
TO LADY WASHINGTON

In Massachusetts, General Washington wrestled with the myriad problems of creating an army. He missed his wife's companionship—and her skills as a hostess. He was living in the house of the president of Harvard College and found himself constantly entertaining important visitors. In October he wrote Martha a note, urging her "to come to me, although I fear the season is too far advanced . . . to admit this can be done with any tolerable degree of convenience." Martha was at Eltham, visiting her favorite sister, Nancy, and Nancy's husband, Burwell Bassett, when George's letter finally reached her. She stayed at Eltham for another week, thinking about the six-hundred-mile journey.

For someone who had a tendency to fear the worst, it was a daunting proposition. Late fall weather could make the roads impassable; numerous rivers and creeks would have to be crossed; roadside taverns were often dirty, inhospitable places. But she sensed George needed her. With Jack and Nelly Custis for company, she was soon on her way. Nelly had recently given birth to a baby who lived only a few weeks, and Martha hoped the trip would raise her daughter-in-law's spirits.

The journey took on overtones of a triumphal procession. Washington made sure one of his aides, Joseph Reed, met them in Philadelphia. It was Martha's first glimpse of the huge expansion of her husband's fame. A troop of uniformed horsemen escorted her into the city, and hundreds of people thronged the sidewalks to get a glimpse of her. In the rooms Reed reserved for them, a veritable stream of congressmen and prominent Philadelphians rushed to welcome General Washington's wife. Martha was gracious and warm, but she did not get carried away. In an amused letter to a Virginia friend, she wrote, "I don't doubt but you have see the figuer [*sic*] our arrival made in the Philadelphia paper . . . and I left it in as great pomp as if I had been a great somebody."[1]

Reed was charmed by Martha and Nelly. In a note, he told Washington that he was sure they would be good company "in a cold country where wood is scarce." In Cambridge, Martha swiftly solved George's hospitality problems. She persuaded him to move to a larger house and was soon a cheerful presence at a bountiful dinner table. She charmed grumpy Yankees such as James Warren, speaker of the Massachusetts legislature, and his formidable wife, Mercy, a fierce intellectual who wrote satires and plays (under a man's name) pillorying the British and would later turn out a three-volume history of the Revolution. Mercy told her friend Abigail Adams how impressed she was by Martha. She praised "the complacency of her manners" as well as her "affability, candor, and gentleness."[2]

II

Martha Washington's journey to Cambridge was the first of many trips she would take from Mount Vernon to join her husband during the next eight years of the War for Independence. Her importance as a wife and human would grow larger and more apparent to everyone—above all to her husband. She was the only person with whom Washington could relax and speak candidly. With most people he had to maintain the role of the confident, decisive commander in chief.

By 1778, Washington was being called "the father of his country" in the newspapers. He did his utmost to avoid acknowledging this tendency to view him as a demigod. At Valley Forge, on the night of February 22, 1778, General Henry Knox, commander of the artillery, sent a regimental band to serenade the commander in chief on his birthday—the first semi-official celebration of the day. Washington sent Martha out into the snowy road to thank the musicians and give them generous tips. She politely informed them that the general had gone to bed and that was why he was not thanking them in person.[3]

The chorus of adulation continued to grow. Poems and speeches hailed

Washington's greatness. In France, Ambassador Benjamin Franklin kept a full-length portrait of him on the wall in his study. But fame seemed to have no impact on George and Martha's loving relationship. If anything, they became even more intimate. Martha began calling him "my old man" and in private often addressed him as "Pappy." No one sympathized more deeply with the commander in chief's travails. More than once she wrote a friend, exclaiming that "the pore general" was looking weary and discouraged, as the war dragged on and on.

General Nathanael Greene, who rose to second in command of the American army, wrote to his wife, Caty, "Mrs. Washington is excessive fond of the general and he of her . . . they are happy in each other." Throughout the war, Martha wore a miniature of Washington by Charles Willson Peale on a chain around her neck; he carried one of her beneath his shirt.

III

In 1776, Jack Custis reached the age of twenty-one and took over the management of his estate. With no apparent prodding from Martha, who was with her husband in New York, he wrote Washington a touching letter. Jack was "extremely desireous . . . to return you thanks for your parental care which on all occasions you have shown to me." He had lost a father at an early age, but "few have experienced such care and attention from real parents as I have done." Jack asked him to "continue your wholesome advice and reprimands whenever you see occasion." He promised they would be "thankfully received and strictly attended to." Meanwhile he would never cease looking for opportunities to testify "to the sincere regard and love I bear you."[4]

Washington never even hinted that Jack should join the Continental Army; he knew that Martha would be prostrate with worry at the mere thought of him going into battle. Jack bought Abington, a handsome 900-acre estate near Alexandria, and demonstrated a modicum of patriotism by serving in the Virginia Assembly. He also invested a considerable sum in a privateer that would, like hundreds of similar warships, attack British merchant vessels. He remained deeply devoted to his Nelly, who gave him four children in the next six years. The fourth, a boy, was named George Washington Parke Custis. But running a plantation bored Jack. He was equally unenthusiastic about traveling to his plantations in the vicinity of Williamsburg to make sure they were being properly managed by his overseers. Such responsibilities interfered with his favorite pleasures—betting on horses and cards, hunting, and partying.

When Jack decided to sell some of his more distant plantations, Washington was so dismayed that he took time he could not spare from the war to write him a long, earnest letter, urging him not to do it. The buyers paid Jack in paper dollars printed by the Continental Congress, money backed by nothing but the hope that the Americans would win the war. As the conflict dragged on, these dollars had begun to depreciate. Washington urged Jack to invest the money in land closer to his home while the dollars still had value. Jack ignored his advice.

Worse was to come. Jack did not gamble in the restrained style of his stepfather, and he was soon deeply in debt. One year, he bought some of Mount Vernon's cattle. Washington assumed he would have them appraised and pay him a fair price. Jack had the appraiser examine only the worst beasts in the herd and applied the price to the rest of them. He then waited months while the dollar depreciated toward virtual waste paper and paid his stepfather in the now almost worthless currency. "You might as well attempt to pay me in old newspapers," a furious Washington told him.[5] He stopped signing his letters to Jack "with love." But he could not do or say more. Jack remained Martha's darling. She would not tolerate criticism of him from anyone, even George.

IV

If Washington had known some intimate details of Jack's home life, he might have changed his mind and risked Martha's wrath. When Nelly's second baby was another girl, Jack expressed his disappointment so vehemently that Martha offered to raise the child. Nelly demurred, but Jack's sulk about not having a son continued. When his oldest daughter, Eliza, was four or five years old, he decided she could entertain him and his friends at dinner parties. She had a good memory and loved to sing. Jack proceeded to teach her the lyrics of several raunchy songs. At the end of the dinner, when the table was cleared and serious drinking began, Jack would order Eliza to be brought to the dining room in one of her prettiest dresses.

Jack would lift the child onto the table and she would prance up and down, singing the salacious lines, while her father and his friends roared with laughter. Nelly Custis protested, but Jack brushed her off, claiming "his little Bet could not be injured by what she could not understand—that he had no boy and she must make fun for him until he had." Nelly gave up and left the little girl "to the gentlemen" and her father's "caresses." For a while, the experience made Eliza "think well of myself." Later in her unhappy life, after a broken marriage and several failed affairs, Eliza would change her mind about this and many other things.[6]

V

The "long and bloody war," as Washington sometimes called it, was in its sixth year with no end in sight when Washington received an anxious letter from an old friend, Benjamin Harrison, who was now the speaker of the Virginia assembly. Seventy-seven-year-old Mary Ball Washington had apparently asked some members of the legislature to propose a bill granting her a pension. She claimed she was "in great want, owing to the heavy taxes she was obliged to pay." Implicit was the accusation that her famous son was allowing her to all but starve. An uneasy Harrison assured Washington that the assembly would be glad to pass the bill, but he thought it might be wise to consult him about it first.

A thunderstruck Washington replied that he had no idea that his mother was having any financial difficulties. He had instructed his cousin Lund Washington, who was managing Mount Vernon, to answer all her requests for money without a moment's hesitation. Moreover, Mary did not have a child "who would not divide the last sixpence to relieve her from any *real* distress." He had told her this repeatedly. The general made it clear to Harrison that he and his brothers and sister would feel "much hurt" at having their mother a pensioner, especially when she had "ample income of her own." He begged Harrison to block the bill and make the whole matter disappear as quickly and quietly as possible.[7]

VI

Not long after this contretemps with Mary, Washington visited Mount Vernon following an absence of more than six years. He was accompanied by General Comte de Rochambeau and other leaders of the French army. Martha, Nelly, and Jack and their four children greeted them joyously. The soldiers spent only three days in the house before pushing on to Williams-burg, where the French and American armies were gathering to assault a British army entrenched in the nearby port of Yorktown. The sweet smell of victory was suddenly in the air, after so many years of discouragement.

Jack Custis asked his stepfather if he could join his staff as a volunteer aide. How could Washington say no? Yorktown was almost certainly going to be a siege rather than an all-out battle. There would be little or no danger for someone on a general's staff. The British soon proved Washington correct; they retired inside their fortifications and tried to hold out, betting on the royal navy to rescue them. But a revived French fleet easily defeated a halfhearted attempt at seaborne relief, led by one of the most inept admirals in British history. The stunned redcoats realized they were trapped.

The siege lasted almost three weeks. For a while Jack enjoyed himself. On October 12, he wrote his mother a cheerful letter, telling her, "The General tho in constant fatigue looks well." Like many soldiers, Jack Custis developed dysentery from eating army commissary food that had grown stale or partially spoiled and drinking water polluted by thousands of men living in the vicinity with only minimal sanitation. A few days later, he began running a high fever. Dr. James Craik, the same physician who had cared for Washington two decades earlier, diagnosed "camp fever"—the disease we now call typhus. He urged Jack to retreat from the battlefield to a house where he could stay in bed, drink good water (or better, liquor), and eat unspoiled food.[8]

Jack shook his head, determined to have his own way as usual. The siege was thundering to a climax. The British fortifications were battered wrecks from the ferocious day-and-night Allied bombardment. On October 17, a British officer waving a white flag appeared outside their works and delivered a letter from his commander, Charles, Lord Cornwallis, asking for terms. Two days later, the British marched out and surrendered their guns. By this time, Jack Custis was so weak that he could barely sit up in a carriage to watch the ceremony.

A distressed Washington joined Dr. Craik in virtually ordering Jack to go to Eltham, his uncle Burwell Bassett's plantation, about thirty miles from Yorktown. The general stayed at Yorktown for the rest of October, dealing with the thousand and one details of the new situation created by the victory. Not until the first days of November did he set out for Mount

Vernon with his staff. On the way, he left his aides at a nearby tavern and rode to Eltham to check on Jack Custis.

Imagine his consternation when he was met at the door of the Bassett mansion by a weeping Martha; Jack's wife, Nelly; and their oldest daughter, Eliza. Jack was dying. Washington raced upstairs to the sickroom, where several doctors were standing around Jack's bed, their heads bowed in defeat. Jack's breath was dwindling in his throat. Washington could only watch as he expired.[9]

The general's first concern was Martha. She was overwhelmed with grief. Washington spent the next five days at Eltham, presiding over Jack's funeral and doing his utmost to console his wife. It was a dismaying interlude at a time when everyone else in Virginia and the rest of the nation was celebrating the Yorktown victory. Her son's death must have triggered an inner struggle for Martha. As her husband ascended to glory, she was plunged into despair. But she was consoled by Nelly and the four adorable grandchildren, and sustained by Washington's strength and love.

Still fighting a war, the general asked Martha's brother, Bartholomew Dandridge, to take charge of Nelly and the children and act as the executor of Jack's estate. Dandridge refused to accept responsibility for the children, and when he got a look at Jack's account books, he was soon moaning that everything was an unholy mess. Some historians have noted that Washington never uttered a word of grief for Jack. This may be true, but that does not mean he did not grieve for him. By now the general had no illusions about human nature. He had seen too many other men with character defects to judge Jack harshly; he could not deny a stepfather's memories of tender moments. As a man who had lost his own father in his boyhood, his heart went out to Jack's four children. He ruefully but willingly made them his responsibility.

VII

The war dragged on for two more years. Most of the time, the general and Martha lived in a house in Newburgh, a few miles from the Continental Army's camp in New Windsor, about forty miles north of New York. Often during these months, Washington reported Martha was "low," suffering from "bilious fevers and colic" and other complaints, probably symptoms of depression. Washington was bored and not a little irritated at Congress's inability to raise money to pay his troops. He, too, had minor physical woes; missing teeth made it difficult to chew his food and his eyes were beginning to fail. He ordered two sets of spectacles from Philadelphia, one for distant vision, the other for reading.

From Virginia came more bad news about his mother. Mary was back at her old game, telling everyone in Fredericksburg and elsewhere that she was penniless and close to starvation. An angry Washington wrote to his favorite younger brother, Jack, begging him to pay their mother a visit. He urged Jack to tell Mary "in delicate terms" that she should not accept money or favors from anyone but her "relations." He had no doubt whatsoever that they would be able to satisfy her "*real* wants." As for her "*imaginary* wants," they were "boundless and always changing."[10]

On his way to Mount Vernon from Yorktown, Washington had stopped in Fredericksburg to visit Mary. She was not at home—an accident that he probably regarded as a stroke of luck, saving him from listening to a litany of complaints. He left her some money and went on his way. Several weeks later, she wrote him a letter lamenting that she had missed his visit. She had taken a trip "over the mountains" that "almost kill'd" her. There she had seen some land he owned that she thought would be perfect for "a little hous of my one [own] if it is only twelve foot squar." George was paying rent for her to live in an elegant house in Fredericksburg, but it was apparently unsatisfactory. Whatever he did for Mary was unsatisfactory. There is little doubt that George's encounters with his mother invariably increased his affection for Martha Custis.

VIII

By the time the War for Independence ended, Nelly Custis had found another husband, an Alexandria physician, Dr. David Stuart. She took her two older girls with her into the new marriage. The Washingtons adopted the two youngest children, four-year-old Nelly and two-year-old George Washington Parke Custis, whom everyone called "Wash." When Washington arrived at Mount Vernon on Christmas eve, 1783, a private citizen once more, the ex-general's trunk was full of toys he had bought in Philadelphia for the "little folks" as well as presents for a beaming Martha.

In the same two years, George and Martha acquired other parental responsibilities. His younger brother Samuel had died at forty-seven.

Samuel had never been much of a businessman, and the deaths of no fewer than four wives had added to his woes. He had left his fifth wife penniless, wondering how she was going to feed her newborn baby, plus three boys and a girl from one of Samuel's previous unions. "In God's name," Washington asked his brother Jack, "how did my brother Saml. contrive to get himself so enormously in debt?"

Jack Washington could only plead with his older brother not to ask him for help with the indigent children. He had his hands full trying to provide for his own family. Their youngest brother, Charles Washington, was a hopeless alcoholic. George wearily saw he again had no choice. He ordered Samuel's three youngest children sent to Mount Vernon. The oldest, a fourteen-year-old boy, and the baby would stay with his brother's widow. The two younger boys turned out to be hell-raisers who drove Washington and several schoolmasters to distraction. But he persisted in paying their tuitions and lecturing them on good behavior, and they eventually graduated from the University of Pennsylvania. Their sister, Harriot, lived at Mount Vernon for almost two decades. Also in residence was Fanny Bassett, daughter of Martha's sister, Nancy, who had died in 1777.

The older children could more or less fend for themselves. George and Martha's chief concern—and pleasure—were Nelly and Wash. When their step-grandfather became president, they traveled with George and Martha to New York, and later to Philadelphia when the capital was shifted there. This ready-made family was enormously important to Martha, whose maternal needs remained intense. Washington was equally involved in the "little folks'" future. He began thinking about a tutor for "Wash," who was so fat he was often called "Tubby." George told one friend he planned to "fit the boy for a university." He was hoping to succeed where he had failed with Wash's father. Alas, he was doomed to another disappointment. As rich as his father, Wash was to prove equally resistant to scholarly effort. He quit or was expelled from no fewer than three colleges.[11]

IX

When Washington became president, Martha once more journeyed north and became a crucial part of his household in the new capital, New York. Many people are under the mistaken impression that because he was elected unanimously, Washington's presidency was a love feast. The opposite is closer to the truth. There were still a substantial number of Americans, loosely called anti-federalists, who feared and disliked the Constitution and the new government it had created. Much of this hostility focused on the presidency, which they regarded as an office fraught with menace to American rights and liberties. Thousands of eyes were on Washington, suspecting him of being ready to turn into an American version of George III.

When Martha arrived in New York on May 27, 1789, these critics were growling because the president, after being overwhelmed with impromptu visitors ten hours a day, announced he was restricting such time-wasting encounters to two hours a week so that he could get some work done. Others carped at his weekly receptions, which were too formal, they thought, and smacked of an audience with a monarch. Still others complained about the poor quality of the dinners he served. Martha took charge of the kitchen, and soon guests were telling friends how deliciously they had dined and wined. Next, she launched her own weekly receptions, at which ladies were welcomed, and everyone was charmed by her relaxed, cheerful style. She also acquired a title that she neither sought nor liked: "Lady Washington." It was the invention of well-meaning people who felt a need for something better than "Mrs. President."[12]

Equally important was the way Martha made friends with the wives of cabinet members and other VIPs, above all, the vice president's wife, Abigail Adams. "A most becoming pleasantness sits upon her countenance," Abigail declared. She was particularly pleased by Martha's insistence that Abigail sit beside her and join in greeting the guests at her receptions.[13]

In little more than a month, Martha discovered another important role: nurse. Washington began complaining about a severe pain in his left thigh. Doctors discovered a growth that swelled and festered, making them fear it was malignant. Another physician suspected anthrax, a disease common among farm animals and sometimes contracted from people who sorted newly sheared wool. The president ran a high fever, and a rumor swept the city that he was dying. Two doctors operated without anesthesia, a discovery still far in the future. The pain was agonizing, and afterward Washington's head ached so intensely he could not tolerate the slightest noise. His secretary roped off the street around his house to eliminate passing carriages and bawling peddlers with creaky carts. It took him the entire summer to recover his strength.

Most of the time the Washingtons enjoyed New York. The city was full of exhibits of exotic animals, sometimes stuffed, often alive. A waxworks on Water Street featured "The President of the United States, sitting under a canopy, in his military dress." Martha seems to have found this an especially enjoyable sight. She took the children and several other "young misses" and persuaded Washington to go for a private viewing.

The Washingtons' favorite recreation was the John Street theater, which they attended so often that the proprietor created a presidential box, with the coat of arms of the United States in gold across its front. They particularly liked plays such as Sheridan's *School for Scandal,* which was considered racy in its day. Their attendance was usually advertised in advance to drum up business. As they entered their box, often with a party of friends, the band struck up "The President's March" and the audience gave them a standing ovation.[14]

X

Politics absorbed most of the president's attention. Congress was torn by wild wrangles over how to create a workable government. Secretary of the Treasury Alexander Hamilton had proposed a financial plan to cure the nation's chronic bankruptcy. The government would assume the wartime debts of both the states and Congress and utilize a new entity, the Bank of the United States, to gradually repay them. Secretary of State Thomas Jefferson and his friend Congressman James Madison violently opposed Hamilton's vision of the United States as an industrial and commercial powerhouse. Their opposition morphed into a detestation of New York as the nation's capital, because the city's numerous wealthy merchants supposedly corrupted Congress. The Virginians wanted a rural capital beyond the reach of big-city temptations. These clashes, which soon spilled into the newspapers, made President Washington a very worried man.

While the politicians called one another vicious names, Washington caught a cold that transmuted into pneumonia. This time, the fear that he was sinking toward death was more than a rumor. In an eerie replay of Jack Custis's demise, four doctors watched helplessly while the president struggled for breath. Another physician was summoned all the way from Philadelphia but had nothing to offer but more hand-wringing. The story spread throughout the nation, causing acute anxiety everywhere. "Every eye full of tears," one senator wrote.[15]

Martha was at George's bedside constantly. On the sixth day, one of the doctors grimly predicted the president's imminent death. About four o'clock that afternoon, George broke into a terrific sweat—a sign the disease had reached its crisis. Within two hours, he was smiling at Martha and speaking in a low voice. Probably at Martha's urging, he was soon taking rides with her in their carriage to escape the perpetual pressures of the presidency.

The realization that Washington was mortal may have influenced the politicians to reach a major compromise. Jefferson agreed to round up southern votes for Hamilton's financial plan and the New Yorker persuaded northerners to support a bill placing the permanent capital of the nation on the Potomac River in a newly created District of Columbia, carved from Maryland and Virginia. In the meantime, the federal government would move to Philadelphia, presumably less corrupt than New York. President Washington signed both bills and political tensions relaxed for a while. A booming market in government bonds and shares in the Bank of the United States began creating prosperity throughout the nation.

XI

There were times when Martha tired of public attention. It was often overwhelming. At one point she told her niece Fanny Bassett that she felt more like a state prisoner than anything else." When Washington left her in New York while he visited the New England states, she almost slipped into a depression. She wrote a remarkably frank letter to Abigail Adams's friend Mercy Otis Warren, in response to a "very friendly" letter Mercy had sent her. Martha said she was pleased by "the demonstrations of respect and affection" that the president had received from the American people. It made the burdens of the presidency tolerable for him—and for her. "You know me well enough to . . . believe that I am only fond of what comes from the heart," she wrote.

But Martha still yearned for Mount Vernon, where she had thought when the Revolutionary War ended she and George would be "left to grow old in solitude and tranquility together." Her problem, Martha confessed, was how acutely she missed her "grandchildren and domestic connections" in Virginia. But she was determined to be cheerful. "Everybody and everything conspire to make me as contented as possible." She had learned from experience that "the greater part of our happiness or misery depends on our dispositions and not upon our circumstances."[16]

After seventeen months on public display, President Washington decided he and Martha could take a vacation. They headed back to Mount Vernon, where grandchildren and grandnieces rushed to join them. At one point there were no fewer than ten young people, from teenagers to toddlers, rampaging around the house. Martha loved every minute of the chaos. Washington rode out regularly to his outlying farms and soon regained his health. He and Martha began discussing a topic that would absorb them for the next year. Should he accept a second term? The answer, they jointly decided, was an emphatic NO. The president asked James Madison to help him write a farewell address to the American people.

XII

The Washingtons were soon forced to change their minds. Frantic letters from Hamilton, Jefferson, and numerous other politicians warned the president that the country would come apart if he did not serve for another four years. Once more, Washington bowed to necessity. He knew Martha was deeply disappointed, but he also knew that she would remain at his side as his loyal partner. By this time they had settled into a comfortable mansion on Market Street in Philadelphia and were enjoying the numerous amenities of this sophisticated city. Close friends such as merchant Robert Morris added to their pleasures. But politics soon soured their lives in a tumultuous new way.

The second term had scarcely begun when news arrived from Paris that King Louis XVI, the monarch who had supported America's revolution, had been guillotined, along with his Austrian-born queen, Marie Antoinette. Within weeks came word that England had joined a coalition of European nations that were determined to crush France's revolution. At first the Marquis de Lafayette, a man for whom Washington had deep, almost paternal affection, had been among the leaders. But the marquis had been forced to flee and was now in an Austrian prison. Radicals known as Jacobins had seized control of France and had launched a reign of terror that sent thousands of people to grisly deaths beneath the guillotine's relentless blade.

Theoretically, the treaty America had signed with France in 1778 obligated the United States to join the war on her side. But Washington decided that the murder of Louis XVI, the man who had signed the treaty, meant France was now a different country. He also knew that the United States could not fight a war without wrecking its fragile economy, which depended heavily on trade with Britain. A grim President Washington issued a Proclamation of Neutrality—and immediately became violently unpopular among the thousands of Americans who saw the French Revolution as a sacred cause in the ongoing struggle for worldwide liberty.

These angry voters soon turned Washington's second term into a nightmare. More than once, thousands of people jammed Market Street in front of the president's house, shouting insults and waving pro-French slogans and banners. Newspapers castigated the president as a pro-English lackey who was betraying the American Revolution. At a public dinner in Virginia, journalist James Thomson Callender proposed a toast to "a speedy death to General Washington."[17] Tom Paine published an open letter to the president in which he raged, "the world will be puzzled to decide whether you are an apostate or an imposter; whether you have abandoned good principles or whether you ever had any."[18] Martha soon acquired an intense dislike for Secretary of State Thomas Jefferson, who was the leader of this new Republican party.

At one point Washington exploded into a gigantic rage at a particularly obnoxious newspaper story written by Benjamin Franklin Bache, Ben's fanatically pro-French grandson. In a rant that left no doubt he was Mary Ball Washington's son, the president pounded his desk and shouted that he was sick of being treated like a "common pickpocket." He swore he would rather be in his grave than put up with another day of this thankless job. For a half hour, he sat at his desk drained and dazed before he regained his equilibrium.

The public never saw—or heard about—this explosion. Nor did Martha. It was witnessed by only a few anxious secretaries. The public saw only the serene, dignified leader, who never showed the slightest evidence that he took seriously the abuse that was showered on him. One day, Washington was sitting for a portrait by Gilbert Stuart. Martha was nearby on a sofa, probably knitting while joining in the conversation. The painter considered himself an amateur psychologist. He remarked that he was fascinated by Washington's physiognomy—his face was that of a man with a terrific temper and violent passions.

Martha was shocked. "You take a great deal upon yourself, Mr. Stuart!" she said.

"Ah but Madam, let me finish," Stuart said with a bow in her direction. "Mr. Washington has these qualities under perfect control."

"He's right," Washington said with a smile.

Ninety-nine percent of the time Stuart's intuition was on target. Otherwise, Washington would never have survived his second term.

XIII

The president found Martha's company and the world of their family a welcome escape from the nation's increasingly rancid politics. He was especially interested in Martha's granddaughters and their thoughts and feelings as they grew to womanhood. His favorite, Nelly, was strikingly beautiful, and suitors thronged from all directions. Nelly told Washington they all left her cold. She had begun to think she would never marry because she could not imagine the callow "youth of the present day" approaching the awesome stature of the man she admired most—her grandfather. She was determined never to give herself "a moment's uneasiness on account of any of them." Washington took her announcement with the utmost seriousness and wrote her an earnest letter:

Dear Nellie. . . . men and women feel the same inclinations for each other now that they always have done, and which they will continue to do until there is a new order of things, and you, as others have done, may find, perhaps, that the passions of your sex are easier raised than allayed. Do not therefore boast too soon or too strongly of your insensibility to, or resistance of, its powers. In the composition of the human frame there is a good deal of inflammable matter, however dormant it may lie for a time. . . .

When the fire is beginning to kindle, and your heart growing warm, propound these questions to it: Who is this invader? Have I a competent knowledge of him? Is he a man of good character: a man of sense? For, be assured, a sensible woman can never be happy with a fool. What has been his walk in

*life? Is he a gambler, a spendthrift, or drunkard? Is his fortune sufficient to
maintain me in the manner I have been accustomed to live?*

Finally, Washington urged Nelly to remember that the declaration
of love must come from the man, without any invitation on her part, in
order "to make it permanent and valuable." Her task was to draw the line
"between prudery and coquetry." He had no doubt that she would do this
and find a good husband "when you want and deserve one."[19]

In 1799, Nelly married Captain Lawrence Lewis, son of Washington's
sister, Betty, on Washington's birthday at Mount Vernon. The man she
admired most escorted her up the aisle.[20]

XIV

At last came the day in March of 1797 when Washington's second term
ended, and he and Martha and Nelly returned to Mount Vernon. Sixteen-
year-old George Washington Parke Custis was studying at Princeton and
wrote the ex-president a charming letter, wishing him a happy retirement.
Martha told one of her friends, Lucy Knox, wife of the secretary of war,
that she and the general [a title she preferred to president] "feel like children just released from school or from a hard taskmaster." It was wonderful to have a home again "after being deprived of one for so long."
Nothing would ever tempt them to leave their "sacred roof-tree again,
except . . . private business or pleasure."[21]

The Washingtons were still unpacking when the ex-president opened a
letter from Eliza Powel, the widow of the mayor of Philadelphia. She had
become one of their closest friends. Behind them in their Philadelphia house
they had left several pieces of furniture they could not use in Mount Vernon.
One of them was a handsome rolltop French desk, which Mrs. Powel had
bought. She informed the ex-president that in a secret drawer of the desk she
had found "love letters of a lady addressed to you under the most solemn sanc-
tion; and a large packet too." She teased him for several lines about her shock
to discover that he, a "votary" of the goddess of prudence, who never made a
single mistake as president, could commit such a blunder. Then she informed
him that she was describing "a large bundle of letters from Mrs. Washington,
bound up and labeled with your usual accuracy." She assured him she had not
read a line of them and they would be "kept inviolable until I deliver them."[22]

Washington cheerfully thanked her for rescuing him from what we would call a senior moment. He assured Mrs. Powel that he would have no hesitation in telling Martha about his lapse. She would be as amused by the discovery as he was. If Mrs. Powel had peeked at them, the letters would have been found to be "more fraught with expressions of friendship than of enamoured love." If they had been discovered by someone with inclinations of "the *romantic order*," he would have had to set them on fire to get any warmth from them.[23]

Though his tone was joshing to match Mrs. Powel's arch prose, those lines about friendship were a significant tribute to Martha Custis. In the eighteenth century, a man could choose to be a husband with all the panoply and power that word implied in a world that gave him virtually absolute authority over a wife. Or he could seek and hopefully find in his life's companion a person with whom he could share his deepest hopes and fears and ambitions. Francis Bacon, the English philosopher whose *Essays* exerted a profound influence on the founders' pursuit of fame, wrote: "A man cannot speak to his son but as a father, to his wife but as a husband . . . whereas a friend may speak as the case requires." For George and Martha, the years had deepened the word "friend" until it became a synonym for happiness.

XV

The final years at Mount Vernon were not as tranquil as Martha had hoped they would be. Washington remained involved in the nation's politics. President John Adams, in an unwise act of deference to Washington, retained the members of his predecessor's cabinet and they wrote regularly to the ex-president, frequently expressing their doubts about Adams's unstable presidential style. For a while, a war with Revolutionary France looked possible, and General Washington was summoned from retirement to head a new 10,000-man army. He appointed Alexander Hamilton as his second in command and let him do most of the work. The "Quasi-War," as it came to be called, was fought mostly at sea, and the U.S. Navy soon forced the French to start talking peace.

Washington firmly rejected several overtures to persuade him to accept a third term as president. He had no enthusiasm for another four years of insults and attacks on his reputation. His fame was beyond the reach of carping critics now, and he had no desire to add anything to it. More and more, he was content to be "Farmer George," the master of Mount Vernon. He still worried about his country's future, of course. There were many problems that had to be solved, if the federal union he had given so much time and attention to constructing was to endure. As a step in this direction, when he made his will, he freed all his slaves. He hoped this example might be a signal to other southerners to consider a similar step. He knew that most northern states had begun programs to gradually abolish slavery within their borders.

More than once, when he rode out to visit one of his distant farms, his route would bring him within sight of Belvoir. A fire had reduced the once handsome house to a charred ruin. In the spring of 1798, Washington learned that Bryan Fairfax, Sally's brother-in-law, was going to England to inherit the dukedom. George William Fairfax had lost his lawsuit and died in 1787, a disappointed man. Washington gave Bryan a letter to deliver to Sally, who was living in Bath.

He told her that "many important events have occurred and changes in men and things have taken place" since they parted in 1773. They were too complicated to discuss in a letter. What he wanted to say to her was simpler—and more important. None of those events, "nor all of them together have been able to eradicate from my mind the recollection of those happy moments, the happiest in my life, that I have enjoyed in your company."[24]

In the same packet, he enclosed a letter from Martha. She assured Sally that "although many years have elapsed since I have either received or written [a letter] to you   my affection and regard for you have undergone no diminution." In fact, now that she was "again fixed (I hope for life) at this place" [Mount Vernon], among her greatest regrets was "not having you as a neighbour and companion."

Martha went on to tell Sally that "the changes which have taken place in this county since you left . . . are in one word, total." In Alexandria, there was not a single family left "with whom you had the smallest acquaintance." In their neighborhood, she reported that Colonel George Mason and many other nearby landowners "have left the stage of human life." But they had been replaced by a younger generation, some of them Fairfax relatives. She described their marriages and in some cases early deaths in womanly detail.

Martha devoted a more cheerful paragraph to her own family, noting the marriages and offspring of Jack Custis's two older daughters and her fondness for their siblings, Nelly and Wash, who were still living at Mount Vernon. Martha signed the letter "Your affectionate friend M Washington."[25]

XVI

Life continued serenely at Mount Vernon for the next two years. The usual stream of visitors, some invited, some little more than intruders, flowed through the bedrooms and dining room. At one point, Washington told a correspondent that "unless someone pops in unexpectedly, Mrs. Washington and I will do what has not been [done] by us in nearly 20 years—that is set down to dinner by ourselves."[26] Martha corresponded cheerfully with friends from Philadelphia, especially witty Eliza Powel. They joked about growing old and what the next generation would think of them. In one letter Martha referred to George as "the withered proprietor" of Mount Vernon. Mrs. Powel said she would prefer to visit Mount Vernon in the winter. She did not want to contrast herself with "all the bloom that will pervade that delightful spot" in the spring.[27]

Except for an occasional cold, the Washingtons were in excellent health. When Nelly Custis married at Mount Vernon, she asked George to wear his Revolutionary War general's uniform when he gave her away. It fit him perfectly. Nelly returned from a five-month honeymoon pregnant and gave birth to a daughter in November 1799. The birth was difficult, and the doctor ordered her to stay in bed for several weeks. The weather was cold and rainy, and she did not object. Martha was constantly at her bedside, cooing over the new arrival.

At about 10 a.m. on December 12, 1799, Washington rode out to inspect some of his more distant farms. He was in the midst of reorganizing Mount Vernon to improve the profitability and efficiency of the entire operation. He had not gone far when the weather became vile. A cutting northeast wind brought swirls of snow, then a shower of hail, and settled into an icy rain that gradually turned to snow again. Washington spent about five hours riding through this storm. His heavy greatcoat gave him some protection, but when he strode into Mount Vernon that afternoon, flecks of snow clung to his hair and neck. His secretary, Tobias Lear, noticed this testimony to the storm. But Washington assured him the greatcoat had kept him dry.

Without changing his clothes, he sat down to dinner and chatted cheerfully with Martha and Lear. The next day, he had a cold and the weather was even more unpleasant, with heavy snow whirling in the wind until about 4 p.m. That left another hour of cold, clear daylight, and Washington ventured out on the lawn to mark some trees he wanted to remove. That night after dinner he was hoarse, and Lear urged him to take something for his cold. Washington waved the suggestion aside. He never took anything for a cold. He would "let it go as it came."

That night, about 3 a.m., Washington awoke Martha to tell her he was very ill. He was having difficulty breathing. She immediately began getting out of bed to call a servant. But he forbade her to do this. She had only recently recovered from a cold and the room, with the fire long since out, was icy. She lay beside him, listening to his labored breaths until 7 a.m., when a housemaid arrived to light the fire. Martha told her to summon Tobias Lear, who swiftly perceived that the disease was serious. He sent a servant racing to the Alexandria home of Washington's close friend and personal physician, James Craik.

The doctor arrived around 9 a.m., and his diagnosis was dire. He thought Washington was suffering from "inflammatory quinsy," a deadly form of sore throat. We now know the disorder was probably acute epiglottis, a bacterial infection of this small structure at the base of the tongue at the entrance to the larynx. Craik immediately called two other doctors who practiced in the vicinity and began putting blisters on Washington's throat, hoping to draw the infection to the surface. Martha sat at the foot of the bed, watching the agonizing process, which accomplished nothing.

For the rest of the long day, Martha seldom took her eyes off the man she loved as he fought a losing battle with death. "I die hard," he said to Dr. Craik at one point. "But I am not afraid to go." The doctors tried everything in their limited repertoires, but the disease was beyond their knowledge as well as their skills. Finally, Washington summoned Lear to the head of his bed and told him he did not want to be buried until three days after his death. Lear was too overcome to do anything but nod. "Do you understand me?" Washington asked.

"Yes, sir," Lear replied.

"'Tis well," Washington said.

Those were his last words. About 10 p.m. his breathing became easier. The doctors did not know it, but this was the final stage of acute epiglottis. Tobias Lear took his hand. Dr. Craik sat by the fire, a study in despair. Washington withdrew his hand and began taking his own pulse. Lear called to Dr. Craik, who rushed to the bedside. The hand fell from Washington's wrist. Lear drew it to his breast. Dr. Craik put his hand over Washington's eyes. Forty years of experience told him he was in the presence of death.

No one said a word for a long, sad moment. "Is he gone?" Martha asked in a voice that was amazingly strong and calm.

The weeping Lear could only make a convulsive affirmative gesture. "'Tis well," Martha said. "All is now over. I have no more trials to pass through. I shall soon follow him."[28]

XVII

That same day, Martha closed off the bedroom she had shared with her husband and moved to a smaller room on Mount Vernon's third floor. She lived for another two years, presiding over the mansion with the same dignity and charm she had so regularly displayed as George Washington's wife. She greeted a continuing stream of visitors with the same generous hospitality. Nelly and her husband, Lawrence Lewis, remained with her, as did the faithful Tobias Lear. She retained a lively interest in politics, and frequently startled visitors with her sarcastic comments on President Thomas Jefferson's administration.

In these two years, Martha burned all the letters she had exchanged with her husband. It was a statement of how deeply she valued the private world of love and partnership they had enjoyed for forty years. She had shared George Washington with the larger world, but there was a limit to the public's claims on their journey together. Only four letters survived the curtain Martha drew across their personal life. They were found in her writing desk by a granddaughter, after her death. Two were trivial notes she may have decided were not worth burning. Two were letters that she must have found it impossible to destroy—the ones Washington wrote to "My Dearest" in 1775, telling her of his appointment to command the Continental Army, and five days later, reporting his imminent departure for the camp at Cambridge.[29]

In the first months of 1802, a perceptive visitor, the Reverent Manasseh Cutler, saw deeply into Martha's spirit in the hours he spent with her. "She frequently spoke of the General with great affection," he recalled. "She repeatedly remarked on the distinguished mercies heaven still bestowed on her, for which she had daily cause for gratitude, but she longed for the time to follow her departed friend."[30]

In May of that year, Martha experienced a stomach upset and summoned Dr. Craik. Although it seemed to be just another case of bilious fever to the doctor, she abruptly asked him to stay with her. He remained at Mount Vernon for three weeks while she slowly slipped away. Nelly, Wash, the older daughters, and their children—all the members of her family—were at her bedside. They wept as Martha received holy communion and said farewell to each of them. But their grief was softened by the knowledge that she was almost visibly eager to join the man she had loved so long and so deeply in an eternity of happiness.

THE OTHER GEORGE
WASHINGTON SCANDALS

*L*ong before Sally Fairfax's letters stirred speculations about George Washington's relationships with women beyond his ostensible devotion to Martha Dandridge Custis, newspapers carried stories that suggested he was a womanizer of epic proportions.

The first of these tales appeared in 1775, soon after Washington reached Cambridge to take command of the impromptu army besieging the British in Boston. He found himself plunged into chaos. The army was closer to a mob, living in crude huts and tents with little or no concern for sanitation. There was a shocking shortage of gunpowder, and the haphazard fortifications erected against a British foray from Boston were next to worthless. Washington was soon bombarding Congress with requests for ammunition and trained engineers and complaining mightily about the undisciplined ways of his New England troops.

One of the general's correspondents was Congressman Benjamin Harrison of Virginia. Older than Washington, he was known as a jovial man whose conversation was frequently racy. On August 17, 1775, *The Massachusetts Gazette; and the Boston Weekly News-Letter,* a newspaper published inside the British-occupied city, reported in gleeful terms that a royal navy warship had captured a letter from Harrison that revealed a side of General Washington that might surprise the public.

The letter began with several businesslike paragraphs about the difficulty of finding engineers and Congress's anxiety to locate ammunition for the army. Then the tone abruptly shifted:

> *As I was in the pleasing task of writing to you, a little noise occasioned me to turn my head round and who should appear but pretty little Kate the washer-woman's daughter over the way, clean, trim and rosey as the Morning: I snatch'd the golden glorious opportunity, and but for that accursed antidote to love, Sukey,* [probably a house slave] *I had fitted her for my general against his return. We were obliged to part but not till we had contrived to meet again; if she keeps the appointment I shall relish a week's longer stay—I give you some of these adventures to amuse you and unbend your mind from the cares of war.*

In a recent book on the sex lives of the presidents, the author gleefully accepted this letter as authentic. He stated that Harrison was Washington's procurer in Philadelphia. He noted that the letter was soon published in England, as if this guaranteed its authenticity.[1]

Fortunately for those who prefer their history unflavored by fiction, the story of the letter's seizure and publication can be explored in depth. We know that Harrison actually wrote it and gave it to a young Massachusetts lawyer named Benjamin Hichborn, who was about to leave Philadelphia for Boston. John Adams added two letters to Hichborn's pouch—one for his wife, Abigail, the other for his friend James Warren. Hichborn decided the quickest route home was by way of Rhode Island. On a ferry from Newport to the mainland, he was seized by a boarding party from a British warship and was soon a prisoner in a cell aboard the flagship of the admiral commanding the British fleet in Boston harbor. The bloody battle of Bunker Hill had recently been fought, and the British assumed all-out war was now inevitable.

Hichborn had foolishly hesitated to throw his letters overboard. Ashore in Boston, the British high command read them with interest. All three made it clear that the Americans were getting ready to declare their independence and fight a war to defend it. John Adams assailed various people in Congress who hesitated to take this plunge. The British published both his letters, hoping to sow dissension in the rebels' ranks. But there was nothing noteworthy in Harrison's letter unless it could somehow be improved.

This turned out to be a fairly simple task. On the admiral's staff was a fluent writer named Gefferini who was able to compose the authentic-sounding paragraph about Kate the washerwoman's daughter and put it into the middle of the letter. Why are we sure of this? Because there are copies of the original letter in the Public Record Office in England—without the fraudulent paragraph. General Thomas Gage, commander of the British army, was an old friend of General Washington, and he forwarded the original to his superiors in London without comment or tampering.[2]

II

When the war shifted to New York, Washington found he was in a very different city and state from Massachusetts, where loyalists were only a handful. A committee of the New York legislature worked full time to detect conspiracies and ferret out spies. In June they discovered a loyalist plot to assassinate and/or kidnap General Washington. A member of his guard, Thomas Hickey, was arrested, convicted of treason, and hanged.

In 1777, the British used this plot to go to work on Washington's reputation again. From London came a pamphlet, "Minutes of the Trial and Examination of Certain Persons in the Province of New York." The printer, John Bew, claimed it was based on documents from the files of the New York Assembly committee that was ferreting out loyalist conspiracies. The records had been captured in New York when the Americans retreated from the city in late 1776.

This claim was true up to a point. A comparison of the actual minutes of the committee makes it obvious that Bew had the records on his desk when he wrote the pamphlet. All the witnesses were involved in the trial of Thomas Hickey.

There was no attempt to win sympathy for that unlucky conspirator. The meat of the pamphlet was testimony from two witnesses, who claimed that it would have been far easier to kidnap or assassinate General Washington than was generally believed. Why? The first witness, William Cooper, said that the general was in the habit of visiting a woman named Mary Gibbons, "late at night in disguise." Mary was a spy who passed along everything Washington told her to a loyalist named John Clayford, who often talked about what he learned while drinking with other loyalists at the Serjeants-Arms Inn. One of the things Washington supposedly told Mary was that "he wished his hands were clear of the dirty New Englanders."

Witness two was a soldier named William Savage. He testified that while General Washington snored the sleep of a sexually satisfied man, Mary Gibbons went through his pockets and extracted numerous letters and documents that she slipped to John Clayford for quick copying and clandestine return.

The testimony of these witnesses was inserted into the overall minutes with the same skill displayed by the writer who altered the letter from Benjamin Harrison. The Mary Gibbons story, told with great sincerity and seeming plausibility, acquired a life of its own. It has inspired novels and nonfiction books portraying Mary as one of dozens of women with whom Washington enjoyed voracious sex. Its believability sinks to zero if we recall that Martha Washington was in New York at the time of George's supposedly insatiable visits to Mary. Morever, there is no record of soldiers named William Savage or William Cooper in the American army. Nor has anyone ever heard of John Clayford, outside the pages of Bew's pamphlet.[3]

III

The British pursued Washington's infidelity as a topic in another pamphlet that John Bew published in 1777, titled "Letters from George Washington to Several of his Friends in the year 1776." These letters were supposedly discovered in a satchel carried by Washington's slave, Billy Lee, who was reportedly captured at Fort Lee, New Jersey, when that Hudson River bastion surrendered without a fight. Most of the letters concentrated on Washington's political foibles. He repeatedly insulted New Englanders. He accused them of leaking his military plans to the British. At another point he told a friend that he considered the struggle hopeless, with such despicable allies. He also confessed that he was still loyal to the king and the war was a terrible misunderstanding.

In one letter, the British veered into Washington's private life. Supposedly writing to Martha, he began it with "My Dearest Life and Love" and closed with "Your Most Grateful and Tender Husband." Both are phrases that Washington never used in any surviving letter to her. The letter is dated June 24, 1776—the same period in which he was supposedly enjoying his midnight visits to Mary Gibbons. It is addressed to Martha as if she were at Mount Vernon, and urges her to go to Philadelphia, where she can be inoculated against smallpox. After this advice, the general expatiates on how intensely he is hoping for negotiations with the "English commissioners" and an early "pacification."[4]

We now know that these letters were probably written by John Randolph, the former attorney general of Virginia, who had remained loyal to the king and retreated to London before the fighting war began. He knew Washington well and was in touch with numerous loyalists who picked up the gossip about Washington's dislike of New Englanders. The pamphlet was reprinted in New York by the loyalist newspaper editor James Rivington. When Washington read the letters, he was with his army at Valley Forge. He wrote an outraged letter to Congressman Richard Henry Lee of Virginia, condemning the British for being "governed by no principles that ought to actuate honest men." In 1778, most Americans seemed to agree with this opinion. The letters were also dismissed and generally disbelieved in England, where Washington's reputation remained high throughout the war.[5]

IV

In Washington's second term as president, these 1777 forgeries underwent a rebirth and acquired a following of remarkably virulent proportions. The letters were reprinted in New York and Boston in 1795 and used to argue that Washington had been a secret enemy of numerous patriots during the Revolution, and in many ways had favored the English. This supposedly explained his declaration of neutrality in the war between England and Revolutionary France and his decision to sign a treaty that John Jay had negotiated with the British, which pro-French Americans regarded as a capitulation to London's hegemony. Soon the letters were collected into a book, *Domestic and Confidential Epistles,* with a preface that solemnly declared they would be "regarded as a valuable acquisition by a very great majority of the citizens of the United States."

At first Washington attempted to ignore these exhumed attacks, but they circulated so widely that on his last day in office, he decided to write a letter to Secretary of State Timothy Pickering and "disown them in explicit terms." He considered this act "a justice due to my own character and to posterity." Along with denying he ever wrote the letters, he pointed out that Billy Lee had never been captured during the Revolutionary War, nor had any part of his (Washington's) baggage fallen into enemy hands. Pickering thought the letter was so effective that he released it to the newspapers.

One might think this would have ended the matter, but many booksellers simply printed Washington's letter in the front of the book, apparently convinced that if he took so much trouble to deny the letters, they were probably authentic. As late as 1872 the book was being sold by a rare-book collector, who noted it had the Washington letter in it and commented, "To this day there are writers who from choice or warped moral vision give credit to lies rather than truth."[6]

V

In 1871, an oil portrait of Thomas Posey was unveiled in the Indiana statehouse and a hitherto unnoticed chapter in Washington's early life suddenly became headline news. Thomas Posey was an authentic American hero. He had volunteered for the Continental Army when the Revolution began and had served for the entire war, rising to the rank of lieutenant colonel. He had repeatedly distinguished himself as a battle leader in many bloody clashes. Afterward, he served in the army of the 1790s as a brigadier general and was named territorial governor as Indiana moved toward statehood. Anything written about him was bound to attract attention, at least in the Midwest.

Not long after the portrait was unveiled, the Cincinnati *Daily Commercial* published an article by an unnamed man from Indianapolis asking, "Was Geo. Washington a father?" The writer declared that "none who are acquainted with the evidence . . . doubts the assertion that Posey was the son of George Washington." The accusation was based on a claim that Posey's parents had been tenants on one of Mount Vernon's farms. After his father died in 1754, Posey's mother had a liaison with the then unmarried George Washington and gave birth to Thomas. Thereafter, Washington supervised the boy's education and saw that he had a decent home life when his mother remarried. General Washington named Colonel Posey to his staff during the Revolution and appointed him a brigadier general in the 1790s and finally territorial governor of Indiana.

Papers all over the Midwest ran the article, never bothering to check its wild divergence from easily ascertainable facts. Posey never spent a day on Washington's staff during the Revolution, and the president was dead when Posey became territorial governor of Indiana. Instead, the *Daily Commercial* dispatched a reporter to Indiana. The newsman interviewed numerous unnamed persons who assured him that in "the generation that has passed away," the tradition that Washington was Thomas Posey's father was frequently discussed and widely accepted.

The kernel of truth in the story was Thomas Posey's connection to Mount Vernon. He was the oldest son of Washington's closest neighbor, Captain John Posey, who lived with his wife and five children at Rover's Delight, just west of Washington's plantation. John Posey had served on the frontier with the young Colonel Washington, courageously leading a company of soldiers who specialized in building fortified camps and roads, often under enemy fire. At home, Posey was always ready to join Washington in a fox hunt, followed by a few drinks at Rover's Delight or Mount Vernon. Unfortunately, Posey's few drinks frequently multiplied into many later in the evening and often continued multiplying for several days.

Captain Posey had a steady income from a ferry that he ran across the Potomac to Maryland, but he had no head for keeping track of his money. He began by borrowing small sums from Washington and soon owed him 750 pounds—the equivalent of perhaps $75,000 today. Worse, he could not even pay the interest on the debt. He owed more money to people in Virginia and in Maryland. Washington wrote him strenuous letters, but kept loaning him money even when it became obvious that he was never going to get it back. Meanwhile, Posey and his wife and children were always welcome at Mount Vernon. Their oldest daughter, Milly, became Patsy Custis's favorite companion.

In 1769, as creditors closed in on Posey, Washington advised him to sell his Potomac lands and the ferry contract, pay his debts, and move west. Land was cheap there, and a man could make a fresh start with very little money. Washington frequently rode to Rover's Delight to repeat this advice, but found it was "no easy matter to find the Captn at home and still more difficult to take him in a trim capable of business." Drunk most of the time, the distraught Posey ignored Washington's advice.[7]

Posey's tall nineteen-year-old son, Thomas, grimly aware that his father was bankrupt, took Washington's advice. After a consultation at Mount Vernon, Thomas headed west to Augusta County in Virginia's Shenandoah Valley, a region where Colonel Washington's name still meant a great deal to many people. Washington may have loaned him enough money to make the journey. Thomas also probably obtained letters from his father, introducing him to friends in Staunton, the principal town in the county. The captain had soldiered there during an expedition against the Cherokees in the early 1760s. In 1772, Thomas married the adopted daughter of Staunton's wealthiest merchant and opened a thriving saddlery shop in a nearby town.[8]

By that time, John Posey was a ruined man. He showed up at Mount Vernon to borrow small sums, which Washington noted in his account book as "Charity to Capt. Posey." The ex-colonel remained deeply sympathetic to the plight of Posey's other sons and paid for the education of two of them. In 1774, with Posey's wife dead and his family scattered, the captain wrote a last pathetic letter to his Mount Vernon neighbor, describing himself as "advanc'd in years" and "really not able to work." He thanked Washington for his "many favors" and apologized for being reduced so "very low." Washington invited him to dinner and gave him twelve pounds. That was the last time Washington had any contact with Captain John Posey. Nothing in this account reveals or even suggests that Washington was a former lover of Posey's wife.

VI

The *Daily Commercial*'s story remained dormant for the next fifteen years. In 1886, the St. Louis *Daily Globe-Democrat* suddenly reported that in Shawneetown, in southern Illlinois, where Thomas Posey was buried, virtually every man, woman, and child was convinced that George Washington was Posey's father. The story was based on statements made by some of Posey's descendants and also by a strong resemblance to Washington noted in one of Posey's sons. The reporter dredged up a physical description of Posey, published in 1824, that stressed his six-foot-two-inch height and muscular appearance—supposed proof of Washington's paternity.

The paper followed this story with an interview with a great-grandson of Thomas Posey, a Missouri banker named George Wilson. He dismissed any and all blood connection to Washington. He analyzed portraits of

Posey and Washington, pointing out numerous dissimilarities. Wilson added that Colonel John A. Washington, a grandnephew of the general, had told him there were at least a dozen other people who claimed descent from a Washington-fathered son or daughter. Not one had withstood investigation. Finally, Wilson pointed to a brief autobiographical statement by Posey, in which he wrote that he was born of "respectable parentage." Would he have written this if he were illegitimate?

The story refused to die. It surfaced in other papers in Wisconsin, Ohio, and Missouri. The St. Louis *Daily Globe-Democrat* returned to the fray in 1898 with a four-part story that drew on seemingly reliable research. The source seems to have been an unpublished article by a director of the Smithsonian Institution that propounded the theory that Thomas Posey had been placed in the Posey family by his real father, George Washington. His mother was from "one of the most distinguished families in Virginia." She was the reason Thomas Posey's real identity was never revealed. Backing the argument was another analysis of the supposedly strong physical resemblance between the two men.

Out of nowhere at this point reemerged George Wilson, the chief repudiator of the story, to announce that he was now a believer. He had discovered the name of the woman who had been the partner in Washington's teenage passion—Elisabeth Lloyd. She died giving birth to Thomas, and Washington had arranged for him to be raised by a widow named Posey, a "woman of culture." In this version, the name Posey was a coincidence—he had nothing to do with John Posey and his family. Wilson's story ran in the Indianapolis *News*. Adding to the confusion, Wilson died not long after the story was published, and no one ever found his research. Nor did he tell anyone where he had heard about Elisabeth Lloyd, who remains a mystery woman to this day. All Wilson left were two paintings, one of Posey, the other of Washington, which purported to prove the strong resemblance between father and illegitimate son.

The story nonetheless convinced editors of encyclopedias and historical compendiums such as Revolutionary Soldiers Buried in Illinois to refer to Posey as "reputedly the natural son of George Washington." By the 1920s, John C. Fitzpatrick, director of the Library of Congress and soon to be editor of George Washington's papers, felt compelled to attack the story in *Scribner's Magazine*, dismissing it as fiction. In his biography of Washington, Fitzpatrick returned to the attack, noting Washington had befriended and supported several of Posey's sons. Did this mean he was also their father? he asked mockingly. He also noted that Washington financed the education of many other young men. "If every child whose education was assisted by Washington were to be stigmatized," Fitzpatrick wrote wryly, "the distinction of being The Father of His Country might take on a new meaning."

Fitzpatrick was particularly hard on those whose arguments were based on "alleged physical resemblance." He called it "the quintessence of inexcusable credulity." He also took aim at a letter Washington wrote to Posey during the Revolution, which according to some people began with "My Dear Son." The letter did not contain another personal reference. It was all military business, ending with the usual "Your most obdt. & humble Servant." When Fitzpatrick examined the original letter, he saw that the claimants were unfamiliar with "one of Washington's pen characteristics." It was sometimes hard to distinguish his "word-ending letters"— understandable because he was often writing in haste. A close look at the letter reveals it began with "My Dear Sir."[9]

This is not the end of the story. A recent biographer of Thomas Posey points out that if he was born, as most people agree, in 1750, it is unlikely that John Posey was his father. Posey did not marry Martha Harrison, Thomas's putative mother, until 1752. This suggests that Thomas may have been born of another woman and taken into the Posey family. How eighteen-year-old George Washington, with very little money and not an iota of fame, could have managed this arrangement remains unexplored. But the secret might explain Washington's "extraordinary liberality" to Captain Posey and his family. The writer stubbornly ignores Fitzpatrick's observation that Washington was frequently a generous supporter of the sons of cash-short friends.

Nevertheless, the biographer, a descendant of Thomas Posey, presses on. He cites Washington's early love letters, in which he moaned about his passion for a "lowland beauty" and for another "agreeable young woman" who added "fuel to the fire." The author recounts Thomas Posey's similar physical appearance, his courage under fire, and his undoubted gift for leadership. "Both [Washington and Posey] died at the same age, of similar causes," he writes. But he is forced to conclude that there is no hard evidence to prove Washington's paternity. The best he can say is that "those who choose to do so may perhaps be forgiven if they continue to believe that Thomas Posey was really the son of George Washington."[10]

VII

After he returned to Mount Vernon in the final days of 1783, Washington devoted most of the next two years to reviving the plantation's commercial vitality. Almost as absorbing was the stream of visitors who took advantage of his hospitality to have a meal with the man who was now considered the greatest living American. Yet in these same years some people maintain that Washington fathered a boy named West Ford by a slave named Venus, who lived at Bushfield, his brother Jack's plantation, ninety-five miles away. The child was born sometime in the year 1784 or 1785.

No mention of the boy as George Washington's son has been found in any letter, diary, memoir, or newspaper before Washington's death in 1799. West Ford never visited Mount Vernon until 1802, when Martha Washington died and the plantation was inherited by Bushrod Washington, Jack's son.

Bushrod brought West Ford and a number of slaves from Bushfield. By that time, Ford was a free man. Bushrod's mother, Hannah, freed him in her will in 1801. From that time until his death in 1863, Ford was a fixture at Mount Vernon, a sort of combination servant and caretaker. In 1850, historian Benson Lossing, who "discovered" the forged letter George supposedly wrote to Martha in 1758, interviewed him. He drew a sketch of Ford, in which some people later found a strong resemblance to George Washington. But Lossing made no claim of paternity in his written account of the interview.

Not until the closing decade of the twentieth century did many people take seriously the claim that West Ford was George Washington's son. Linda Allen Bryant and Janet Allen, descendants of Ford, launched a media campaign that won widespread attention in newspapers and on television. Stories ran in *USA Today,* the *Chicago Tribune*, the *St. Louis Post-Dispatch*, and the *New York Times,* quoting them and discussing the validity of their claim. The sisters created a website, and in 2001 Bryant published a book, *I Cannot Tell A Lie, The True Story of George Washington's African American Descendants.* In her preface, she explained why the book is a historical novel rather than a nonfiction work backed by documents and footnotes: "This format allowed me to relay my heritage in the way it was passed down through the generations by the Ford chroniclers."[11]

In the narrative's first chapter, seventeen-year-old Venus recalls the night at Bushfield that "Master John"—Washington's brother—told her that Master George "needs comforting and has asked for you."

Venus replied that she would "go and light the fire and warm some bricks for his bed."

"Ah, Master George needs warming of another kind," Jack replied.[12]

After that first fateful night, Venus supposedly became Master George's preferred bedmate whenever he visited Bushfield. When she accompanied Hannah Washington, Jack's wife, to Mount Vernon, Master George enjoyed her there, too, until she became pregnant. Thereafter he "left her alone." When the child was born, Venus decided not to christen him George. "Master George Washington was too politically important to have scandal attached to his name," she thought. Also, in her mother's words, "the responsibilities of commanding an army had made him sterner, almost unapproachable." So Venus called the baby "West."[13]

Later in the book, Mrs. Bryant describes how Washington took a special interest in West. At the age of four, he became Washington's "personal attendant" when he visited Bushfield. He "would fetch and carry and do all kinds of small errands" for him. He sat beside the general on "wagon rides" around the countryside and even accompanied him to church.

One day, after spending some time with Master George, West asked his mother, "Mamma, is the old General my papa?"

At first Venus was panicky. She wanted to know "who done told you that?" West said Bushfield's cook and her helpers talked about it all the time. They said he looked like the General. Venus realized "the [whole] slave population at Bushfield and Mount Vernon knew. No news could escape the slave telegraph."

Venus studied her son's chestnut-colored hair, put her arms around him, and said, "The Old General be your papa." But she warned him to tell no one, for the time being. "One day you can tell your children but for now it be our secret."[14]

It is a touching story, but facts show it to be fatally flawed. The assumption that Washington paid many visits to Bushfield in the years after he returned from the war is refuted by the detailed information we have for his whereabouts almost every day, thanks to his diary and account books. In fact, there is no documented evidence that he visited Bushfield even once in the years between his return from the war in 1783 and Jack Washington's death in 1787. Hannah Washington visited Mount Vernon once during this period, and Venus may have accompanied her. But the notion that George would ask for Venus at Mount Vernon, where Martha shared his bedroom and the house was filled with visitors eager to ogle the most famous man in the country, is difficult to accept. Also, the Ford family oral tradition clearly identifies Bushfield as the site of the supposed tryst.[15]

There would seem to be little doubt that West Ford had Washington blood in his veins. But it was probably inherited from Jack Washington or one of his three sons, Bushrod, Corbin, or William Augustine. The latter died around the time West was conceived. Ford's emancipation and the consideration with which he was treated by Bushrod Washington, who left him over a hundred acres of land in his will, is not untypical of how many southern planters attempted to provide for their own or their family's mulatto slave children. Equally familiar is the tendency of a slave mother to tell her mulatto son or daughter that their real father was "old master" or someone even more distinguished, rather than the overseer or a temporary white workman, or some nameless white guest to whom the master had given access to his slaves, in what some considered the great tradition of southern hospitality.[16]

VIII

The same faux historian who described Congressman Benjamin Harrison as Washington's procurer claims that the sixty-eight-year-old ex-president caught the cold that led to his death while jumping out a back window with his trousers in his hand after an assignation with an overseer's wife. This is a story that John C. Fitzpatrick labeled "the most nebulous of all the slanders," but over the years it "gathered its strength from mere repetition." It is an extreme example of the things people have believed—and some continue to believe—about George Washington's imaginary love life.

Ultimately, the George Washington pseudo-scandals are a cautionary tale about the way fame attracts this sort of defamation from those who want to believe the worst about a great man. Their reasons run the gamut from envy to political partisanship to revenge for perceived or imagined wrongs.[17]

BOOK TWO

 Benjamin Franklin

THE SINS OF THE FATHER

A young man named Daniel Fisher, who worked as a clerk for Benjamin Franklin and lived in his house on lower Market Street in Philadelphia, later told a revealing story about his employer's marriage. Fisher was relaxing in a downstairs room, perhaps reading a copy of Ben's newspaper, *The Pennsylvania Gazette*. Ben's wife, Deborah, was sitting a few feet away. William Franklin, Ben's illegitimate son, walked through the room. He did not even glance at his stepmother or say a word to her. As Fisher put it, he did not make "the least compliment to Mrs. Franklin."

As William left the room, Deborah Franklin exclaimed, "Mr. Fisher, there goes the greatest villain on earth." Fisher could only stare at her in amazement. He thought of William as a relatively harmless young man. Deborah proceeded to denounce William in "the foulest terms I ever heard from a gentlewoman," Fisher said. In a few months, the clerk quit his job and went elsewhere. There were too many times when Deborah's "turbulent temper" hurled volleys of abuse at him as well as at William.[1]

Deborah was also violently jealous of the long hours Franklin devoted to the politics of Pennsylvania. She once bitterly complained to Fisher, "All the world claimed a privilege of troubling her Pappy." She grew even more unhappy when Franklin plunged into the worldwide fad for studying electricity and made a host of original discoveries, climaxed by proving lightning was a form of electricity and could be tamed by the judicious use of lightning rods. Franklin proved his case by reporting that he had flown a kite in a thunderstorm and was able to measure the electricity that flowed from a metal key attached to the string. The discovery won him worldwide celebrity. Few people knew that the person who had risked his life to fly the kite was William Franklin. The triumph cemented the already strong bond between father and son and further inflamed Deborah Franklin's anger. The discovery accelerated the Franklins' pursuit of fame in a sphere that both Benjamin and William found fascinating and Deborah acrimoniously continued to dislike: politics.[2]

II

The oldest of the founding fathers, Benjamin Franklin was born in 1706 in Boston. His father, Josiah Franklin, had emigrated from the mother county in 1683 after becoming disillusioned with the flaccid Christianity of the Church of England. He was fifty years old when Ben was born. Josiah earned his living with his hands, making soap and candles. Between Josiah and Benjamin stood fifteen older children by two wives. Two more children followed Benjamin. As a boy, Franklin recalled thirteen people at the supper table.

Benjamin's mother, Abiah Folger Franklin, seems to have managed this large ménage with skill and warmth. His later letters to her, addressed to "Honoured Mother," glow with affection. In his autobiography, he remarked that she had an excellent constitution and suckled all ten of her children. Benjamin was extremely proud of her Folger ancestors, who had settled Nantucket—especially Peter Folger, who wrote a long verse essay defending liberty of conscience in religious matters. Benjamin seems to have inherited his sunny disposition from his mother. More important, she inspired his lifelong fondness for women.

At the age of seventeen, after several years as a restless apprentice printer to an older brother, Ben decamped from Boston to Philadelphia. The city delighted Franklin from the day he arrived. Its Quaker rulers were religious men, but they did not impose their beliefs on everyone in the rancorous style of Boston's puritans. Philadelphia was equally delighted with young Franklin. His engaging personality charmed everyone he met. In this atmosphere of easygoing bonhomie, Ben soon lost touch with his pious parents and the moral and spiritual creed they had taught him.

In Boston, Franklin had devoted his spare time to books and became a freethinker, liberated from the religion of his boyhood. He gave little time to another important subject, women. In Philadelphia, Franklin was attracted to Deborah Read, his landlord's buxom daughter. "I had a great respect and affection for her and had some reason to believe she felt the same for me," he later recalled. But her widowed mother objected to an early marriage. Franklin was still only a journeyman printer, working for weekly wages.

Stung, Franklin accepted an offer from William Keith, the governor of Pennsylvania, to go to London and buy a printing press and other equipment to start a newspaper. The governor, also seemingly charmed by the clever young Bostonian, would provide the credit. Deborah and Benjamin "interchanged some promises" before he sailed. This semi-engagement was all that her mother would tolerate. She may have known enough about Governor Keith to make her doubt that they would see Benjamin again.

In the imperial capital, Benjamin discovered that the governor had no credit and had sent him on a fool's errand. Keith was one of those men who wanted to please everybody and was locally famous for making promises he could not fulfill. Benjamin proceeded to break Deborah's heart by informing her that he might never return to Philadelphia. Finding work as a printer's assistant in London, he and a Philadelphia friend, James Ralph, a would-be poet who had sailed with him, proceeded to enjoy themselves in a city where women of pleasure swarmed the streets and some shopgirls were ready to give themselves to a man who bought them a drink and dinner.

Over the course of a year of dissipation, Benjamin loaned James Ralph almost thirty pounds—the equivalent, today, of about $2,500. Benjamin decided this generosity entitled him to enjoy Ralph's mistress. When the lady informed Ralph, the yet unpublished poet angrily told Ben that he was never going to see his money. Around the same time, it dawned on Franklin that he was unlikely to become more than a printer's helper in London, where the business was controlled by men of wealth. His only hope of independence was a return to Philadelphia. Grimly, he settled down to daily toil at low wages, saving a few shillings a week to pay his passage back to America. He ate only the simplest food and avoided all amusements.

When he returned to Philadelphia eighteen months later, Benjamin found more disarray. Deborah Read had made a bad marriage to a potter named John Rogers, a spendthrift who ran through her dowry and then mistreated her. Rogers had turned out to be a bigamist in the bargain, with a wife in London. He capped matters by fleeing to the West Indies to escape his creditors. Deborah was living forlornly with her mother, too disconsolate to face anyone but her family.

Franklin's conscience bothered him acutely, but there was little he could do about the situation as he struggled to make a fresh start as a printer. He also found it difficult to control what he later called "that hard to be governed passion of youth." His "intrigues with low women" cost him money and were frequently "inconvenient"—he meant embarrassing. A man struggling to start a business needed a good reputation. There was also the danger of catching a venereal disease, a possibility that repeatedly filled Franklin with dread.

The more Ben thought about his life, the more he began to suspect that his pious parents had some worthwhile ideas after all. He also grew critical of his freethinking friends—and of himself. He noted that James Ralph, whom he had converted to a freethinker, had felt no compunction about cheating him out of his money—and had incidentally abandoned a Philadelphia wife and child to stay in London (where he eventually became the first American-born professional writer). Several other friends who had joined Benjamin in irreligion had also welshed on their debts to him. He himself, he ruthlessly concluded in this spasm of insight, had behaved with equal lack of decency toward Deborah Read.

What did it mean? Benjamin could only conclude that "truth, sincerity and integrity" in dealings between people were of the "utmost importance" to a man's happiness. While he could not join his father and mother in practicing these virtues as commandments of God, he vowed in his journal to observe them as long as he lived. He had learned the hard way that virtues were good in themselves.[3]

III

Taking on a partner and borrowing money from his father, Ben set up as an independent printer. After more vicissitudes with the partner, who turned out to be a drunk, Ben shed him, started a newspaper, and seemed on his way to success. But he still found it difficult to control that "hard to be governed passion of youth." Deciding he needed a wife, he at first tried to find one with a dowry. But printing was not considered a prosperous trade, and he met with several humiliating rebuffs.

Meanwhile, Franklin's ungovernable sex drive had presented him with a problem that threatened to be extremely inconvenient. One of the lower-class women with whom Ben satisfied his desires presented him with a son. Here the record grows murky—and Franklin the autobiographer evasive. Scholars have spilled a lot of ink trying to identify this woman and what happened to her. In 1779, an American loyalist wrote a hostile sketch of Franklin for a London newspaper. He claimed the woman was "an oyster wench in Philadelphia whom he left to die in the streets of disease and hunger." Another hostile sketch, written by a political enemy in 1764, described her as "his hand maid Barbara" whom he allowed his wife and daughter to mistreat abominably. But most scholars have concluded it would have been impossible for Ben to have hired his illegitimate son's mother in the glare of Deborah's hostility. The most probable description comes from a letter written by a Franklin friend around this time to an acquaintance in England: "Tis generally known," he wrote, that the mother "is not in good circumstances." But the "report of her begging bread in the streets" was untrue. "Some small provision is made by him (BF) for her, but her not being one of the most agreeable of women prevents particular notice being shown, or the father and son acknowledging any connection with her."[4]

One thing is certain: at some point in the year 1730 Benjamin turned to Deborah Read, who was still "generally dejected and seldom cheerful." He explored with her the possibility of becoming man and wife. There were serious problems to be solved on her side of the equation. If her husband, John Rogers, returned from the West Indies—where rumor had him killed in a drunken brawl—Deborah could be charged with bigamy. That crime carried a life sentence in 1730 Pennsylvania. Rogers still owed money to several people in Philadelphia and if Benjamin and Deborah married in a formal way, Ben could be sued for those debts. Franklin declared himself ready to accept a common-law arrangement, whereby Deborah would simply move in with him and set up a household. That would keep the debtors at bay—and if Rogers showed up they could claim they were committing nothing more heinous than adultery.

It would seem more than probable that, at this point, Franklin displayed the first sign of those skills that later made him a master diplomat. Having declared himself ready to risk bigamy and lawsuits on her behalf, Benjamin now wondered whether Deborah would be willing to do something for him. It must have been hard for the grateful young woman to imagine anything she was not ready to do to please this ingenious young man who was rescuing her from a dismal future as an abandoned wife.

Raise my illegitimate son as your own child, Franklin said.

How could Deborah Read say no? She may have been momentarily dismayed to discover that Franklin was not a paragon of virtue or a lovelorn swain who had been pining for her with anguished fidelity all these years. But her encounter with John Rogers and her observations of street life in Philadelphia must have left her with few illusions about the perfectibility of mankind. Deborah agreed to the bargain. She and the infant, soon named William Franklin, began sharing Benjamin's house on lower Market Street.

IV

Deborah's dislike of William Franklin probably began on the day her husband persuaded her to take the boy into their house. But her jealousy became acute when their only son, Francis Folger Franklin, died of smallpox in 1736, at the age of four. She later bore Franklin a daughter, Sarah, who had a strong resemblance to her big, bustling mother. Daniel Fisher reported that in one of Deborah's rants she accused Franklin of having "too great an esteem for his son" and far less warm feelings for her and their daughter. There is no doubt that Sarah (soon nicknamed Sally) was never able to compete with William for their father's affections. William inherited his father's brains and was remarkably handsome in the bargain.

William's importance grew larger for Ben after he achieved wealth as a newspaper publisher and fame as a scientist. He began envisioning a distinguished family, a line of Franklins who would play a dynamic role in their nation's history. William helped fuel this dream by displaying talent as a politician and a soldier. Ben's pride in his accomplishments only added to the tension between husband and wife.

Ben did his best to cope with his short-tempered, resentful spouse. He repeatedly paid tribute to the way she had helped him survive his precarious early years, when he was struggling to launch his newspaper and the print shop and general store attached to it. She "assisted me cheerfully in my business, folding and stitching pamphlets, tending shop, purchasing old linen rags for the papermakers etc etc." Thanks to Deborah's skill with a needle, for a while he was "clothed head to foot in woolen and linen of my wife's manufacture."

In the store, where Deborah presided, she did a profitable business selling everything from spectacles to sealing wax to an ointment made by her mother that was guaranteed to cure "the most inveterate itch." Another profitable item was Crown Soap, made in Boston by Franklin's older brother, John. Even more of a moneymaker was the annual edition of *Poor Richard's Almanac,* a book in which Ben mingled wise and witty aphorisms and weather predictions. Deborah called it "Poor Dick."

Deborah kept the books for the business, a task Franklin admitted was beyond him. In spite of his fondness for telling everyone that a penny saved was a penny earned, frugality did not come naturally to his expansive nature. In his autobiography he remarked: "We have an English proverb that says, 'He that would thrive must ask his wife'; it was lucky for me that I had one as much disposed to industry and frugality as myself . . . We kept no idle servants." He was soon declaring that his wife "became a fortune to me." There was not the slightest doubt in his mind that a money-wise wife was the key to a prosperous marriage. "What we get the women save."

For eighteen years Ben and Deborah were partners in a constantly expanding business. The store's inventory eventually included goose feathers, Rhode Island cheese, Franklin-designed stoves, and lottery tickets. Even slaves flowed in and out of its busy doors. Quakers were only beginning to question the morality of slavery during these years. Like other shopkeepers in Philadelphia, Deborah opened at 5 a.m. and did business until darkness fell. The sales of *Poor Richard's Almanac* topped ten thousand a year—the equivalent in today's vastly more populous America of a million and a half copies annually.

Without Deborah's help Ben could never have combined success as a newspaper and almanac publisher with politics, which enabled him to acquire the plum job of postmaster of Philadelphia. That entitled him to send copies of the *Pennsylvania Gazette* through the mail free. More important, hundreds of people came to the print shop and store, because it was the city's post office.

During these years, the limitations as well as the advantages of their marriage slowly became apparent to both Franklins. Deborah remained the same poorly educated shopgirl who had married Ben in 1730. Her letters to him were a blizzard of misspellings. Ben was becoming a political and scientific thinker of world-class proportions. When he wrote a letter to Deborah, he addressed her as "My Dear Child." There was affection in the words, but also more than a hint of paternal superiority.

In this unspoken drama, Ben played both a villain and a savior's role. If he had remained faithful to his original promise to Deborah, she would not have married Rogers and undergone the public humiliation of discovering he was both a bigamist and a thief. She also undoubtedly knew, thanks to the gossip that pervaded the small business world of Philadelphia, that Ben had shopped around for a better marriage deal before retreating to her. These memories would not contribute to any woman's peace of mind—especially one born with a hot temper.

By 1748, the Franklins were earning 2,000 pounds a year—the equivalent of about 300,000 modern dollars. Only very successful merchants and prominent lawyers made that much money. Franklin decided he and Deborah could relax. They closed the shop and moved to a bigger, more comfortable house. Ben handed the *Pennsylvania Gazette* to his well-trained assistant, David Hall, who agreed to pay him 650 pounds a year as his share of its profits.

It was the beginning of a new life for Ben, with ever-widening intellectual and political horizons. He resigned as postmaster of Philadelphia, making William his successor, and became deputy postmaster general for America, a job that required him to travel throughout the thirteen burgeoning colonies. Deborah never went with him. She did not share Ben's exuberant delight in meeting new people and exploring distant colonies and towns. She preferred familiar Philadelphia, with the same friends and relatives she had known since girlhood.

We have only one portrait of Deborah, the work of an unknown Philadelphia painter. She looks prosperous; her dress is expensive and she has an ornament in her hair. Her fleshy face has not a hint of refinement, but she looks fiercely determined to be herself and deal with life's problems on her own terms. It does not take much effort to imagine her brow furrowing, her thick lips curling, and angry words exploding from them. Her figure verges on the plump—which did not displease her husband. When Ben's penchant for travel took him to England again, he sent Deborah a large jug with a note explaining that he thought it resembled "a fat jolly dame, clean and tidy . . . and just put me in mind of—somebody."

Franklin never faulted Deborah for her stubborn refusal to change her ways or broaden her interests. "Don't you know a wife is always right?" he wrote wryly to a friend. In *Poor Richard's Almanac* he published a song that paid tribute to his hardworking helpmate. It also inadvertently summed up their marriage:

My Plain Country Joan

Of their Chloes and Phillises poets may prate
I sing of my plain country Joan
Now twelve years my wife, still the joy of my life
Blest day that I made her my own.
My dear friends
Blest day that I made her my own.
Not a word of her face, of her shape, or her eyes
Or of flames or darts you shall hear:
Tho' beauty I admire, tis virtue I prize,
That fades not in seventy year.
My dear friends . . .
Some faults have we all, and so may my Joan
But then they're exceedingly small
And now I'm so us'd to 'em, they're just like my own
I scarcely can see them at all.
My dear friends . . .
Were the finest young princess, with million in purse
To be had in exchange for my Joan
She could not be a better wife, mought be a worse,
So I'll stick to my Joggy alone
My dear friends
I'd cling to my lovely ould Joan.[5]

V

Franklin's trips as deputy postmaster general often took him to Boston, where he stayed in the mansion of his brother John, who had grown wealthy as a soap maker. There, in 1754, Franklin met a slim, flirtatious twenty-three-year-old brunette named Catherine Ray. She had grown up on isolated Block Island, the child of older parents, and was hugely excited by this opportunity to visit bustling Boston and meet an already famous American.

Catherine took delicious pleasure in tormenting men with her dancing eyes and low-cut gowns. Simultaneously she proclaimed her pride in her virginity, which she was determined to yield only to a man of surpassing charm and ability. The forty-nine-year-old Franklin found himself mesmerized. For the first time, Ben met a woman whose beauty and seeming availability made him realize a dimension of love that he had never encountered in his relationships with the lower class women of his early amours—or in his practical, self-interested marriage to Deborah.

This deeper attraction, known today as romantic love, was (and is) ancient and forever modern. It was just beginning to emerge in Europe as an experience that plumbed the depths and heights of human emotions and sometimes involved the surrender of a man or woman's soul.

Ben plunged into a half real, half make-believe dance of desire and mutual delight. He and Katy, as he began calling her, talked about sex and love with a candor that came naturally to their era. Playing the sorcerer, a role that came naturally to him, Ben invented a game that required Katy to tell him every detail of her previous loves. He warned her that if she held anything back, he had powers that would enable him to penetrate her deception. Katy's eyes gleamed with delight; she poured out her youthful heart to Franklin.

One night the sorcerer revealed he was human. He suggested that he could become a lover who would send Katy into transports beyond imaginative shivers. Katy recoiled and reminded the sorcerer that he was a married man. With a sigh, Franklin retreated to a bittersweet affection in which words remained ardent but actions were firmly within the bounds of propriety. His rational head had prevailed over his wayward heart.

After they parted, and Ben returned to Philadelphia, he wrote Katy a wry letter describing the experience from his side. For a while, on his return journey, he "almost forgot I had a home." He was lost in romantic visions of an ecstatic sojourn with Katy in some little-known inn or forest bower. But as he drew closer to his adopted city, he began driving his horse faster and faster until "a very few days brought me to my own house and to the arms of my good old wife and children."[6]

For a while their letters continued to glow with suppressed passion. Katy sought his advice on new affairs of the heart, and Franklin responded with warnings that she had better not appear "too knowing." When Katy told him she would send him kisses on the wind, Ben replied that northeast storms, which traveled south from New England—one of his early scientific discoveries—were often blamed for depressing people. Now their gray clouds and freezing winds would no longer bother him. "Your favors come mixed with the snowy fleeces which are as pure as your virgin innocence, white as your lovely bosom—and as cold."[7]

Ben added that Deborah was aware of their correspondence and threatened to leave him to Katy in her will. "She is willing I should love you as much as you are willing to be loved by me." The sorcerer was wryly reminding Katy that in the arithmetic of ordinary love, this added up to very little. But in the emerging world of romantic love, there were exquisite gradations of emotion that approached but did not reach physical expression yet could mean a great deal on both sides. Both Ben and Katy sensed that they had experienced feelings neither would forget.[8]

VI

Absorbed in politics and business, Ben Franklin did little to protect William from his stepmother's antagonism. At the age of fifteen, the boy advertised his unhappiness by trying to run away to sea. Ben heard about it before the ship sailed, rushed to the docks, and obtained William's release from the captain. He protested in a letter to his sister Jane that "no one imagined it was hard use at home that made him do this." The remark suggests that the busy father was not paying much attention to William's trials with Deborah. But the attempted escape troubled him nevertheless. One of Ben's older brothers had run away to sea and had never been heard from again.[9]

Ben obtained a commission for William in a Pennsylvania regiment formed to help fight the French in Canada during a colonial clash called

King George's War. William liked being a soldier and won a promotion to captain, but the war ended before he saw any battlefield action. Back home, he persuaded his father to let him move out of the house to a nearby set of rooms beyond the reach of his stepmother's harangues.

William's military bearing and aura of adventure, plus his good looks, enabled him to circulate with ease in the upper echelon of Philadelphia society. When Ben was elected to the Pennsylvania Assembly, he made sure William was appointed to his former job as clerk. He also made him controller of the North American postal system. No one disapproved; in the eighteenth century it was customary to appoint relatives to political jobs. But Ben had larger ambitions for his son. He arranged for him to begin studying law with Joseph Galloway, one of Philadelphia's best younger attorneys and a staunch political ally. He also wrote to friends in England, asking them to register William at the Inns of Court, where the elite of the British legal profession studied the intricacies of common law.

Meanwhile, William had become a member of Philadelphia's exclusive Annual Assembly, which staged balls that attracted the wealthiest and most attractive young women in the city. Soon he was in love with Elizabeth Graeme, the vivacious, intelligent daughter of one of the city's elite families. Their sumptuous country house, Graeme Park, was one of the showplaces of the Middle Colonies. Unfortunately, her pompous father violently opposed the match. Dr. Graeme was not put off by William's illegitimate birth; he had sired two such "byblows" in his youth. But he did not like Ben's Pennsylvania politics. Franklin invariably opposed the greedy, self-interested policies of the governors the Penn family appointed to run the colony. Elizabeth tended to side with her father, further complicating matters.

William soon revealed he was a romantic, eager to play the game of longing sighs and barely restrained passion. He wrote letters to his "dear Tormentor" and thanked her for her "agreeable vexatious [and undoubtedly flirtatious] little billet." At one point, things grew so heated, they almost eloped. But William still had no serious means of support, and the lovers decided not to risk her father's wrath. Ben apparently watched the burgeoning romance with bemused interest. His handsome son was enjoying an approach to love and marriage that had been beyond the reach of Ben's cash-strapped youth. It did not bother him that William was spending his father's money on fashionable clothes and books such as *The True Conduct of People of Quality.*[10]

Ben's politics remained strongly populist. He began calling for the dismissal of the heirs of William Penn as the colony's proprietors. The Penns repeatedly ignored the resolutions of the assembly, and their appointed governors frequently vetoed them. Franklin's stance made him even more obnoxious to the Graemes.

The assembly voted to send Franklin to England to plead their case with the British government. He took William along as his secretary, telling him, as an added inducement, that he could complete his law studies in London. Ben may have been hoping to cool the romance with Elizabeth Graeme, but before he departed, William became secretly—and perhaps defiantly—engaged to his dear tormentor.

Ben urged Deborah to accompany them to Britain. Her response was a vociferous *NO*. Like many people of the time, she had an obsessive fear of sea voyages. The thought of spending six weeks or two months on a ship with her detested stepson no doubt also played a part in her emphatic refusal. Ben warned her that they might be gone for several years, but Deborah still refused. It was the beginning of the end of their marriage. By this time, Deborah may have sensed that she was still Ben's wife in name only. No longer a partner in their day-to-day business, she had little or nothing to share with him.

VII

Both Ben and William were delighted with London. The imperial capital was twice the size and ten times as wealthy as the city Ben had visited as a teenager in the 1720s. It spread along the north bank of the Thames for three miles, wrapped in sooty smoke that gave it, on cloudy days, an infernal appearance. On the south bank of the river, the town of Southwark was rapidly matching its parent; it would eventually become part of Greater London.

The warm welcome Ben received from scholars and newspapermen with whom he had corresponded pleased him immensely. Universities gave him honorary degrees and members of the Royal Society, the elite of British science, rushed to shake his hand. Newspapers published his articles. This surge of admiration confirmed Ben's conviction that England and America were one country, part of the triumphant British empire, which was on its way to ruling the globe. Father and son rented rooms in the house of a charming well-to-do widow, Mrs. Margaret Stevenson, on Craven Street, just off the busy Strand, in the heart of the city. Her household included a pretty teenage daughter, Polly.

William reveled in London's "infinite variety" and exulted in accompanying his father to dinners and receptions with "politicians, philosophers and men of business." Within a few months, his letters to Elizabeth Graeme dwindled to brief scribbles. Elizabeth sent William an enraged letter, breaking their engagement. She accused him of dumping her because of their fathers' political antagonism. William seems to have felt no more than a tremor of regret. He had enrolled at the Inns of Court and was studying law while working closely with his father in his political war with the Penns.

Not long after he arrived, Ben fell seriously ill with a cold that seemed to verge on pneumonia. London's polluted air probably worsened his condition. For eight weeks he wheezed and coughed and often struggled for breath. Mrs. Stevenson nursed him day and night, winning his heartfelt gratitude. William was left in charge of their political operations, and impressed everyone with his energy and intelligence. He represented Ben at Court, where he made his bow to the king and his courtiers and ministers. When the Penns unleashed hired hacks to smear Franklin and the assembly in the newspapers, William wrote a blazing reply that was printed in three papers and in the prestigious *Gentleman's Magazine*. Wealthy publisher William Strahan, Franklin's closest friend in London, praised William's "solidity of judgment" and declared that he was one of the most gifted young Americans he had seen in years.

The admiring Strahan remarked that William had become his father's "friend, his brother, his intimate, and easy companion." Over the next few years, this intimacy grew even more intense. Father and son traveled together to Scotland and to France. They visited the home village of the English Franklins and discovered an aged relative who knew Ben's father before he emigrated. They found a Franklin of the previous generation who was a local leader with an uncanny resemblance to Ben. Their pride in their English heritage expanded exponentially.[11]

Unaware of Deborah Franklin's hatred of William, the jovial Strahan launched a campaign to persuade her to join Ben in London with her daughter, Sarah. Along with enthusiastic descriptions of William's warm relationship with his father, the publisher mentioned that he had a son around Sarah's age whom she might marry. He also warned Deborah that the "ladies of London" liked her "amiable" husband as much as his men friends. He described how Mrs. Stevenson had nursed Ben during his "severe cold . . . with an assiduity, concern and tenderness which perhaps only yourself could equal." Strahan urged Deborah to risk the Atlantic to protect her "interests." The printer assured her that Ben was "as faithful to his Joan as any man breathing." But "who knows what repeated and strong temptation may in time . . . accomplish?"[12]

VIII

On Craven Street, the first stage of what would become Ben Franklin's transatlantic divorce was already in progress. If he was not yet Mrs. Stevenson's London husband, he was moving in that direction. He took over her house, setting up his electrical apparatus and giving demonstrations of his scientific prowess to dozens of friends and acquaintances. He invited people to musical evenings at which he played the harp or violin in a quartet. Then there were his air baths. Every morning Ben rose at dawn, shed his calico bed gown and flannel night trousers, and sat around for an hour totally naked. He did this summer and winter; occasionally he strolled the Stevensons' secluded garden in the altogether. He was convinced that fresh air was good for his health. Not many landladies would have tolerated such a routine.

Deborah declined William Strahan's invitation to London, as her husband knew she would. Groping for a compliment, Ben wrote he was "pleased" with her answer. She had not succumbed to Strahan's "rhetoric and art"—even though her stubbornness left her three thousand miles away. Ben made no secret of Mrs. Stevenson's presence in his life. He praised her as "very diligent" when he was "indisposed" but added that he still wished Deborah were with him, along with "my little Sally." Only "sincere love" inspired a woman to nurse a sick man with "tender attention." Meanwhile, more and more London friends were treating Ben and Mrs. Stevenson as a couple. They were regularly invited to dinner parties together.

Ben did everything in his power to reassure Deborah of his continuing affection. He found a Book of Common Prayer in large type that would save her the trouble of wearing spectacles in church. He shipped her thousands of dollars worth of clothes, carpeting, china, silverware, bedding, and tablecloths. The goods arrived by the crate in ship after ship. But when Ben's sojourn in England reached its third year, Deborah wrote him a plaintive letter, complaining that she was lonely and kept hearing rumors about him enjoying other women. Franklin smoothly assured her that "while I have my senses," he would "do nothing unworthy of the character of an honest man, and one that loves his family." This was an enigmatic guarantee, at best. If he and Margaret Stevenson were living as man and wife, as seems probable by this time, she was the last woman in the world who would betray him.[13]

Mrs. Stevenson's fatherless daughter, Polly, soon all but worshipped Franklin. She was a remarkably brainy young woman, far more intellectual than Catherine Ray, and Franklin became her semi-father, teacher, and at times, soul mate. At one point she told him she was inclined to remain single and devote herself to the study of science, but Franklin insisted every woman had a responsibility to have children. At times, he was playfully erotic in his letters to her, pretending they were lovers. These remarks have prompted some scholars to speculate that they conducted a secret affair. However, it is hard to believe Margaret Stevenson would have tolerated a star boarder who seduced her only daughter.[14]

Instead, Ben and Mrs. Stevenson both hoped that Polly might fall in love with William Franklin, and vice versa. But the two young people showed no inclination to comply with their elders' wishes. On the contrary, toward the close of their third year in London, William informed his father that he, too, had found it difficult to control the passions of youth; an affair with an unnamed woman had produced an illegitimate son. Ben could hardly censure him for this lapse; all he could do was insist that William follow his example by giving the boy the family name and taking responsibility for him. They named the child William Temple Franklin, and Mrs. Stevenson found a temporary foster home for him. Ben promised to pay all his expenses.

William had completed his studies at the Inns of Court and had been "called to the bar." He was now entitled to practice law in London or Philadelphia. But the Franklins did not depart for home, even though their war with the Penns had ended in a frustrating stalemate. They spent another two years in the imperial capital, pursuing a possibility, faint at first but growing stronger with the passing months, that William could obtain an appointment as an official somewhere in the empire—perhaps as judge of an admiralty court or deputy governor of one of the colonies. This prospect soon became something even more attractive: through a series of improbable firings and refusals, the governorship of a colony opened up.

Ben went to work; one of his closest friends was an intimate of Lord Bute, tutor and confidant of the new king, George III. The negotiations were conducted in deepest secrecy. By now the Penns hated both Franklins and would have exerted all their influence to block the appointment. Not until the announcement was made in the newspapers did anyone learn that William Franklin was about to become the royal governor of New Jersey. William could justifiably feel that he had played a part in winning this prize; he had demonstrated to his father's friends that he had the brains and good judgment that were essential to handling this important job.

IX

His self-confidence soaring, William proposed to a woman he had been courting for years, Elizabeth Downes, the pretty daughter of a wealthy Barbados sugar planter. She was sweet, dependent, and deeply religious. The contrast between her and outspoken, independent Polly Stevenson suggests that William had not entirely disposed of his inner insecurity about his illegitimate birth. On the other hand, marrying Elizabeth Downes may have been William's way of declaring a limited independence from his role as his father's alter ego. Above all Elizabeth was not from Philadelphia, which made her immune to the vicious local politics that had ruined his romance with Elizabeth Graeme, as well as nasty behind-the-back whispers about his illegitimate birth.

Ben barely managed to conceal his disappointment. He gave the match his "approbation"—he told a friend that Elizabeth was "a very agreeable West Indian lady"—but his enthusiasm was minimal. He had never stopped hoping William would choose Polly Stevenson; their combined brainpower would almost certainly have produced a brood of geniuses. Moreover, Elizabeth was thirty-four, an age at which she was unlikely to have more than one or two children, who might not survive the precarious first years of life. That was a potentially fatal wound to Ben's dream of founding a famous American family.

Ben cooperated with William in telling Elizabeth Downes nothing about the existence of William Temple Franklin, but he abruptly decided that it was time for him to go home without waiting for the wedding ceremony or William's formal appointment as royal governor. Part of his decision made sense: it was already September, and it would be foolish for a man his age to venture onto the cold, stormy Atlantic in the winter. But there was also more than a hint of disapproval in this departure that a sensitive, intelligent woman like Elizabeth Downes would discern—and remember.

X

Five months later, there was only pleasure and pride on Ben's face when he journeyed to New Jersey to watch William Franklin take office as New Jersey's royal governor. Ben's presence helped to promote the cordial welcome William and his British wife received from the colony's leading citizens, despite vicious efforts by followers of the Penns in Philadelphia to poison people's minds against "Ben's bastard," as they sneeringly called the new governor. William swiftly demonstrated he had inherited his father's political skills, extracting a generous budget from the assembly and persuading them to vote a small increase in his salary, a rare accomplishment for royal governors.

Back in Philadelphia, the Penns and their supporters made an all-out effort to break Ben's power in the assembly. They succeeded in defeating him and his right-hand man, Joseph Galloway, in the annual election. But Ben's party retained a narrow control of the assembly, and under his guidance, they rammed through twenty-six resolutions condemning the Penns and calling for an end to their government. The legislators climaxed this blast by resolving to send Ben back to England to press their case. Once more he invited Deborah to come with him, and once again she said no.

At the age of fifty-eight, Franklin began his third crossing of the Atlantic. He had no way of knowing that he was parting with his Plain Country Joan for the last time. Nor could he foresee that the next time he returned to America, his dream of a triumphant British Empire uniting Britain and America would be in ruins. Instead, he would be a leader of a revolution that would inflict terrible wounds on his own heart and the hearts of those he loved.

THE OLDEST REVOLUTIONARY

*B*y the end of December 1764, Benjamin Franklin was sitting in Margaret Stevenson's parlor on
Craven Street. He had arrived without warning
her in advance. He wanted to make her wonder, if only for an instant,
whether he had magical powers that had transported him across the wintry Atlantic. When Margaret returned from a shopping trip, her astonishment delighted him. Polly Stevenson was visiting friends in the country;
Ben dashed off a letter telling her how much he enjoyed the surprise on
her mother's face. Polly hurried back to London and they resumed their
happy family life on Craven Street.

The house was soon thronged with Franklin relatives from England
and America as well as numerous English friends. Mrs. Stevenson never
lost her good humor and managed to keep everyone happy, especially her
star boarder. Her importance to Franklin—and the intimate depth of
their relationship—was visible in a letter Franklin wrote to her when she
took a much needed rest in the country. He reminded her of an old adage
from *Poor Richard's Almanac,* "fish and visitors smell after three days," and
urged her to "return with the stage tomorrow."

With deadpan humor, Ben added that he was by no means confessing
that he could not manage without her. In fact, he would be perfectly happy
if he could get rid of "Nanny" [a difficult servant] and the cat. Then he
would have an empty house, which he would enjoy immensely. However,
such happiness was "perhaps too great" to be given to anyone except saints

and holy hermits. "Sinners like me, I might have said US, are condemned to live together and tease one another."[1]

At one point, during another Stevenson sojourn in the country, Ben published a newspaper, *The Craven Street Gazette,* to keep her in touch with his doings. In its pages she became "Queen Margaret" and her departure left the Great Person ("so called because of his enormous size") grumpy at the "new administration" even though they promised him one of his favorite dishes, roasted shoulder of mutton, for dinner. Another edition reported that in spite of an "order in council" requiring everyone to attend church, on Sunday "The Great Person's broad-built bulk lay so long abed that breakfast was not over until it was too late." Other editions referred to the Great Person as "Dr. Fatsides" and reported he "begins to wish for Her Majesty's return."[2]

Clearly, Mrs. Stevenson was a far more sophisticated woman than Deborah. Her intellectual daughter, Polly, did not acquire her brainpower and passion for science by accident. Margaret circulated comfortably in London's upper-middle-class business and artistic world. Among her close friends were the American-born painter Benjamin West and his wife. She had excellent taste and helped Ben select the china, silver, and other presents he sent Deborah and Sally. Thanks to her tact and good humor, Franklin was able to convince himself he was keeping both women happy.

II

Politics suddenly transformed Franklin's mission from a war with the Penns into a struggle to preserve his dream of a British empire in which Americans were treated as equals. To pay off the huge debt created by the Seven Years' War with France, Parliament decided to tax the Americans directly rather than request grants from the various colonial assemblies. Their first attempt was a Stamp Act, which would require a royal stamp on every conceivable public document. American reaction was violently hostile, but it took Ben several months to realize this was an issue that transcended his feud with the Penns.

The law did not cause serious pain to Franklin's ample personal exchequer. He advised friends in America to accept it. Frantic letters from Philadelphia told him he was being slandered as one of the sponsors of the act. Deborah described how she had defended their house against an angry mob with a loaded pistol in her hand. Faced with a choice between Britain and America, Ben joined the opposition. His facile pen—and his shrewd, carefully rehearsed testimony before Parliament—played a key role in persuading the legislature to repeal the Stamp Act. But at George III's insistence, Parliament added a clause that affirmed its right to tax Americans "in all cases whatsoever." An ugly wedge of hostility had been inserted between England and America.

This new situation made Ben Franklin a semipermanent boarder on Craven Street for the next ten years. From Philadelphia, Deborah's letters became more and more plaintive, as age and illness darkened her days. Again and again Franklin promised to return home, only to apologize because a new political crisis required his presence in London. He gradually became the spokesman not only for Pennsylvania but for all the American colonies in their growing antagonism to the mother country.

In the midst of this turmoil, Ben and William did not forget William Temple Franklin. When the boy was about four years old, Ben began bringing him to Craven Street. William approved of this decision and a few years later wondered if "Temple" could be brought to America to live with him. William had no intention of acknowledging him as his son; he intended to describe him as the "son of a poor [English] relative" that he was raising as his own child. Ben's repeated postponements of his return to America and William's second thoughts slowly let this proposal slide into oblivion. The royal governor realized there was a good chance that people would start whispering that the boy was the bastard father of a bastard son. On Craven Street, the boy was known only as William Temple—he was not told that William was his father and Ben his grandfather. But Ben assured him he was a member of the Franklin family. He was sent away to school, returning only for holiday visits.[3]

III

Back in America, William had assumed leadership of the Franklin family, dealing with quarrels and pleas for help from some of his father's aging brothers and sisters, and even offering Deborah advice and assistance when she asked for it. But his modest income severely limited his largess, and he got into a full-fledged quarrel with Deborah when she asked him to investigate a merchant named Richard Bache, whom Sally

Franklin wanted to marry. William found Bache had gone bankrupt several times and dismissed him as a fortune hunter. His father agreed with him at first, but when Deborah backed her daughter, Ben wisely decided he was too far away to play a role and agreed to the match. William was hurt by the way Ben disregarded his advice; father and son exchanged angry letters before jointly deciding to drop the subject. But William's hard feelings persisted; he claimed he was too busy to attend his half sister's wedding.

As the quarrel between England and America grew more complex and intense, William had to deal with a restless and often defiant New Jersey assembly. It did not make him happier to know that his father was being viewed in some circles as an agent provocateur, urging ever greater resistance. Crown officials viewed William with suspicion and hostility, and at one point there was a serious attempt to harass him into resigning. William was proud of the way he repeatedly soothed New Jersey's angry voters. He was doing a good job, but Ben's role as the spokesman for America's complaints deprived the governor of the credit he felt he deserved.

More and more, William began to wonder whether his father had lost his perspective on what was happening in America. The complainers and defiers were a noisy minority, many of them bankrupts like Sam Adams of Boston, who were trying to blame their failures on the king and Parliament. The best people, men of wealth and judgment, with whom William associated, did not share their views. Ben was also getting old, reaching a time of life when it was difficult for a man to change his opinions.

William shared these thoughts with Elizabeth Downes. She had not given him a son or daughter, and as the years passed it became obvious that she was not going to do so. This lack of a family only drew them closer together as a couple. Soon, William became a member of the Church of England and shared her fervent faith in a Christian God. With that step, remarkable for a son of Ben Franklin, William crossed a spiritual divide. A devout Anglican saw the king as the incarnation of God's plan for an ordered, peaceful society. Faith reinforced reason to make men and women deeply committed to the world that had emerged from the Glorious Revolution of 1688—an era in which Britain had achieved global power and wealth.

IV

Meanwhile, Ben was changing his mind about the British empire. In 1773 he told William that he now thought Parliament had no right to make laws for the colonies; the Americans could and should govern themselves. The king was the only figure to whom Americans owed allegiance. He casually added that he knew William did not see things this way. "You are a thorough government man," he wrote, "which I do not wonder at, nor do I aim at converting you."[4]

Upheavals in Boston and London soon disrupted this precarious harmony. Yankee radicals dumped tea in Boston harbor, and the British reacted with mindless anger and brutality. In London, crown officials turned on the man they considered the source of much if not all of America's defiance: Benjamin Franklin. They converted a hearing before the king's Privy Council on a petition Franklin had submitted to them on behalf of Massachusetts into an impromptu trial. No less a personage than the king's attorney general assailed Ben Franklin with sarcasm and insults for over an hour, while courtiers crowding the chamber laughed and applauded. Franklin never said a word.

The next day, the government stripped him of his job as deputy postmaster general for America. Ben wrote a letter to William, advising him to resign. Two weeks later, he changed his mind. Another letter told William to sit tight; better to wait until the American-haters in George III's cabinet fired him as they had dismissed Ben. "One can make something of an injury but nothing of a resignation," he wrote.

Less than a year later, he received dismaying news from Governor Franklin: Deborah was dead. William had struggled through snow drifts to attend her funeral on December 19, 1774. She had suffered a stroke in 1773 that left her enfeebled in both mind and body; a second stroke had carried her into eternity. She had been a pathetic sight in her last days. Even William referred to her as "my poor old mother." Deborah lamented that Ben had not come home immediately after his humiliation before the Privy Council. She had often wept and sighed that she would never see her Pappy again.

The drift to war continued, with Ben still in London desperately trying to negotiate a settlement. But the British were intransigent, and he finally decided to sail for home. On his last day in England, tears streamed down his face as he said goodbye to Mrs. Stevenson, Polly, and numerous friends. He promised them he would come back as soon as possible to continue his political struggle with the king and his parliamentary allies. But in his heart he doubted he would ever return to this imperial city, where he had found love and admiration.

Ben was also weeping for the death of his dream of a united British empire. With him to the port of embarkation went William Temple Franklin. Ben had decided he would take the sixteen-year-old with him to America. It was a silent signal of his intention to center his hopes for the future on an independent America. For the first time, he told Temple that he was not merely a Franklin—he was an American Franklin.

On shipboard, Ben spent much of his time writing the longest letter of his life—it eventually totaled ninety-seven pages. It began "Dear Son." It was a narrative of his final, hitherto secret negotiations with the British government. He thought it might be used to justify America's decision to declare independence. But those opening words made it clear that Governor William Franklin was the one person Ben wanted to read it. His chief purpose was to persuade William he had done everything in his power "to preserve from breaking that fine and noble China vase, the British Empire." [5]

When the ship reached Philadelphia on May 5, 1775, Ben heard news that did not entirely surprise him: war had begun. Blood had been shed on Lexington Green and on Concord Bridge. Pennsylvania appointed him a delegate to the Continental Congress, but he found that many people eyed him with uneasy suspicion. William was still the royal governor of New Jersey. Where did his father's loyalty lie?

Ben quickly sent a messenger to William and arranged a meeting. He was thunderstruck to discover that William was still loyal to the king and Parliament. The governor denounced extremists in both London and America, but he felt "obligations" to the ministers who had trusted him in spite of his father's opposition to their policies. William was stunned to hear his father say—or at least strongly intimate—he favored American independence. He angrily warned Ben against trying to "set the colonies in a flame"—words he had already used with some success in a speech to the New Jersey legislature. [6]

Only William Temple Franklin prevented a violent quarrel. William was almost pathetically overjoyed to claim the young man as his son. He took

Temple with him when he returned to Perth Amboy, New Jersey, where the colony had recently built him a splendid new house. Mournfully, Ben could only remember a letter he had written to William a year earlier:

> *I don't understand it as any favour to me or to you, [your] being continued in an office by which, with all your prudence, you cannot avoid running [financially] behindhand if you live suitably to your station. While you are in it I know you will execute it with fidelity to your royal master, but I think independence more honourable than any service, and in the state of American affairs, you will find yourself in no comfortable situation and wish you had disengaged yourself.*

That prophecy would soon be fatefully fulfilled.[7]

V

Ben quickly learned that independence was not a popular idea in the Continental Congress or in many homes in Philadelphia. When Ben chose to say little or nothing in Congress, some of his former enemies in Pennsylvania's politics wondered aloud whether he might be a spy, sent by the ministry to identify the leaders of the rebellion. "He wants to discover our weak side," these Franklin-haters said. Richard Henry Lee, leader of the Virginia delegation to Congress, declared himself "highly offended" by Franklin's reticence.[8]

In distant England, a drama of longing and eventual heartbreak was unfolding. Margaret Stevenson was hoping that Deborah's death meant that she and Ben could now marry. She wrote fervent letters to him, looking forward to the "hapey day" when he would return to England as he had promised and they would become man and wife.

A letter from a mutual friend, Dolley Blunt, leaves little doubt of the nature of Margaret's relationship with Ben. Dolley told him Margaret was hoping desperately for a letter that would tell her when he would return. Mrs. Blunt urged him to send "the strongest assurances" of an early arrival, adding, "I am firmly persuaded that without the animating prospect of spending the remainder of her life with you, she [will] be very wretched indeed." It was not that Margaret lacked friends, but "all of us are less to her than you." With the kind of irony that history seems to favor, Mrs.

Blunt wrote these words on April 19, 1775, the day the shooting war began at Lexington and Concord.

Franklin wrote a warm letter to "My dear dear Friend," but he offered Margaret only financial advice—to take her money out of government bonds and put it into private investments. He did not respond to her suggestion that she and Polly might join him in Philadelphia. He could not explain that marrying an English wife would be tantamount to political suicide for him in the atmosphere of doubt and suspicion swirling through the city.[9]

VI

When British and American soldiers clashed at Bunker Hill, leaving over a thousand men dead and wounded, Franklin wrote an angry public letter to his friend William Strahan, with a mordantly witty closing line: "You are now my enemy and I am yours, B Franklin." In Perth Amboy, William Franklin became more and more isolated as New Jersey's allegiance shifted to the rebel cause. His peril only increased William's stubbornness. In August 1775, Ben visited him and again tried to change his son's mind. Ben spoke without reserve, telling William it was now or never. The swing to independence was growing stronger every day. If he joined his father, William could become one of the leaders of a new nation.

Governor Franklin remained intransigent. One suspects that deep in William's heart smoldered a hidden anger for his illegitimate birth—and his father's failure to protect him from Deborah's nasty tongue when he was a boy. A more immediate motive may have been a reluctance to trust his fellow Americans, who were all too ready to fling the epithet "bastard" at him. Some weight must be also be given to the influence of Elizabeth Downes Franklin, who remained devoted to the church and the king. Ben went back to Philadelphia a very disappointed man.

Meanwhile, William was exerting his formidable charm on Temple. Elizabeth was equally persuasive. William had told her that Temple was his son and asked her to accept him. She was soon calling herself Temple's "mamma" and urging him to admire his father's courage. In the fall of 1775, Temple returned to Philadelphia to begin studying at the College of Pennsylvania. Early in 1776, musket-wielding rebel militiamen surrounded the governor's mansion in Perth Amboy and pounded on the door at two a.m. to demand his surrender. Elizabeth was terrified by this possibly murderous intrusion. William wrote his son an emotional letter aimed at winning Temple's sympathy by dwelling on his stepmother's distress:

> *I hope you will never be wanting in a grateful sense of her kindness to you. I should be able to keep up my spirits in struggling through all the present and expected difficulties if it was not on her account. Her constitution is naturally weak and delicate and the late brutal treatment she has received, and her anxious concerns for me, have nearly deprived her of her life. . . . Her spirits continue so agitated that the least sudden noise almost throws her into hysterics . . . She has no relations of her own in this country to whom she can resort, or from whom she can receive any comfort in a time of distress and she cannot but take notice that mine do not at present seem disposed to give themselves any concern about her.*[10]

In June 1776, New Jersey's rebel government arrested Governor Franklin. After a hasty trial before two leaders of the rebel assembly, during which one of them, John Witherspoon, the president of the College of New Jersey (later Princeton), called him "a base born brat," the Continental Congress ordered him deported to Connecticut. William wrote Temple a farewell note, urging him to be "dutiful and attentive to your grandfather to whom you owe great obligations." But the governor also begged his son to "love Mrs. Franklin for she loves you and will do all she can for you if I should never return more. If we survive the present storms we may all meet and enjoy the sweets of peace with greater relish." He signed this incompatible advice, "Your truly affectionate father."

VII

Elizabeth wrote pathetic letters to Temple and to Sally Franklin Bache, describing her isolation in the governor's mansion surrounded by unruly American soldiers who insulted her to her face if she ventured outside. All the wealthy loyalist friends in the vicinity had been arrested and shipped to the interior of New Jersey. "I can do nothing but sigh and cry," she told Sally Bache in July 1776. "My hand shakes to such a degree I can scarcely hold a pen." Sally tried to assure her sister-in-law of her continued friendship. When she gave birth to a baby in December 1775, she asked Elizabeth to be the godmother.

A distraught Temple persuaded his grandfather to let him go to Perth Amboy to protect his stepmother. Soon he was writing strident letters, protesting the treatment she and his father were receiving from the rebel government.[11] Elizabeth had not heard a word from William, nor did she have any address to which she might send a letter. She wrote a plaintive letter to her father-in-law, begging him to intercede with the Continental Congress on William's behalf. He did not answer her.

Next, Temple asked permission to take a letter from Elizabeth to his father. Ben refused to let him risk his life. The British and Americans were fighting furious battles in and around New York, and the rebels were losing most of them. The whole region was a war zone. Temple replied with an angry letter that made his grandfather realize he was becoming a loyalist like his father.

Political decisions in Philadelphia flowed inevitably from those unnerving military defeats. Congress decided that the Americans could not win the war without foreign aid. They voted to send Benjamin Franklin, the one American with international fame, to France to plead for it. Ben made a fateful decision. He summoned Temple to Philadelphia to discuss "something offering here that will be much to your advantage." That something was an appointment as Benjamin Franklin's private secretary in France. Ben told Temple it was a chance to launch a diplomatic career that would make him famous in America. He would also have unique educational opportunities in Europe, if he wished to pursue them. Ben was taking his daughter Sally's oldest child, seven-year-old Benjamin Franklin Bache, with him to send him to school in Geneva, Switzerland.

We do not know what Temple thought and felt during this discussion. But it is not difficult to imagine his inner torment. To accept his grandfather's invitation meant he was leaving in Perth Amboy the fragile, loving woman he had learned to call mother. Yet his father had told him to obey his grandfather. On October 26, 1776, Franklin, Temple, and Benny Bache boarded the American warship USS *Reprisal* and sailed for France. A year later, Elizabeth Downes Franklin died a lonely, penniless refugee in British-occupied New York without seeing her "dear persecuted prisoner" (William) or Temple again.[12]

Abandoning Elizabeth Downes Franklin was the cruelest act of Benjamin Franklin's life. He never offered her the shelter of his house in Philadelphia or urged his daughter, Sally, and her husband, Richard Bache, to look after her. He never interceded on her behalf with the leaders of the Revolution in New Jersey. It was grim evidence of how deeply William Franklin's defection from the Revolutionary cause had wounded his father—and how bitterly Ben Franklin felt that Elizabeth shared the responsibility for this demoralizing pain.

MON CHER *PAPA*

*I*n France, seventy-year-old Benjamin Franklin began the third phase of his extraordinary life. His fame as a scientist and philosopher blended with the huge excitement he generated as the spokesman for the embattled new republic, the United States of America. With consummate shrewdness, Franklin wore the simple clothes of an American Quaker, an imaginary character created by savants such as Voltaire and Jean-Jacques Rousseau. The French wanted to believe that in the new world a new kind of man was emerging, free of the corruptions and infirmities of their decadent old world. Franklin was more than ready to encourage this illusion. One excited Parisian wrote: "Everything about him announces the simplicity of primitive morals . . . The people clustered about him as he passed and asked: 'Who is this old peasant who has such a noble air?'"

The old peasant, whose primitive morals had enabled him to maintain wives on both sides of the Atlantic without a hint of scandal, was soon displaying his gift for backstairs diplomacy. He began by charming France's foreign minister, the Comte de Vergennes. With the help of several American victories on the battlefield, Franklin persuaded this cautious veteran of twenty-four years' service in Europe's capitals to back the United States, first with secret aid and then with a formal alliance in 1778. This was only the beginning of Franklin's French accomplishments. He secured over $40 million in loans and gifts from the French treasury—the equivalent of perhaps $600 million today—money that kept the bankrupt American government functioning. He supervised the shipment of tons of supplies and weapons to America. He armed and equipped American sea captains, such as John Paul Jones, who preyed on British shipping in their home waters with spectacular success. He raised money and aroused sympathy for American captives in British jails. He wrote letters and gave interviews that encouraged opposition in Parliament to George III's determination to smash the rebellion.

One might think these achievements would have made Franklin's fellow American diplomats among his most ardent admirers. But the Americans of the eighteenth century were almost as complicated and unpredictable as their descendants of the new millennium. In the contorted souls of several colleagues, Franklin's performance stirred envy and suspicion rather than admiration. They accused him of gross sexual misconduct and a dangerous subservience to France. Back in America, members of Congress echoed these charges—or read them aloud from the diplomats' letters. The torrent of abuse became so vitriolic that in May 1781, Franklin resigned as ambassador to France. A panicky Congress persuaded him to withdraw the letter of resignation—but it remains a testimony to the industry if not the accuracy of Franklin's enemies.

Franklin's seeming lack of system and orderliness as a diplomat irritated his critics. But it was his tendency to spend so much time with women that drove those with puritanical tendencies into paroxysms. They were appalled by the way the ladies of France swarmed to exchange kisses with the ambassador. This was visible proof that the septuagenarian Franklin was a libertine with sexual appetites of gargantuan proportions. Franklin's fellow diplomat, Arthur Lee of Virginia, told his brother, Congressman Richard Henry Lee, that Franklin was "a wicked old man" who had made his headquarters in France "a corrupt hotbed of vice." John Adams chimed in with equally savage letters. By 1780, Richard Henry Lee was asking Samuel Adams, "How long should the interests of the United States be sacrificed to the bad passions of that old man under the idea of his being a philosopher?"[1]

In a cheerful letter to a grandniece in America, Franklin had a different explanation for his dalliances: "Somebody gave it out that I loved ladies; and then every body presented me their ladies (or the ladies presented themselves) to be embraced, that is to have their necks kissed. For as to kissing on the lips or cheeks it is not the mode here, the first is reckoned rude, and the other may rub off the paint. The French ladies have however 1000 other ways of rendering themselves agreeable; by their various attentions and civilities, & their sensible conversation. 'Tis a delightful people to live with."[2]

There were times when female enthusiasm almost overwhelmed Franklin. At one point three hundred ladies surrounded him and placed a laurel wreath on his bald pate. Then they chose the prettiest among them to kiss him—and of course, to be kissed in return. A French neighbor who enjoyed watching Franklin in action told of women talking to him "for hours on end, without realizing that he did not understand much of what they said because of his scant knowledge of our language. But he greeted each one of them with a kind of amiable coquettishness that they loved. Occasionally, one madam or mademoiselle asked him if he cared for her more than the other pursuers. With a smile Franklin would reply in his limping French, 'Yes, when you are closest to me, because of the power of the attraction.'"

The remark combined flirtation and a reminder of his fame as a scientist. He was comparing the lady's impact on him to the way an electrified piece of metal drew iron filings to it. Behind these amorous games lay the goal Franklin never forgot—persuading the French to back the faltering American Revolution. He knew—and cheerfully approved—the passion for politics among upper-class French women. He hoped their enthusiasm for his amiable American ways would be transmitted to their influential husbands or lovers.[3]

II

During his first years in France, Franklin's favorite Frenchwoman was his neighbor in the suburb of Passy, Madame Brillon de Jouy. At thirty-five, she was exquisitely beautiful—and was also one of the best amateur pianists in Europe. Leading musicians dedicated compositions to her, and she wrote a number of sonatas for the harpsichord and the piano. She first attracted Franklin with her music. He began visiting her house to hear her play and her two pretty daughters sing. Franklin was soon calling these visits "my opera."

The ambassador also met Monsieur Brillon, who was twenty-four years older than his wife and rather brusque and self-satisfied, as became a powerful treasury official. Madame Brillon's beloved father had recently died, which left her depressed and vaguely discontented with her husband. She began discussing religion with Franklin—one of his favorite topics. One night, sitting on the terrace of the Brillon mansion, Ben remarked that if the Catholic religion was the true one, he feared he was damned. He was constantly committing one or another of the seven deadly sins—pride, covetousness, lust, anger, gluttony, envy, or sloth. Madame Brillon said he needed a spiritual director. Franklin asked her if she would take the job.

Madame Brillon accepted the appointment and solemnly insisted she now had to hear his confession. The ambassador annotated his sins, placing special emphasis on his fondness for beautiful women. She announced she would have to think about this overnight and decide on his penance. The next day a note arrived on Franklin's desk, informing him that Madame Brillon would grant him absolution. She forgave him for all his past sins and even his future ones, and would tolerate his weakness "as long as he loves God, America and me above all things, and I promise him a paradise where I shall lead him along a path strewn with roses."[4]

Amorous banter became the order of the day whenever Ben visited Madame Brillon. At one point he confessed there was a sin he could not stop committing—coveting his neighbor's wife. On another visit he wondered whether a certain long-dead church father might have been right when he advised people that the best way to get rid of temptations was to yield to them.

Madame Brillon coolly replied that she would have to discuss that question with "the neighbor whose wife you covet." She stormily rejected his argument that the friendship a man has for women could be divided among many recipients like the music from her pianoforte. "This is something I shall never put up with," she imperiously announced. Her heart, which was "capable of great love," had chosen "only a few objects," and Franklin was at the head of the list.

Franklin protested that his "poor little love" was "thin and ready to die of hunger" for want of the "nourishment that his mother inhumanly refuses!" Now she wanted "to cut his little wings so that he cannot go seek it elsewhere." But Madame Brillon was unrelenting. She insisted on a special place in Franklin's heart—and she knew exactly what it should be. She began calling him "*Mon Cher* Papa" and Franklin, ruefully at first, began calling her his daughter.

Franklin knew too much about women to abandon these amorous games. He continued to welcome Madame Brillon's kisses, often delivered with fervor while she sat on his lap. Occasionally they played chess together while Madame Brillon was in her bath. This was not nearly as erotic as it sounds to modern ears; eighteenth-century bathtubs were covered with boards that prevented indecent exposure.[5]

In mid-1779 came a crisis. Madame Brillon discovered what almost everyone else in Passy already knew: Monsieur Brillon was having an affair with their children's governess. She fled to Franklin for solace and advice. "My soul is very sick," she cried. "You are my father." Here was the moment when Franklin, if he had only seduction on his mind, could have satisfied his desire. Instead, he did his utmost to restore Madame Brillon's spiritual strength.

He urged her to forgive her husband—to continue to be a "good mother, good wife, good friend, good neighbor, good Christian (without forgetting to be a good neighbor to your Papa) and to neglect and forget, if you can, the wrongs you may be suffering at present." He advised her to practice one of his favorite maxims: "Doing an injury puts you below your enemy, revenging one makes you but even with him, forgiving him sets you above him." If the advice made Franklin feel his age, so be it. Like a caring father, he watched and encouraged and approved as Madam Brillon slowly regained her equilibrium.[6]

III

Ambassador Franklin grew more and more harassed by America's bankruptcy and his enemies' assaults on his reputation. This may explain an explosion of bad temper that could easily have alienated another woman in his life, his daughter, Sally Franklin Bache. She had not enjoyed the two years that followed Ben's departure to France with Temple and her oldest son, Benjamin. Twice she and her husband and younger children had to flee into the countryside when the British army threatened Philadelphia. In the fall of 1777 the enemy army captured the city, and Sally had spent some very unpleasant months living in a town near Lancaster. Her husband, Richard Bache, was postmaster general of the new United States, a job his father-in-law had obtained for him. But the Continental Congress seldom paid him, and the paper money was depreciating so fast that it did not come close to feeding the family. They had to live on credit. When the British fled Philadelphia at the news of the French alliance, the Baches returned to find their house had been wrecked and looted by the British officers who lived in it. Nonetheless, they were in a mood to celebrate, like the rest of the city. Everyone assumed the alliance would swiftly end the war.

Sally wrote to her father early in 1779, asking him for some French "finery" that she might wear to the dinner parties and balls that were proliferating in Philadelphia. "There was never was so much dressing and pleasure going on," she remarked. She particularly wanted something to wear for a ball in honor of General Washington. She saw herself telling people that her father had sent her the dresses. They would be an illustration of his good taste. It took months for the letter to reach Franklin. By that time, the French alliance was turning into a very complicated and disappointing affair. The fleet that the new ally sent to America was too small to have any impact on the war. The French were more interested in joining their Spanish ally in an invasion of England, which turned into a humiliating fiasco.

A depressed Franklin reacted with uncharacteristic nastiness to Sally's request. Lecturing her as if she were ten years old, he declaimed that her letter "disgusted me as much as if you had put salt in my strawberries." Instead of making her own cloth at the spinning wheel, she had to be "dressed for the ball!" He wondered why she did not know that "of all the dear things in this world, idleness is the dearest, except mischief." She had told him that wartime inflation and the depreciating dollar were driving up the price of everything. But instead of adding that she and everyone else were becoming "frugal and industrious," she wanted him to spend money on feathers and lace! Ben told her to get her feathers from the end of a cock's tail.

Reading this rant, Sally must have wondered whether the ghost of parsimonious Poor Richard had acquired power over her famous father. But she refused to lose her temper. She told her father that she was sure he never would have written such a nasty letter if he had stopped to realize it would hurt her. She asked him if he wanted her to stay away from the French ambassador's or General Washington's entertainments. In fact, the general and "his Lady" had asked her to spend the day with them. The Benjamin Franklin she knew was the "last person" who would want her to accept such an invitation wearing "singular" clothes. She had never insisted on looking "particularly fine," but she would never go out "when I cannot appear so as to do credit to my family and husband."[7]

Still grumpy—the war news continued to be terrible—Franklin replied that he would send Sally everything she wanted. But he could not restrain a needless comment that he was rewarding her for "being a good girl" because she had told him she was spinning cloth and knitting stockings for her family. Another reason for Franklin's hostility may have been Sally's insistence on remaining in contact with her half brother, William. She told Temple in a separate letter that she still loved William and was not going to let their "political difference" change her mind. She urged Temple never to let that happen to him. Temple may have mentioned Sally's stance to Ben, and the wound inflicted by William's defection bled again.

Undeterred by her father's black mood, Sally launched a project that required Franklinesque political skills and energy. As 1780 began, the war was looking interminable and American morale was drooping. Washington's troops were in rags. Sally organized a group of Philadelphia women who raised thousands of dollars for the army. They wanted to give the cash directly to the soldiers. General Washington demurred and suggested they give it to him, as the army's commander in chief. Sally politely but firmly refused. She wanted the soldiers to know how much the women cared about them. She bought cloth, which the women sewed into more than two thousand shirts. Her chastened father sent her a warm letter of praise and published the story in French newspapers.

IV

As the war dragged on, Franklin found spiritual refreshment in the company of a remarkable woman who lived in the nearby village of Auteil. Her baptismal name was Anne-Catherine de Ligniville d'Autricourt. Distantly related to Queen Marie-Antoinette, she was simultaneously bohemian and regal, frivolous and frank, reckless and reserved. Almost sixty when Franklin met her, she was the widow of Claude-Adrien Helvetius, a wealthy financier and freethinker whose books had been banned by royal decree. Madame Brillon called her "my charming and redoubtable rival." Franklin dubbed her "Notre Dame d'Auteil"—Our Lady of Auteil. Madame Brillon was a moody and gifted bourgeois, full of the insecurities of her class. Madame Helvetius was an aristocrat with a hauteur that came as naturally to her as breathing. Once Franklin realized that Madame Brillon wanted him only as a platonic "Papa," he began spending more time with Madame Helvetius.

In her charming country house with its three blooming acres, Madame Helvetius maintained two unfrocked abbots and Pierre-Jean-George Cabanis, a sometime medical student and poet of twenty-two. They all worshipped Madame H. and were equally delighted with Franklin's company. Dinners were a feast of fine food and even finer wit and bursts of poetry.

Through Cabanis, Franklin dashed off notes such as this one:

> *If Notre Dame is pleased to spend her days with Franklin, he would be just as pleased to spend his nights with her; and since he has already given her so many of his days, although he has so few left to give, she seems very ungrateful in never giving him one of her nights, which keep passing as pure loss, never making anyone happy but Poupon [Madame's dog].*[8]

At first, Madame Helvetius did not take Franklin's invitations and intimations seriously. She teased him about a pain he had developed in his shoulder: "Would it be, by any chance, a rheumatism caught under the windows of one of my rivals? Surely you are young enough to go and spend all clear nights playing the guitar, while blowing on your fingers!"[9]

Suddenly, Franklin moved beyond jokes and solemnly, seriously proposed marriage to Madame Helvetius. Notre Dame d'Auteuil was nonplused. She sought the advice of her old friend, the famous economist Anne-Robert Turgot, whose proposals she had rejected innumerable times. Turgot curtly informed her that she—and Franklin—were much too old to be swooning like Romeo and Juliet. Madame mournfully informed Franklin that she could not accept his proposal. She had resolved never to marry again, to honor the memory of her husband.

A few mornings later, one of Franklin's wittiest masterpieces was on Madame Helvetius's breakfast table. He told how he had stumbled into bed after Madame's rejection, a forlorn discouraged man. Suddenly he found himself "in the Elysian Fields." Assuming he was dead, he asked to meet some famous philosophers. He was soon chatting with Socrates and Helvetius. Knowing no Greek, he could not make sense of Socrates, but he chatted with Helvetius in his makeshift French and discussed "religion, liberty, the government of France." Finally he wondered why the philosopher had not inquired about his wife.

Helvetius informed Franklin he had taken another wife. She was not quite as beautiful as Madame but she had "just as much common sense, a little more wisdom, and she loves me infinitely." At this point, in strolled the second wife. It was Deborah Franklin! Her astonished earthly husband claimed her as his spouse. But Deborah was not even slightly interested, saying, "I have been a good wife to you for forty nine years and four months; almost half a century; be content with that. I have formed a new connection here that will last for eternity." The thunderstruck Franklin "resolved there and then to abandon these ungrateful shadows, and to come back to this good world, to see the sun again, and you. Here I am! Let us *revenge* ourselves."[10]

Madame remained adamant. To make his rejection seem part of a game, Franklin published the little essay on a private printing press he had set up at Passy. It was circulated through the salons of Paris, and irony blended with romance to enable Franklin to return to Auteuil, where he was soon back to cheerful fooling. One day he asked Madame to help him solve a mathematical puzzle: "Usually when we share things, each person gets only one part; but when I share my pleasure with you, my part is doubled. The part is more than the whole."[11]

Madame of course had no need to reply to this nonsense. She understood that Franklin was telling her she no longer had to worry about importunate demands for marital bliss. Soon Franklin was cheerfully telling Cabanis, who was absent for a few weeks, that now and then "I offend our good lady who cannot long retain her displeasure but, sitting in state on her sofa, graciously extends her long handsome arm and says: '*la, baisez ma main; je vous pardonne*' (there, kiss my hand, I forgive you) with all the dignity of a sultaness. She is as busy as ever, endeavoring to make every creature about her happy, from the abbes down thro all ranks of the family to the birds and Poupon."[12]

V

Two other women still on Franklin's horizon were Margaret Stevenson and her daughter, Polly. In 1770 Polly had married a doctor named Hewson and had three children in four years. When her husband contracted septicemia while dissecting a corpse and died in agony, Polly had returned to her mother's house. She was soon writing to Franklin, discussing the war, politics, science, and her aging mother.

After Ben negotiated the alliance with France, and he was a recognized and feted diplomat, Mrs. Stevenson began expressing the hope that she could join him in Passy. But by this time the Stevensons had run into shoal waters financially. When an inheritance from an aunt was tied up in litigation, they had to give up their house on Craven Street and retire to a country village. Mrs. Stevenson was miserable. She was a city woman and she yearned for the lively social world that Franklin had woven around her for almost two decades. Only Polly and her three children occasionally lifted her spirits.

Polly put the children in boarding school and moved back to London, but Mrs. Stevenson's health began to decline. She suffered from dropsy, which affected one of her legs. Yet she was still determined to come to France and live with Franklin. He was more than willing, even though she did not speak a word of French. In that careful style they used with each other in letters (which were too often opened by inquisitive governments), Franklin assured her that he, too, yearned for her "faithful tender care and attention to my interests, health and comfortable living." When Polly hesitated, unsure about bringing her children to France, Mrs. Stevenson proclaimed her readiness to come with a friend. "Don't be surprised if I pack myself up and . . . pipe upon you," she playfully warned.

On January 13, 1783, only a few weeks after Franklin and his fellow negotiators had signed a preliminary treaty of peace with England, Polly wrote Franklin a tearful letter: "I know you will pay the tribute of a sigh for the loss of one who loved you with the most tender affection." Her mother had died on New Year's Day. A saddened Franklin replied that the loss reminded him of how many of his English friends had died recently, but this news "strikes the hardest." He somberly speculated that it was the way life loosened old people's attachments to the world and made them more willing to follow their friends.[13]

A few months later, when his spirits rose with the prospect of peace, Ben wrote to Polly in another vein, recalling his friendship with her and Margaret Stevenson as "all clear sunshine without the least cloud in the atmosphere." Here, in a single sentence, was the difference between Franklin's English wife and his American wife.

VI

After the British defeat at Yorktown it took another year and a half to negotiate the peace treaty that ended the struggle. As this often torturous diplomacy unfolded during the year 1782, William Franklin's name abruptly intruded on the scene. In New York, William was being accused of the brutal murder of an American militia officer in New Jersey. The actual crime had been committed by one of the officers in a band of guerillas he had organized, the Associated Loyalists, but the officer claimed he had only been following William Franklin's orders.

The victim had been hanged in revenge for the supposed murder of one of William's loyalist raiders. An infuriated General Washington had chosen by lot a young British officer captured at Yorktown and was ready to hang him if William was not surrendered for trial and certain conviction. The British officer's mother had appealed to King Louis XVI to intercede with Congress to spare her son, and the king had acquiesced. William had decided he might be safer on the other side of the Atlantic and retreated to London.

This messy affair was a huge embarrassment for Ambassador Franklin. He soon learned that William had been made the London spokesman for the hundreds of loyalists who were seeking compensation from the crown for their losses. In the peace negotiations, the British fought desperately to persuade the Americans to agree to join them in paying these people for their confiscated lands and houses. Franklin stonily opposed giving a penny to these ruined men and women, even when his fellow negotiators wavered and admitted that some sort of palliative gesture might be agreeable. Finally the British begged them to accept a clause in the treaty, stating that Congress would simply appeal to the states to compensate loyalists. Franklin saw to it that the clause excluded loyalists who had "borne arms against the United States." William Franklin was, of course, his target.

In the summer of 1784, almost a year after peace between England and America became official, William wrote his father a letter, wondering if they could "revive that affectionate intercourse and connexion which till the commencement of the late troubles was the pride and happiness of my life." The defeated royal governor would not admit he was wrong; "If I have been mistaken, I cannot help it," he wrote. "I verily believe were the same circumstances to occur again tomorrow, my conduct would be exactly similar." Not even the "cruel suffering, scandalous neglect and ill-treatment we poor unfortunate loyalists have in general experienced" would change his mind. He did not mention Elizabeth Downes Franklin's lonely death in New York, but it had to be an unspoken part of his "cruel suffering."[14]

Franklin's reply was a study in torment. He began by all but confessing that he, too, remembered those years of "affectionate intercourse and connexion" with deep pleasure. Reviving them would be "very agreeable to him." But in the next sentence the wound spoke: "Indeed nothing has ever hurt me so much and affected me with such keen sensations, as to find myself deserted in my old age by my only son; and not only deserted but to find him taking up arms against me, in a cause where in my good name, fortune and life were at stake."

Desperately, the father struggled to summon some shreds of forgiveness: "I ought not to blame you for differing with me in sentiment in public affairs. We are men, all subject to errors. Our opinions are not in our own power; they are form'd and govern'd much by circumstances . . . Your situation was such that few would have censured you for remaining neuter . . ." This was as close as Franklin could come to mentioning Elizabeth Downes Franklin's influence.

But the bitterness, the rage, could not be denied. Franklin underlined the next sentence: *"Tho there are natural duties which precede political ones, and cannot be extinguished by them."* He was telling William that his loyalty to his father should have transcended his loyalty to his wife.[15]

William had asked to come to France to discuss family business. Franklin curtly informed him that a visit would be inconvenient for the time being. Instead, he would let William Temple Franklin go to London and find out what was on William's mind. He warned William not to introduce Temple to his loyalist friends, who might damage his standing with patriotic Americans. But the visit still ignited the fears that had impelled Benjamin to take Temple to Paris with him in 1776. He knew how persuasive William could be.

Ben harassed Temple with a stream of letters, insisting that he tell him virtually every move he made. At one point Temple, now twenty-four, had to ask his grandfather's permission to go to the seashore with his father. By the time Temple returned to France, even a pretense of affection between Benjamin and William had evaporated. Ben never wrote his son another letter, and William plunged into the political struggle to win compensation from the crown for the loyalists.

During the peace negotiations, Temple, revealing a lack of judgment appalling even in a young man, covertly advised one of the British negotiators that they ought to do something handsome for William, saying it might make Ambassador Franklin more amenable to their point of view. Of all people, Temple certainly knew this was a total fiction. It was a sign of how gravely Temple, too, had been wounded by the brutal clash between his father and grandfather and the sad death of the woman who had offered him a mother's love. His future years would reveal the depth of his trauma.[16]

VII

With peace restored, Ambassador Franklin had one more decision to make. Should he stay in France and continue to enjoy Madame Brillon's adoration in Passy, Madame Helvetius's effervescent charm in Auteuil, and the devotion of a half dozen other ladies in Paris? He was the most famous American in Europe. Distinguished men and women from a half dozen countries visited him to pay their respects. For a while Franklin wavered. He was not at all sure how he would be welcomed in America.

His hesitation was grim evidence of the damage inflicted by the smear campaign Arthur Lee, John Adams, and his other enemies had waged against him. William Franklin's defection was another reason to avoid mocking or accusatory eyes. How could a man who failed to persuade his only son to be a patriot hope for praise from his countrymen?

As we shall see, Franklin was not the only founder who felt that his fame had been diminished by the vagaries and rivalries of the Revolution. Today most historians are ready to say his diplomatic achievements in France place him just behind George Washington in the pantheon of those who contributed most to winning independence. But this was by no means apparent in 1784. Few Americans understood or appreciated what he had accomplished in his years as ambassador to France.

VIII

"The moment that an American minister gives loose to his passion for women, that moment he is undone; he is instantly at the mercy of the spies of the court, and the tool of the most profligate of the human race." John Adams struck this lowest of low blows against Franklin on September 3,

1783, the very day that a definitive peace treaty with England was signed in Paris.[17]

More than two hundred years later, has any worthwhile evidence emerged that substantiates this vicious portrait of Franklin's love life? Scholars who have devoted decades to studying Franklin in France have not found a single diary entry, memoir, or letter from anyone who claimed to have had an affair or even a one-night liaison with him. In a country where love has always been a topic openly and almost continuously discussed, this surely is evidence that nothing even approaching Adams's innuendo occurred.

This conclusion has not prevented playwrights, novelists, and screenwriters from portraying Franklin as, in the words of one reviewer of a recent TV portrayal, "somewhere between a phony and a dirty old man."[18] Our current view of sex as an irresistible drive to which everyone succumbs has endorsed this mythical Franklin to the point where, in the words of the best historian of this aspect of his career, it has "become part of our national heritage." This fable deprives Franklin of his "rich and complex humanity."

The French understood and still understand the variety of love that Franklin was practicing in his old age. They call it *amitié amoureuse*. It is sometimes described as a game; a better term would be art. It is a branch of love that has added spice to the lives of men and women of all ages—and it is by no means confined to France.[19]

Perhaps the best insight into Franklin's French love life comes from an anecdote told by Thomas Jefferson. Not long after he arrived in France to replace Franklin as ambassador, the Virginian watched while a group of French women showered their favorite American with kisses. With a smile, Jefferson asked Franklin if he could arrange to transfer "these privileges" to him when he took over his post. "You are too young a man," Franklin said.[20]

VIII

Another factor in Franklin's struggle to decide whether to stay in France or return home was his health. He suffered repeated attacks of gout and had developed painful bladder stones. He wondered whether he could survive a two- or three-month ocean voyage. But he found himself yearning for his Philadelphia family, his daughter, Sally, and her husband and children. "Who will close my eyes if I die in a foreign land?" he asked Madame Brillon, who replied, "I will."

Finally, Franklin made up his mind. He told Madame Helvetius he wanted to be buried in his own country. One is tempted to wonder whether he also wanted to make amends to Sally Franklin for the harsh words he had written to her. More probable was a desire to see the four grandchildren Sally had borne since he left America. She had sent him a stream of lively letters full of their sayings and antics.[21]

Temple was also on Franklin's mind. In the letters he sent to Congress asking for a replacement, he had added numerous pleas for a government appointment for his grandson. The ambassador argued that Temple's fluency in French and his seven years' experience as embassy secretary would make the young man an excellent *chargé d'affaires* of an American legation in Europe. Congress ignored him.

Temple had become a handsome young man, with a passion for expensive clothes and an eye for pretty women. When he went to England to visit his father, he took with him a list of the best tailors, bootmakers, and hatters compiled by a French friend. Appended was another list that Temple also undoubtedly put to good use: "And when lewd, go to the following safe girls, who I think are very handsome."

In Passy, Temple had fathered an illegitimate son by the daughter of a French neighbor. The baby died of smallpox a few months after it was placed in the care of a country family. Instead of sympathizing with the heartbroken mother, Temple blamed her for the child's death and broke off the affair. This was not the sort of conduct that would persuade Congress's numerous puritans to approve Temple as a representative of the United States in France or any other European country. Temple's plight played an important part in Franklin's decision to go home. If he could not find a place for the young man in Europe, he would do his best to help him in America.

Also in the picture was Sally Franklin Bache's oldest son, fourteen-year-old Benjamin Franklin Bache. He had spent most of his time in Europe at school in Geneva, Switzerland, where he had often been miserably homesick. Franklin had brought him to Passy when peace arrived and found Benny a startling reproduction of himself. He had his grandfather's intelligence and wry humor, and he became an ardent and expert swimmer.

Swimming had been Ben's favorite hobby in his Boston boyhood. Franklin decided to train him as a printer and set him up in the newspaper business in America.

So the day of departure arrived. The queen of France, Marie Antoinette, sent a special litter pulled by snow-white Spanish mules to carry Franklin to the seacoast. His bladder stones made riding in a jolting carriage agony. King Louis XVI sent a miniature of his royal personage, encircled by more than four hundred diamonds. On July 12, 1785, Franklin climbed into his traveling bed while his Passy neighbors gathered to say farewell. "A solemn silence reigned around him, interrupted only by sobs," Benny Bache wrote in his diary. The leave-taking at Auteuil had been equally mournful. "Many honorable tears were shed on both sides," wrote the young doctor, Cabanis.

Madame Helvetius found the separation almost unbearable. A day later, she dashed off a frantic letter and sent it in pursuit of her rejected lover. "I picture you in the litter . . . already lost to me and to those who love you so much . . . I fear you are in pain . . . If you are, mon cher ami, come back to us."[22]

But the decision was irrevocable. In a week Franklin was in Le Havre, waiting for 128 crates of baggage to be loaded onto a ship from London. The ex-ambassador's thoughts were of Madame Helvetiuis. "I am not sure I will be happy in America," he wrote. "But I must go back. I feel sometimes that things are badly arranged in this world when I consider that people so well matched to be happy together are forced to separate.

"I will not tell you of my love. For one would say there is nothing remarkable or praiseworthy about it, since everybody loves you. I only hope you will always love me some . . ."[23]

IX

In Southampton, many of Franklin's English friends gathered to say goodbye to him. Refreshed by the voyage across the channel, Franklin partied and joked and drank with them for four days at the Star Tavern. Then came a visitor who dampened everyone's spirits: William Franklin. In the privacy of Ben's room, father and son faced each other and both saw there would be no forgiveness. There was not even an attempt to achieve it. The only subject they discussed was money.

Ben had decided that Temple might be happiest living on the six-hundred-acre farm that the ex-governor still owned in New Jersey. The state had never confiscated it, probably because William was Benjamin Franklin's son. There may also have been some twinges of conscience for New Jersey's role in Elizabeth Downes Franklin's death. Ben announced he wanted to buy the farm for Temple—and drove the hardest imaginable bargain. He paid less than William had paid in the 1760s. Then he presented William with a bill for fifteen hundred pounds—money he had advanced to him during his governorship. With Parliament still sitting on his claims, William was virtually penniless. He was forced to sign over to Temple valuable lands he had purchased in northern New York. Another parcel of good farmland in New Jersey went to Ben in return for cancelling the remainder of his debt. It was Ben's bitter way of severing William's last connections with America. In a letter to his sister Sally a few days later, William wrote: "My fate has thrown me on a different side of the globe."

The voyage across the Atlantic was smooth and pleasant. The ship glided up the Delaware River to Philadelphia, where Ben's son-in-law, Richard Bache, came aboard to greet the travelers. Another ship had outrun Franklin's vessel and informed the city of his imminent arrival. There was a cheering crowd on the dock as they came ashore. Church bells rang and cannon boomed. Joy coursed through Franklin's aged heart. His enemies' smears about his love life had not ruined his fame, as he feared. An enormous crowd lined the streets and applauded until Sally Franklin Bache embraced her father in the doorway of his house. The greeting was "far beyond my expectations," he admitted in a letter a few days later.

Even better news awaited him. Within six weeks his friends had nominated and elected him chief executive of Pennsylvania. "This universal and unbroken confidence of a whole people" was immensely pleasing to him. It was some compensation for the ambassador's secret wounds.

X

The next three years added fresh achievements to Franklin's fame. When the new nation's rickety constitution, The Articles of Confederation, proved inadequate, he joined the call for a constitutional convention. The conclave met in Philadelphia in the summer of 1787, with George Washington as the presiding officer. Franklin played a key role in working out compromises that persuaded deadlocked delegates to agree on thorny issues such as a strong presidency and equal representation for small states in the senate. In a witty speech at the close of the convention, Franklin admitted there were some things he disliked in the final version of the charter, but he planned to sign the document and urged everyone to do likewise. This spirit of compromise played a crucial role in persuading the states to ratify the Constitution and launch the new government in 1789, with George Washington as the first president.

Alas, this added fame could not prevent the slow decline of Franklin's body. The bladder stones that tormented him in France became more and more painful. Soon he was confined to his bed most of the time. His daughter, Sally, was constantly at his bedside. Polly Stevenson Hewson had followed him to America with her three children, and was another hovering presence. Temple frequently visited from his farm in New Jersey, concealing from his grandfather his unhappiness with rural life.

After Franklin's death, Temple sold his farm and returned to England, where he tried to become reconciled with his father. William had married again, and Temple seduced the second wife's sister, who became pregnant. After the child, a girl, was born, William and Temple quarreled. Temple deserted mother and child and retreated to his favorite city, Paris, for the rest of his troubled life. William raised the abandoned child as his daughter.

In Paris, Temple lived for most of his later years with an Englishwoman whom he married on his deathbed, to give her a claim on his meager estate. More than once, he expressed his detestation of marriage as a source of unique misery. Was he haunted by guilt for deserting Elizabeth Downes Franklin? When Franklin's sister, Jane, heard about Elizabeth's death, she had remarked, "Temple will mourn for her much."

XI

Toward the close of these final years, Franklin said farewell to the women he had loved. To Madame Helvetius, he wrote: "I cannot let this chance [to send a letter] go by, my dear friend, without telling you that I love you always . . . I think endlessly of the pleasures I enjoyed in the sweet society of Auteil. And often, in my dreams, I dine with you, I sit beside you on one of your thousand sofas, or I walk with you in your beautiful garden."

Madame Helvetius spoke of her love for him with the same frankness: "I am getting old, my dear, but I don't mind it, I am coming closer to you, we will meet again all the sooner."

He did not forget the woman who had first stirred romantic yearning in his soul. Catherine Ray had married William Greene, a member of one of Rhode Island's most distinguished families. After the Revolution, Greene served as governor of the state for eight years. Catherine had six children, but neither motherhood nor a busy social life prevented her from writing occasional letters to Franklin. She signed them "your friend who loves you dearly." In one letter she told him, "I impute [a] great part of the happiness of my life to the pleasing lessons you gave me." Franklin responded by telling her, "Among the felicities of my life I reckon your friendship."[24]

XII

Polly Stevenson Hewson was at Franklin's bedside when he died. For her he was an incandescent combination of lost father and almost lover, the man she wished she could have married. After his death, she became disenchanted with America. Without Franklin, the country repelled her. She thought women could look forward to only "insignificance or slavery" here. The political brawls of the 1790s convinced her that the nation was only a step removed from anarchy.

Polly told one of her children that "I . . . repent I ever brought you to this country." But the three young Hewsons differed emphatically with their mother. They all married Americans. Even the son who went back to England to study medicine returned to Philadelphia, declaring that Dr. Franklin had inspired an "enthusiasm for America" that became the lodestar of his life. Margaret Stevenson would have been pleased by this denouement of her troubled love story.

XIII

Of all the women who loved Benjamin Franklin and tried to express the complexity of their affection, Madame Brillon said it best. When Franklin was eighty-three, she wrote to him from Passy about her family's happiness, which had only one troubling aspect: his absence. But her regret had "a certain sweetness," because Franklin had told her he was happy in America. "To have been, to still be, forever, the friends of this amiable sage who knew how to be a great man without pomp, a learned man without ostentation, a philosopher without austerity, a sensitive human being without weakness, yes, my dear papa, your name will be engraved in the temple of memory but each of our hearts is, for you, a temple of love."[25]

BOOK THREE

—

John Adams

AN AMOROUS PURITAN
FINDS A WIFE

*F*riday, June 21, 1775, was hot and muggy, a typical Philadelphia summer day. But the weather did not prevent excitement from coursing through America's largest, wealthiest city. General George Washington, the newly appointed commander in chief of the American Continental Army, was leaving to take command of the thousands of New England militiamen who had rushed to Boston after British troops clashed with Massachusetts men at Lexington and Concord on April 19. The choice of Washington was a brilliant political move. It made Virginia, the largest colony, a visible, unmistakable partner in the confrontation with Great Britain.

Numerous Philadelphians and not a few congressmen such as Thomas Jefferson were waiting on horseback, eager to accompany the new general out of town to testify to their enthusiastic approval of him and his willingness to assume the leadership of the nation's embryo army. The Philadelphia Light Horse, cavalrymen drawn from the best families, were acting as Washington's official escort. They were wearing their expensive uniforms—light-brown jackets, white breeches, gleaming high-topped black boots, and flat round hats bound with silver. Bugles blared, drums beat, and Washington, looking magnificent on a big bay horse, began his journey with cheers from hundreds of spectators.

One of the escorting congressmen was having very different thoughts. John Adams had been the man behind Washington's appointment. He had persuaded his cousin Samuel Adams and other New England delegates to accept the Virginian, in spite of vehement objections from several men. Samuel Adams's emphatic approval—he had seconded John's nomination speech in Congress—was decisive. Most men would have felt a quiet exultation, watching the results of this shrewd, enormously important politicking. But John Adams felt nothing but a gnawing mixture of depression and disappointment.

Back in his hot, dingy room after his brief ride, Adams gazed in the mirror at his pudgy middle-aged torso, with its potbelly and stumpy legs, and tried to accept the fact that he would never be a soldier. He sat down and scrawled a letter to his wife, Abigail. He described the departure of Washington and his subordinate generals, stressing how almost everyone, even officers of the militia, was wearing a uniform. "Such is the pride and pomp of war," he wrote. "I, poor creature, worn out with scribbling, for my bread and liberty, low in spirits and weak in health, leave others to wear laurels which I have sown; others to eat the bread which I have earned. . . ."[1]

This anguished sense of inferiority, which became more and more tinged with raw envy as the Revolutionary War lengthened, is the little understood leitmotif of John Adams's life. Earlier in 1775, when Ben Franklin returned to America and was immediately elected to the Continental Congress, Adams morosely noted that "from day to day [he was] sitting in silence, a great part of the time asleep in his chair." But John was sure the already famous sage would get most of the credit for everything achieved in Congress. Adams was too self-absorbed to appreciate that Franklin's policy of remaining silent was a shrewd tactic aimed at disarming the numerous congressman who suspected he might be a British spy because his son William was loyal to the king.

II

"Honest John" Adams, as he liked to call himself, was born on October 30, 1735, three years after George Washington. His father, John Adams Sr., was a small farmer and leatherworker in the town of Braintree, ten miles south of Boston. A sober, industrious puritan, the elder John traced his lineage back to Pilgrims John and Priscilla Alden. He served the town in various local capacities, including lieutenant of the militia.

Behind his solemn facade lay a warm and loving parent, who paid close attention to the personalities of his three sons and loved to play with them. One of his favorite stunts was rapping out reveille on the kitchen table with a pair of drumsticks he had learned to use in the militia at the age of fifteen. His sons loved it, but his wife, Susanna Boylston Adams, violently objected, claiming it hurt her ears—and may have left a few nicks in her table, which, as a compulsive housekeeper, she would feel obliged to wax smooth.

John's mother was a far more turbulent spirit than his steady, methodical father. Susanna Boylston Adams had a fiery temper that exploded at anyone who crossed her. She and her husband had numerous arguments about money. John called them "rages and raves" that made him fear "all was breaking into flame." Often he fled the house or retreated to his room to try to lose himself in a book.

From other glimpses of her in John's diary, Susanna was a woman whose mood swings were wild and unpredictable. She would fly through the house, dusting, polishing, and scrubbing, scolding her sons to be neater and more careful with their toys and clothes—and then relapse into fits of blank gloom and lassitude in which housework was neglected and the slightest question could trigger tears and disproportionate rage.

John Adams inherited this temperament. In one of his earliest diaries, he wrote: "ballast is what I want. I totter before every breeze. My motions are unsteady." Several biographers have suggested that he—and Susanna— were manic depressives. Whether or not he or she deserves that diagnosis, John's unstable emotions would cause him—and his country—not a little political turmoil in years to come.

III

In his voluminous youthful diary, John admitted he had "an amorous disposition" and as early as eleven was "very fond of the society of females." Growing up on a farm, he had no need for sex education. But he retained his self-control with his many "favorites" throughout his Harvard years. Thereafter his "disposition" toward the opposite sex "engaged [him] much."[2]

As a new attorney, John found it especially hard to concentrate on his law books whenever his thoughts wandered to a certain house in nearby Germantown, where Hannah Quincy lived. Hannah was beautiful—and not at all shy. At twenty-three, she had matrimony on her mind and was determined to get it into John Adams's head. She teased him relentlessly about his studious habits, flatly declaring he would make a very poor husband in her opinion.

"Suppose you was in your study," she asked him, "and your wife should interrupt you accidentally and break off your chain of thought?" Would he snarl at her? Rebuke her? Or welcome her?

"Should you like to spend your evenings at home in reading and conversing with your wife rather than to spend them abroad in taverns or with other company?" Hannah demanded.

By the time he replied to these challenges, John was drenched in perspiration. He found it extremely difficult to discuss love and marriage with Hannah without asking her to be his wife. Among her other charms, Hannah could discuss Alexander Pope's translation of Homer or John Milton's *Paradise Lost* with a perspicacity that John Adams had seldom heard, even at Harvard. Whenever he found Hannah's lovely head bent over a book, he was so overwhelmed with desire that he almost asked the fatal question.[3]

But the law was a sterner mistress—in theory at least. Practice was another matter. After a weekend with Hannah, John desperately lectured himself in his diary: "Here are two nights and one day and a half spent in a softening, enervating, disapating [*sic*] series of bustling, prattling poetry, love, courtship, marriage . . ." A portion of those nights was almost certainly spent strolling with Hannah in "Cupid's Garden," a lover's lane not far from her house. His inability to resist the opportunity to steal a kiss led him into gloomy doubts about his future. He yearned to play a part on a larger stage, to win fame, to become a rich lawyer and great man. How could he hope to do it when he wasted so much time with a mere woman?

Frantic, John tried to deny the reality of Hannah's charms to her face. He lectured her and her equally pretty cousin, Esther Quincy, who lived across the road, on the folly and futility of love. He announced that he despised the whole idea of submitting to such a petty passion. The Quincy women laughed in his face. He reeled home to spend another night denouncing himself in his diary.

One day in the spring of 1759, John found himself in the drawing room of the Quincy home, alone with Hannah. With her usual skill, she turned the conversation to the intricacies of love and marriage. She began talking about her cousin Esther Quincy's engagement to Jonathan Sewall, murmuring in her sibilant, knowing way about how much love had changed Esther.

All semblance of resistance collapsed in John's tormented psyche. Alone in the silent house, the entire world seemed to shrink to the dimensions of the couch on which they were sitting. John leaned toward Hannah, breathing her delicate perfume, lost in the liquid depths of her tantalizing eyes. The words of love and commitment were on his lips . . .

The door crashed open and upon them burst Hannah's cousin, Esther, and her fiancé, Jonathan Sewall. The almost lovers recoiled to opposite ends of the couch, and John rode back to his family's farm in Braintree feeling as if he had narrowly escaped falling off the edge of the earth. For the next month he avoided Hannah's company.

To his astonishment, John next heard that Hannah was engaged to one Bela Lincoln, a handsome militia captain who had begun calling on her in his uniform. Adams found himself incapable of looking at a law book without seeing Hannah's face on the page. He spent half his nights awake thinking about her and the other half asleep dreaming about her. Less amusing were the "chagrin and fretfulness and rage" this highly emotional man struggled to contain. He began to wonder whether he was going mad.

IV

John Adams slowly recovered from his heartbreak over Hannah Quincy and found new preoccupations for his restless mind and unruly imagination. One diversion was the beginning of a serious quarrel between the province of Massachusetts and the rulers of the British Empire over their assumed right to raise revenue from import duties and taxes. He was also slowly advancing up the ladder as a lawyer, winning cases and admittance to the Superior Court as a barrister.

A Harvard friend, Richard Cranch, was courting Mary Smith, the oldest daughter of the Reverend William Smith in nearby Weymouth. Like many other suitors, Cranch felt more comfortable with a friend for support, and he inveigled John into several visits to the Smith parsonage. Here he encountered slim, dark-haired Abigail Smith, age seventeen. Two years earlier, he had met her and her two sisters and had disliked them all. They were "wits," he conceded, but he doubted whether they were "frank or fond or even candid."[4]

John was surprised to find what a difference two years had made in his impression of Abigail Smith. He liked her because she was direct and frank, unlike the fond but devious (in retrospect) Hannah Quincy. Moreover, Abigail was interested in politics and philosophy as well as the arts. She could even read French, a remarkable achievement for someone who had received little formal education. She was not shy about telling him and other people that she considered this failure to educate women a great mistake. She argued that women had an important role to play in the "theater of life" and it was folly not to prepare them for this "trust."

John Adams found himself thinking more and more about Abigail Smith and paying repeated visits to her father's Weymouth parsonage. Sickly Abigail, barely five feet tall, had been a semi-invalid for much of her life. She had despaired of ever having a "spark"—a boyfriend. She found herself unexpectedly attracted to this stocky, somewhat pompous lawyer who treated her numerous strong opinions with remarkable respect. Did she divine how badly he needed someone with whom he could share his perpetual inner agitation? Perhaps.

The courtship was powered by mounting ardor—especially on John's part. One day he showed up at the Weymouth parsonage with a letter written in solemn legalese, ordering Abigail "to give the bearer as many kisses and as many hours of your company after six o'clock as he shall please to demand and charge them to my account." More than once they came close to abandoning their puritan restraint. When a violent rainstorm prevented John from making the trip from Braintree to Weymouth, he told Abigail that it was "a cruel but perhaps blessed" downpour, because it forced him to "keep my distance."[5]

They were often separated by the demands of John's growing law practice. In the fashion of the time, they adopted pen names. She was Diana, the virgin huntress; he was Lysander, a Spartan general who scored a famous victory over Athens and was also a noted diplomat and politician. As her confidence in their love grew, Abigail went to work on reshaping Lysander into a more sociable creature. She told him he was much too intimidating in company, with a haughty, oaklike reserve—"an intolerable forbidding expecting silence which lays such a restraint upon moderate modesty that 'tis impossible for a stranger to be tranquil in your presence."

John replied by sending Abigail a list of her faults: 1. She did not play cards well, holding the deck in a very "uncourtly" way. 2. She was too bashful. 3. She refused to sing. 4. She had a bad habit of hanging her head like a bulrush—undoubtedly caused by another fault that was all but inexcusable in a lady—"reading writing and thinking." 5. She sat with her legs crossed, a habit that would broaden her hips. This, too, was caused by too much thinking. 6. She walked with her toes turned inward. He wanted to see her with a "stately strut and divine deportment." These were all the faults he could find after three weeks of deep thought while they were separated by his inoculation against smallpox—a complicated procedure in their day. "All the rest is bright and luminous."[6]

Abigail's reply was saucy. She told him she was "so hardened as to read over most of my faults with as much pleasure as another person would have read their perfections." This was hardly surprising. All of them demonstrated that Lysander spent most of his time scrutinizing Diana with an intensity that could only be the product of love. She defended her failure to sing by warning him that her voice was as harsh as the screech of a peacock. She promised to stop crossing her legs, though she tartly suggested "a gentleman has no business to concern himself about the legs of a lady." As for her parrot toes, they could be cured only by a dancing master.[7]

Their wedding was set for October 25, 1764. As the date approached and Abigail vanished in a whirl of shopping and visits to friends and relations, John's yearning spilled out in his letters, along with his growing sense that he had found a woman who would gave him the inner balance he needed. "My soul and body have both been thrown into turmoil by your absence," he told her. "A month or two more [of bachelorhood] would make me the most insufferable cynic in the world—I see nothing but faults, follies, frailties and defects in anybody lately . . . But you who have always softened and warmed my heart shall restore my benevolence as well as my health and tranquility of mind."[8]

Perhaps this daunting task caused Abigail to recoil. As the date loomed, she took to her bed under doctor's orders. She was "extremely weak . . . low spirited . . . hardly myself," she told John. He was not much better, reporting from Braintree a catalogue of hypochondriacal complaints. But they both underwent miraculous cures, and the wedding took place on schedule in the Reverend Smith's parsonage in Weymouth.[9]

V

Without Abigail to listen to his explosions of self-pity, one shudders at what might have happened to John Adams. By 1775, the couple had abandoned the nicknames of their courtship days, Diana and Lysander. Abigail had begun signing her letters "Portia"—a name John had chosen, in tribute to her similarity to the heroine of Shakespeare's *Merchant of Venice*. Abigail wrote to John as "My Dearest Friend." There is little doubt that both partners knew the significance of the word "friend" as proof of ultimate intimacy between a man and wife.

No other founding father was as tormented by a hunger for fame as John Adams. It was a subject he never stopped discussing with himself in his diary and in his letters to Abigail. At one point, he pondered the dilemma of this desire and the high standards he inflicted on himself in pursuit of it. "If virtue was to be rewarded with wealth," he wrote, "it would not be virtue. If virtue were to be rewarded with fame, it would not be virtue of the sublimist kind."

Adams was in the grip of an ideology that often drove George Washington and other realists to near distraction and frequently complicated the struggle for independence. There were, according to this doctrine, many kinds of revolutionary patriotism. But the only men that deserved admiration were the "True Whigs"—those who were patriots without a smidgen of self-interest in their motivation. Almost everything John Adams said and did had to meet this impossibly exalted standard. As a result, he was in a constant state of inner turmoil. Without Portia to assure him that he was an admirable but unappreciated man, he would have inevitably slipped into permanent depression and despair.[10]

VI

There was much more to Abigail's role in their marriage than psychological nursemaid. Thanks to John's encouragement, she soon emerged as a thinking woman in her own right. Her most remarkable moment came in 1776, the year when John's leadership in Congress won him his chief claim to fame. Portia wrote to her dearest friend soon after General Washington drove the British army from Boston, restoring a sense of safety and tranquility to Massachusetts. She began by telling him how lighthearted she felt but swiftly moved to a subject that troubled her almost as much as the departed redcoats.

> I long to hear that you have declared an independancy—and by the way in the new Code of Laws which I suppose it will be necessary for you to make I desire you would Remember the Ladies, and be more generous and favourable to them than your ancestors. Do not put such unlimited power into the hands of the Husbands. Remember all Men would be tyrants if they could. If perticuliar care and attention is not paid to the Ladies we are determined to foment a Rebelion, and will not hold ourselves bound by any Laws in which we have no voice, or Representation.
>
> That your sex are Naturally Tyrannical is a Truth so thoroughly established as to admit of no dispute, but such of you as wish to be happy willingly give up the harsh title of Master for the more tender and endearing one of Friend. Why then, not put it out of the power of the vicious and Lawless to use us with cruelty and indignity with impunity. Men of Sense in all Ages abhor those customs which treat us only as the vassals of your Sex. Regard us then as Beings placed by providence under your protection and in immitation of the Supreem Being make use of that power only for our happiness.

Abigail's letter is justly famous today. John's reply two weeks later is much less well known. Basically, he told her: *Equality for women? Forget about it.*

> As to your extraordinary code of laws, I cannot but laugh. We have been told our struggle has loosened the bands of government every where. That children and apprentices were disobedient—that schools and colleges were grown turbulent—that Indians slighted their guardians and negroes grew insolent to their masters. But your letter was the first intimation that another tribe more numerous and powerfull than all the rest were grown discontented.— This is rather too coarse a compliment but you are so saucy, I won't blot it out.
>
> Depend upon it, We know better than to repeal our masculine systems. Altho they are in full force, you know they are little more than theory. We dare not exert our power in its full latitude. We are obliged to go fair, and softly, and in practice you know we are the subjects. We have only the name

of masters, and rather than give up this, which would compleatly subject us to
the despotism of the peticoat, I hope General Washington, and all our brave
Heroes would fight. . . .[11]

Abigail was less than pleased with this reply. She wrote to her friend
Mercy Otis Warren, whose protofeminist views were even more advanced,
and told her about John's flippant dismissal. "I will tell him I have only
been making trial of the disinterestedness of his virtue, and when weigh'd
in the balance have found it wanting," she vowed.[12]

If Abigail embarked on such an exchange of verbal barbs, we have
no record of it. The larger drama of whether "independency" would be
declared in spite of stubborn resistance among the congressional delegates
of several colonies, notably New York and Pennsylvania, soon absorbed
both letter writers. John was at the center of the struggle but even here,
in the midst of the greatest achievement of his life, he found time to let
Abigail know how important she was to him. On May 22, while others
were orating on the floor of Congress, John told Portia he was seated "in
the midst of forty people, some of whom are talking, and others whispering." He told her that the news from England made it look as though
all-out war would erupt this summer, which meant there was no hope of
him getting home.

He was thinking of asking permission to bring her and their four children (Abigail, called "Nabby," John Quincy, Charles, and Thomas) to
Philadelphia, saying, "I am a lonely forlorn creature here." He yearned
to see her and "those babes, whose education and welfare lies so near my
heart." But he could not let anything divert him from "superior duties."
In a previous letter, Portia had told him how much she yearned for him.
But "all domestick pleasures and injoyments are absorbed in the great and
important duty you owe your country."

Those words drew from John another confession of the central role Abigail played in his life: "Among all the disappointments and perplexities,
which have fallen to my share . . . nothing has contributed so much to support my mind, as the choice blessing of a wife whose capacity enabled her
to comprehend and whose pure virtue obliged her to approve the views of
her husband. This has been the cheering consolation of my heart, in my
most solitary and most disconsolate hours."[13]

What could Portia—or any other woman—say in response to such a compliment? Abigail reduced her planned assault on Honest John's disinterested virtue to a mild rebuke. She wondered why he and his cohorts were "proclaiming peace and good will to men [and] emancipating all nations, [but] you insist on retaining an absolute power over wives."

V

Soon, Abigail had the immense pleasure of reading John's July 3, 1776, letter which has become one of the most famous documents in the nation's history. "Yesterday the greatest question was decided, which ever was decided in America, and a greater perhaps never was or will be decided among men." A resolution was passed without one dissenting colony "that these united colonies are and of right ought to be free and independent states." A full-fledged "Declaration of Independency" would be forthcoming in a few days.

John thought July second, when independence was voted, would be the day everyone would remember. But Thomas Jefferson's blazing prose made July fourth, when the Declaration was approved, the nation's birthday. Nonetheless, John's letter should be read as a companion piece to the great document. No other founding father had his special blend of wary hope and stubborn faith. It was a philosophy he shared wholeheartedly with Abigail. Flowing from their New England heritage, with its pessimistic view of human nature, it has a surprising relevance in twenty-first-century America:

> It may be the will of heaven that America shall suffer calamities still more wasting and distresses, yet more dreadful. If this be the case, it will have this good effect, at least: it will inspire us with many virtues, which we have not, and correct many errors, follies and vices, which threaten to disturb, dishonour and destroy us. The furnace of affliction produces refinements, in states as well as in individuals. . . .[14]

VI

Throughout these letters that chart the nation's zigzag progress toward independence are heart-wrenching cries of endearment. "I read and read again your charming letters, and they serve me in some faint degree as a substitute for the company and comfort of the writer," John wrote. "I want to take a walk with you in the garden . . . I want to take Tom in one hand and Charles in the other and walk with you, Nabby on your right hand and John [Quincy] on your left, to view the cornfields, the orchards . . ."

From Abigail came: "I long earnestly for a Saturday evening and experience a similar pleasure to that which I used to find in the return of my friend upon that day after a week's absence." She told John she was doing her best with "our little ones whom you so often recommend to my care." But the "precepts" of virtue and probity she preached to them would be "doubly inforced" if there was "a father constantly before them."[15]

In spite of these endearing sentiments, John Adams stayed in Congress for thirteen months after independence was declared, with only one visit home in the fall of 1776. In August 1777, Abigail was wistfully telling him it was "eight months since you left me." She reported a dream in which John returned but treated her "so coldly that my heart ackd [ached] for a half hour after I awoke." Portia was revealing that she had begun to wonder about her dearest friend's perpetual absence. Was there some less noble motive besides patriotism involved? Even if patriotism was his motive, why was he bearing so much of the burden? Other lawyers in Boston were making fortunes while he debated in Congress about how to win—or to not lose—the war. Why didn't he ask them to spend some time in Congress?

John's salary fell laughably short of meeting his expenses, and the money issued by Congress was depreciating rapidly, sending prices soaring. Abigail had to deal with no less than six tenants on the Adams farm, all of whom owed John money when he went to Congress in 1774. He huffed and puffed about forcing them to pay up, but he left the job of collecting the cash to Abigail. She coped—and then some. She was soon telling John: "I hope in time to have the reputation of being as good a farmeress as my partner has of being a good statesman."

As the months passed, however, the multiplicity of details and the recalcitrance of some of her tenants drained Abigail's patience and pride. She needed to hire laborers, but there were none to be found. Many young men were being lured into the army with $100 enlistment bounties and others into privateering by the hope of bigger but far more risky money. She finally hired one of her tenants, named Belcher, because he was a hard worker, even if he might occasionally "purloin a little."[16]

VI

Complicating matters for the first seven months of 1777 was Abigail's pregnancy. The child had been conceived during John's visit, and his almost immediate return to Congress had left her feeling doubly bereft and morose. All her pregnancies had been difficult, but this one proved to be especially traumatic. "I am cut off," she wrote with uncharacteristic gloom, "from the privilege which some of the brute creation enjoy, that of having their mate sit beside them with anxious care during their . . . confinement." She was tormented by premonitions of death for herself or the baby in her womb. Political quarrels in Massachusetts added to her discouragement. The countryside had rediscovered its longstanding animosity to Boston, and people were insulting and accusing each other of every fault imaginable. "Avarice, venality, pride, dissipation" prevail, she told her absent friend.

In Philadelphia, John grew almost frantic as Abigail's time approached. "Oh that I could be near to say a few kind words, or show a few anxious looks or do a few kind actions. Oh, that I could take from my dearest a share of her distress, or relieve her of the whole." These heartfelt phrases were too frail to withstand Abigail's mounting anxiety. In July, she was seized with paroxysms; her heart pounded wildly, her whole body trembled. Soon she was convinced that the baby was dead, in spite of her doctor's reassurance.

Her labor pains began as she was writing a letter to John. She told him that she had to lay down her pen "to bear what I cannot fly from." When the pangs ceased for a few minutes, she wrote, "I have endured it and resume my pen." She proceeded to give John a detailed account of the condition of the farm—the promising wheat and corn crops, the need for manure, the exorbitant cost of hiring a man to do some mowing. After another twenty-four hours of sporadic labor, Abigail gave birth to a baby girl, stillborn—to the doctor's great chagrin. Abigail's dark intuition had been correct. John's former law clerk, John Thaxter, a Braintree neighbor, told the congressman it was "an exceeding fine looking child."[17]

Abigail recovered with a rapidity that amazed her. Both she and John seemed to accept the baby's death as the will of God—and perhaps a covert blessing. She had more than enough for one woman to do—raising four children and managing the farm. In Philadelphia, a ferocious heat wave reduced John to utter misery. He was "wasted and exhausted in mind and body," he told Abigail. It had been three long years since he had departed for Philadelphia "in search of adventures." Looking back, all he could see now was the toll it had on taken his spirits. He was thoroughly tired of being a legislator. "The next time I come home it will be for a long time," he vowed.

With the courage that seemed to come naturally to her, Abigail told John she felt "my sufferings amply rewarded" by the "tenderness" he expressed for her and—though she did not put it in words—by her own passionate patriotism. The sense that she was sustaining—and sharing—this complicated man's contributions to their country had huge meaning for her.

Behind these brave words lay a growing sense of abandonment. Abigail was not helped by a correspondence with Massachusetts delegate James Lovell, a glib flatterer who left his wife and six children home for five years without a single visit while he surreptitiously enjoyed several women in Philadelphia. When Abigail asked for a map of Pennsylvania so she could follow the progress of the war if the British attacked the American capital, John asked Lovell to send her one. It arrived with a covering letter that suggested Mr. Lovell had shed whatever vestiges of puritanism he may have inherited from his ancestors. Addressing it to "Portia," he told her that knowing she had given her heart to such a man as John Adams is "what most of all makes me yours." More letters would follow, full of even more leering double entendres.

VII

In November 1777, after Washington's army lost three battles and was forced to surrender Philadelphia to the British, John and his fellow legislators fled to York, a town settled by German immigrants, where only a handful of people spoke English. The war seemed to be veering into a stalemate. In October, the Americans had scored a victory at Saratoga, capturing an entire British army. But the British seizure of the American capital was a blow that suggested the struggle might last for years. John decided it was time to retire from Congress and acquire some of the money other lawyers were making.

Back in Braintree by the end of November, he was joyously reunited with Abigail and the children. Soon the revived attorney was in Portsmouth, New Hampshire, handling a case in admiralty court, where privateering captains sold the British ships and cargoes they had captured and frequently quarreled with their officers and crews about how to share the spoils of war. In Braintree, a packet arrived from Congress that Abigail opened. She read with horror and astonishment that John had been appointed a "commissioner" to negotiate an alliance with France, replacing Silas Deane of Connecticut, who had been recalled for purported incompetence and corruption.

Along with a stern letter from the committee on foreign affairs practically demanding John's acceptance was a personal note from James Lovell. He hoped John would accept, "however your amiable partner may be tempted to condemn my persuasions of you to distance yourself from her."

Abigail exploded. She seized her pen and asked Lovell how he could "contrive to rob me of all my happiness." How could he expect her to "consent to separate from him whom my heart esteems above all earthly things? My life will be one continual scene of apprehension and anxiety." Did he really expect her to "cheerfully comply with the demand of my country?"[18]

When John returned from Portsmouth, he accepted the appointment. Lovell had made it sound as if he would be the only capable member of the American diplomatic team. Benjamin Franklin was too old, and the other commissioner, Arthur Lee of Virginia, was erratic. John saw it as his duty. A distraught Abigail announced she would go with him. They would take the two oldest children, Nabby and John. But John regretfully demurred. There was a very good chance that their ship might be captured by British men-of-war. A winter crossing of the icy Atlantic was harrowing, even without this risk. Then there was the expense of living in Paris, which would far exceed his meager salary. Without Abigail at home, there would be no one to run the farm; the crops made money they badly needed. John decided he would take ten-year-old John Quincy with him to relieve Abigail of some responsibility and hope to be home in a year with a treaty signed and the Revolution rescued.

On one level, John's decision—and Abigail's reluctant acceptance—is moving evidence of their commitment to their country's cause. On another level it underscores how ready Portia's dearest friend was to sacrifice her happiness to his pursuit of fame. John did not even try to find out if the accusations against Silas Deane were true. Nor did he ask himself why it was necessary for him to rush to France when Benjamin Franklin was already on the scene.

Adams knew Franklin well, having served with him in Congress in 1775–1776. His envy of the sage had rapidly abated when Franklin became an outspoken supporter of independence. When the seventy-year-old patriot undertook a winter trip to Montreal to try to persuade the French-Canadians to side with the American army that had invaded Canada in 1775, Adams had praised him for his "masterly acquaintance with the French language, his extensive correspondence in France, his great experience in life, his wisdom, his prudence, his caution."

Why did John take the word of a second-rater like James Lovell that Franklin was now too old to deal with the French in Paris without the assistance of John Adams, who was totally inexperienced in European diplomacy and did not understand a word of French? John's hunger for fame had become a passion that overwhelmed his judgment and common sense. Abigail shared his ambition, but she had begun to wonder how much loneliness she could endure.

PORTIA'S DUBIOUS DIPLOMAT

*F*rom the start, the omens for John Adams's venture to Europe were bad. Father and son boarded their ship, the U.S. Navy frigate *Boston,* on Friday, the thirteenth of February, 1778. An Adams relative, who had a habit of reading worrisome portents in nature, noted how "the clouds roll and the hollow winds howl" as the Adamses departed. The portents were on target. A week earlier in Paris, Benjamin Franklin and his fellow commissioners, including the supposedly untrustworthy Silas Deane, had signed a treaty of alliance with France. When John Adams finally reached Paris after a series of terrifyingly close calls from pursuing British men-of-war and ferocious winter storms, he was a superfluous diplomat. There was nothing for him to do but sit around and wait for Congress to ratify the compact, which was as certain as balmy spring breezes succeeding the wintry winds that had almost deep-sixed the USS *Boston.*

Even more dismaying must have been Adams's discovery that the accusations against Silas Deane were anything but open-and-shut. His accuser, Arthur Lee, was also feuding vituperatively with Benjamin Franklin. It did not take Adams long to conclude that Lee was a mental case. As a partner in vitriol Lee had an even more unstable Franklin envier/hater, Ralph Izard of South Carolina. A friend of the president of Congress, Henry Laurens, Izard had been appointed ambassador to the Grand Duke of Tuscany, who declined even to permit him into his domain. So he stayed in

Paris and made trouble for everyone. Izard was a paradigm of Congress's gross ineptitude in the business of foreign policy.

Thus began the worst years of John Adams's career—and of his marriage. His hopes of wider fame withered in Benjamin Franklin's shadow. "There was," he complained, "only one American name on everyone's lips: Franklin. His name was familiar to government and people, to foreign courtiers, nobility, clergy and philosophers, as well as plebians to such an extent that there was scarcely a peasant or a citizen . . . who did not consider him a friend."

Almost as upsetting were the numerous Frenchmen and women who asked Adams if he were "*le fameux* Adams" and were puzzled to learn he was not named Sam. Worse, John's French was so poor—even nonexistent at this point—that he could not explain that he and Sam were cousins and partners in the glorious struggle for independence.

The French readiness to discuss sex in casual conversation shocked John's puritan sensibility. He was barely ashore in Bordeaux when he was invited to a dinner at which he found himself seated beside an attractive young Frenchwoman. Through an interpreter, she asked him if she might conclude from his name that he was descended from the first man and woman in the Garden of Eden. She hoped that he had imbibed from his family's ancient traditions an answer to a question that had long troubled her: "How did the first couple learn the art of lying together?"

Adams felt his face grow hot, but he controlled himself and decided to "set a brazen face against a brazen face." He coolly assured the smiling young lady that he was in fact descended from the first family and was glad to answer her question. There was a "physical quality in us resembling the power of electricity or of the magnet." That was why when men and women came within "striking distance of each other" they flew together like objects in electrical experiments. (John would have been mortified if he knew that Benjamin Franklin was using the same metaphor to charm the ladies of Paris.) The young lady was delighted with the explanation and replied that until now she had never understood the reason for it, but she knew it was "a very happy shock." Adams remained unamused.[1]

Adams's dislike of the French did not lessen when he reached Paris. He told Abigail in an early letter that "if human nature can be made happy by any thing that can please the eye, the ear, the taste or any other sense," France would be a "region for happiness." But he would exchange all the elegance in dress, the magnificence in architecture, even the "handsome and very well educated" women for "the simplicity of Braintree."[2]

Benjamin Franklin welcomed John as a friend and colleague from the glory days of his fight for independence in Congress. He invited him and John Quincy to live rent free in the villa he was using in Passy and introduced John to powerful and influential Frenchmen and women. Adams struggled to be fair and cordial at first. He wrote to Sam Adams, telling him that there was no need for three diplomatic "commissioners" and Congress should appoint Franklin the sole ambassador to France. But his endemic envy and his anger at himself for succumbing to the lure of fame—and at Congress for sending him on this pointless mission—slowly soured his disposition.

Honest John began finding fault with the casual, seemingly unsystematic way Franklin operated as a diplomat. Although Adams enjoyed French food, he grew weary of Franklin's habit of dining out with the rich and powerful almost every day and spending his evenings in visits to Madame Brillon, Madame Helvetius, and other women friends, enjoying music, chess, and card games. "The life of Dr. Franklin was a scene of continual discipation [*sic*]," Adams groused.

Franklin's popularity with the ladies of France grew harder and harder for Adams to endure. In one of his first letters to Abigail, John tried to be lighthearted about it. "My venerable colleague enjoys a priviledge [*sic*] that is much to be envied. Being seventy years of age, the ladies not only allow him to embrace them as often as he pleases, but they are perpetually embracing him." Adams was shocked by Madame Brillon's household, with her husband sleeping with the governess and Madame herself, in Adams's fevered imagination, taking lovers. John was even more appalled by Madame Helvetius, who had the young doctor (and poet), Pierre-Jean-George Cabanis, and two unfrocked abbots living in her house and dining nightly at her table. "These ecclesiastics . . . I suppose have as much power to pardon a sin as they have to commit one. Oh Mores! said I to myself," John muttered in his diary.

II

While Franklin was inadvertently torturing John in Paris, back in Braintree, Abigail Adams was in a continual state of anxiety and dread. It was months before she heard that her dearest friend and John Quincy had reached France safely. She had to survive a rumor that they had been captured and John would soon be on trial for treason in London. The British propaganda machine even circulated the name of the frigate that had captured the *Boston*.

Then came another rumor, that Benjamin Franklin had been assassinated. Abigail immediately thought that if the perfidious British were ready to murder "that great philosopher [and] able statesman," they would not hesitate for a moment to kill John Adams. It took two months for this story to be altered from an accomplished fact to a dastardly plot that had failed. Abigail told her dearest friend that she could only hope "this shocking attempt will put you upon your guard." Without her faith in "a superintending providence" she would be "overwhelmed by my fears and apprehensions."[3]

Meanwhile, she sent letter after letter across the Atlantic into an apparent void. This experience of almost total isolation blended with her sense of abandonment to ferment emotions Abigail never imagined she would feel as a wife. She warned John in one of her first letters that he "must console me in your absence by a recital of all your adventures." When he failed to do so, she became more and more unhappy. His letters were much too short, and as time passed they seemed cold and almost perfunctory. She did not realize that he was struggling to find some purpose in his superfluous diplomatic role. Congress seemed to have forgotten he even existed.

Throughout 1778 and 1779, Abigail grew more and more angry. By November 1778 she was becoming blunt. She "could not help complaining to my dearest friend" that she had received only "3 very short letters" in the previous nine months. By this time John was slipping into a serious depression, and he replied in kind. Her letters, he wrote, gave him something close to a "fit of spleen." In a tone of almost brutal dismissal, he told her "you must not expect to hear from me so often. . . . I have too much to do." He proceeded to list all his duties—keeping track of the embassy's correspondence with Congress, the Navy, and the Court of France, all without the aid of a clerk, which he could not afford. Meanwhile, rumors warned him that he was likely to be recalled, or sent to a foreign capital such as Vienna, where he would be treated as a nonperson. He was being ignored and/or mistreated because he was too independent, he would truckle to nobody. He was ready to accept his fate and "preserve my independence at the expense of my ambition."

Once Abigail would have tried to soothe this outburst of self-pity. But she was too unhappy to deal with it, especially when the moans arrived three or four months after they were uttered. In the last weeks of 1778, John denounced her recent letters as "complaints [that] gave me more pain than I can express." He wrote three angry answers to them but "burnt them all." Grimly he warned her, "If you write me in this style I shall leave of[f] writing intirely. It kills me." What was wrong with her? His letters professed his esteem. "Can protestation of affection be necessary? Can tokens of remembrance be desired? The very idea of this sickens me." Was he not "wretched enough, in this banishment?"[4]

Abigail's mood veered from desperate loyalty to sarcasm to anger. Two days after Christmas 1778, she told John that the Atlantic divided "only our persons for the heart of my friend is in the busom of his partner." A week later, she wryly wondered how she could have lost *every letter* of the fifty John claimed he had written to her. She must be "the most unfortunate person in the world." Perhaps the best solution was to adopt "the *very concise* method of my friend" and send him "billits containing not more than a dozen lines, especially when paper has grown so dear."[5]

Still too depressed to change his mind or tone, John turned to eleven-year-old John Quincy for help. He of course took John's side of the quarrel. At the end of February 1779, he told his mother he had read one of her letters in which she complained "of my Pappa's not writing." John Quincy scolded Abigail for not understanding that Pappa had "so many other things to think of" and had time to write only short notes because they usually heard about a ship's sailing at the last minute and there was no time to write a long letter. But Abigail complained "as bad or worse than if he had not wrote at all and it really hurts him to receive such letters."[6]

This communication must have made Abigail wonder about her dearest friend. Why was her would-be statesman asking an eleven-year-old to be his spokesman?

III

In this atmosphere, a desperate Abigail turned to another correspondent—sneering, leering James Lovell. She was not looking for love or even a substitute for it; she wanted information, and he was in a position to give it to her. He was secretary of the congressional committee on foreign affairs.

She had written to Samuel Adams but he had not even bothered to answer her—a nice glimpse of that worthy's flinty style. Abigail was also aware of another motive in corresponding with Lovell. In one of her first letters, she told him it was "a relief to my mind to drop some of my sorrows through my pen."

Lovell began sending her weekly installments of the journals of Congress, a privilege afforded to few Americans. The deliberations of the national legislature were supposed to be a state secret. But Lovell could not resist injecting erotic innuendos into almost every letter. "Call me a savage," he wrote in April 1778, "when I inform you that your alarms and distress have afforded me *delight*." There was little doubt that in Lovell's imagination, he held a sobbing Abigail in his arms.

To make money on the side, Abigail had become a merchant, selling items John sent her from France. Often these shipments arrived in Philadelphia, where Lovell forwarded them to Boston. Sometimes he opened them to make sure they were not damaged. "I feared moths," he wrote about one consignment. "Have opened your goods—aired and shook the woolens—added tobacco leaves and again secured them for transportation." This assistance in making money Abigail badly needed soon made Lovell difficult to resist.

But she was not enthused by his personal comments and questions. "How *do* you do, lovely Portia, these very cold days, mistake me not willfully, I said *days,*" he wrote in one letter. Remarks like this prompted Abigail to call him a "dangerous man," and to remind him that she was a married woman who had irrevocably given her heart to John Adams. But she signed her letter "Portia."

Lovell responded with a quotation from the Scottish poet Allen Ramsay, "Gin ye were mine ain thing, how dearly would I *love* thee." Next he told Abigail that he was glad her husband's *rigid patriotism* had not produced another pregnancy before his departure to France. He also hinted that John might be doing more to entertain himself in Paris than visiting museums.

Abigail rebuked Lovell for these innuendos—but she continued their correspondence. There is more than a little evidence that she enjoyed Lovell's flirtatious ways. What wife approaching forty could resist being told that she was desirable, especially when the "flatterer"—a term she often threw at Lovell—was three hundred miles away in Philadelphia? But she did not allow him unlimited verbal license. She told him he could call her "amiable" and "agreeable," but "lovely" and "charming" were banned.

Lovell did not relent at first. "Amiable and unjust Portia," he cried. "Must I write to you in the language of the gazettes?" In another letter he protested that he would not be able to resist "all covetousness." He would "still covet to be in the arms of Portia's [here he reached the end of a page; on the next page he continued] friend and admirer, the wife of my busom." Whereupon he sniggered that he hoped Abigail did not stop before turning the page because "a quick turnover alone could save the tenth commandment." He urged her to consult Ecclesiastes 4:11: *Again, if two lie together, they have heat; but how can one be warm alone?* In another letter he teased her as "one of the ____est and ___est and ____est women" he had ever known.

Their relationship changed abruptly when one of Lovell's smutty letters to a fellow Massachusetts delegate, Elbridge Gerry, was intercepted by the British and reprinted in several loyalist newspapers. At first Abigail denounced him. "If what I heard is true, I cannot open my lips in defense," she told Lovell. Calming down in her next letter, she apologized. Obviously she wanted to keep Lovell's friendship for several reasons. Meanwhile, he was being condemned by numerous proper Bostonians. Abigail urged him to come home and spend some time with his wife and children. That would end many of the nastiest rumors.

The vulnerable one now, Congressman Lowell bitterly confessed that he had no profession to support his family. His salary as a congressman would cease the moment he arrived in Boston. Before the war he had been a schoolteacher. He had denounced former students who had stayed out of the war and made fortunes. Did Abigail want him to go groveling to one of these people and ask to become a clerk? In a rage, he assailed her condescension about his "choice of words" and even wondered if she were a genuine American patriot.

Abigail hotly replied that she was only trying to tell Lovell the truth about what was being said about him in Boston. She quoted some of the cruel gossip she had heard from her friend Mercy Otis Warren. Didn't he realize he was becoming an object of derision? Lovell crumpled under this counterattack. "Why do you strive to make me vile in my own eyes?" he whined. He claimed to be true to his wife and even asserted he was "one of the most religious men in the world."

Abigail, sensing she was now in control, relented and assured him they were still friends. "In truth . . . thou art a queer being," she told him. But Lovell "took an interest in my happiness," and she had discovered that "I can make you feel. I hate an unfeeling mortal." She told him he could "laugh and satirize with her 'as much as he pleased.' I laugh with you." But she still felt it would be "necessary to keep a watchful eye on you." Lovell assured her of his renewed respect and friendship, but felt compelled to suggest that "you also Madam are a queer being."

Over the course of the war, Portia and her literary lover exchanged more than ninety letters. But when Lovell finally returned to Boston, he never even tried to visit Abigail. One historian has called their correspondence "a virtuous affair." In many ways that is an excellent description. James Lovell gave Abigail Adams a spiritual and emotional release that helped her survive her dearest friend's political and emotional collapse in Europe.[7]

IV

In an ironic replay of the appointment of Washington as commander in chief, Congress's decision to make Benjamin Franklin the sole ambassador to France deepened John Adams's black mood. He had recommended the change but had apparently hoped that his friends in Congress might turn to Honest John, the one commissioner who had stayed more-or-less aloof from the quarrel. It was another proof, if he needed it, that he could not compete with Franklin's fame. A brooding John told Abigail that "the scaffold is cut away, and I am left kicking and sprawling in the mire." He was not "in a state of disgrace but rather of total neglect and contempt." He could only wonder why he "deserved such treatment."[8]

Arthur Lee had been named ambassador to Madrid. There was not even a hint of an appointment, not a word of either commendation or blame, for John Adams. He vowed to his friend James Warren that he would never again let Congress make a fool of him this way. He told Richard Henry Lee, chairman of the committee on foreign affairs, that he was glad to become a private citizen again.

John's mood was not improved by more letters from Abigail, describing her loneliness and her resentment of his short letters and long months of silence. In a reply that was virtually a snarl, John wrote: "This day, the

Chevalier D'Arcy, his lady and niece, Mr. Le Roy and his lady, dined here. These gentlemen are two members of the academy of sciences. Now are you the wiser for all this? Shall I enter into a description of their dress—of the compliments—of the turns of conversation and all that —? For mercy sake don't expect of me that I should be a boy till I am seventy years of age."

He told Abigail to "consider your age" and acquire some "gravity of . . . character." She had given birth to six children, including a daughter who was "grown up." She should be setting an example to Nabby instead of whining about his short letters. He closed with an oblique apology. "I . . . am grown more austere, severe, rigid and miserable than ever I was. I have seen more occasion, perhaps."[9]

Without asking anyone's permission, John decided to go home. His mood slowly brightened at the prospect, and he again began addressing his letters to "My dearest friend," telling Abigail she would soon be seeing him.

By the summer of 1779, John was back in Braintree, regaling Abigail and the children with Parisian memories. John Quincy delighted everyone with his fluent French and charmed his older sister by volunteering to teach her the language. The gloom that permeated the eighteen-month separation vanished as John and Abigail rode out to visit friends and relations and invited others to join them in cheerful dinners in their farmhouse.

V

John's friends were aware of how deeply he had been hurt by the way Congress had treated him in France. Meanwhile, he represented Braintree at a state convention that gave him a leading role in writing a constitution for Massachusetts. It is considered one of his greatest achievements, second only to his 1776 role as "the Atlas of Independence." In Philadelphia, Congress was grappling with large and unfamiliar challenges. The alliance with France had plunged America into global politics. The new French ambassador informed Congress that Spain had volunteered to mediate the conflict between France and England that had erupted when the French signed the treaty of alliance with the United States. Congress needed to send an ambassador to Europe with power to negotiate a treaty of peace with England, if this diplomacy succeeded.

John Adams's friends decided he was the perfect choice for this large task. At the end of October 1779, a messenger arrived in Braintree informing him that he had been appointed "Minister Plenipotentiary" to explore and hopefully achieve a peace treaty. To underline their seriousness, Congress gave him a salary of $2,500 a year and money for a private secretary, a secretary of the mission, and a servant. John felt enormous satisfaction—even a kind of vindication—in this appointment. His acceptance was a foregone conclusion from the moment he read the document. He was especially pleased that Congress had chosen him unanimously.

Abigail, simultaneously proud and forlorn, could not disagree with him. She made no protest when John decided to take John Quincy with him again, along with his nine-year-old towheaded brother, Charles. They soon boarded the French frigate *Sensible* and were on their way across the Atlantic.

Once more, John did not pause to explore the realities behind his appointment. Congress's unanimous choice of John was a compromise between two political parties that had emerged from the brawl over Silas Deane's recall. Deane's backers won a tacit acquittal of the charges against him, got rid of Arthur Lee and Ralph Izard, confirmed Franklin as the ambassador to France, and won their ally, John Jay of New York, an appointment as ambassador to Spain. Conceding the peace negotiation to John Adams was a minor matter, and making it a unanimous vote was an attempt to heal wounds. The chances of John doing any serious negotiating were dubious, at best. It depended on two very large questions to which no one had the answer: Was Spain serious about playing the mediator? Was Britain interested in peace?

When John, his two sons, and the rest of his entourage reached Paris in mid-February 1780, the answer to both questions became cruelly obvious. Spain's offer to mediate was never sincere. Even before the British rejected the proposal, the Spanish signed a secret agreement with France, committing them to a war that included an invasion of the British Isles by their combined fleets. Worse, the Spanish plan was totally unacceptable to the United States. It called for a peace that would permit the British to retain their grip on New York City and Long Island, Rhode Island, most of Georgia, and a huge swath of what was then called the Northwest, in the center of the continent. In the light of John's previous experience with Congress's ineptitude in foreign policy, one wonders why he did not at least inquire into the terms that Spain was proposing. Once more, his hunger for fame had overwhelmed his judgment and betrayed him into a pointless journey.[10]

VI

With typical amateurishness, Congress had not given John instructions to return to the United States if there was nothing for him to negotiate. He decided to stay in Europe and see whether the French-Spanish armada succeeded in invading England. The attempt was a fiasco. The French admiral in command was a bungler and a coward, and the Spanish ships took forever to reach the English Channel. Meanwhile, disease ravaged the crews of both fleets. The invaders crept back to their home ports without firing a shot.

Seemingly oblivious to this disaster, John Adams asked the French foreign minister, the Comte de Vergennes, when America could open peace negotiations with the British. Struggling to be polite, the comte informed him the timing was inappropriate. In America the British had completed their conquest of Georgia and installed a royal governor. They were about to invade South Carolina. In the West Indies their fleet and army were winning more victories. Vergennes may well have wondered whether Adams was secretly loyal to the British and was trying to make France and Spain look ridiculous. He was getting reports from his ambassador in America about the nasty things John's friends, James Lovell and Sam Adams, were saying about France in Congress. In a scorching reply, Vergennes asked why Adams was trying to become "the laughing stock" of Europe.[11]

The only positive notes John struck in these demoralizing months of flailing in a diplomatic vacuum were in his letters to Abigail. He had regained her love and was determined not to endanger it again. He closed his first letter to her from Paris with: "Yours, yours yours, ever ever ever yours." No matter how discouraging his situation became, he avoided telling Abigail any bad news. He said nothing about his disastrous meeting with Vergennes.

He wrote cheerily of how much the boys were enjoying Paris. Abigail had been dubious about letting him take Charles, who had a sweet dependent nature that made him something of a momma's boy. "Your delicate Charles is as hardy as a flynt," John assured her. "He sustains every thing better than any of us, even than the hardy sailor, his brother." This ebullience did not last long. Charles was soon yearning for his mother's comforting arms.[12]

VII

Having nothing better to do, John also wrote letters to Congress—as many as two or three a day. He reported on British and French politics, naval news from European waters, and rumors that Russia was sending troops to America to help the British. He wrote ninety-five of these epistles, without getting a single reply from Congress. It never occurred to him that he was making himself look silly. "More letters from Adams" became a wry joke among the bombarded legislators.

At Versailles, John's relationship with Foreign Minister Vergennes slid steadily downhill. Adams sent him long letters arguing that France should be doing more to help the Americans. Vergennes soon grew so annoyed that he informed Minister Plenipotentiary Adams that he no longer wanted to hear from him. Henceforth he would consider Ambassador Benjamin Franklin the sole spokesman for America in Europe.

Vergennes handed John's letters to Franklin and told him to send them to Congress. Franklin forwarded them with a covering letter that reported Vergennes's "extreme displeasure" with Adams. The elder statesman spelled out the difference between his and Adams's approaches to France: "He thinks . . . America has been too free in expressions of gratitude to France . . . she is more obliged to us than we to her." Franklin thought this was a mistake. The French should be treated "with decency and delicacy." Franklin closed by asking Congress to let him know which policy they wanted. The ambassador told Adams he had forwarded the letters and urged him to write a reply, defending himself. John ignored him.

Adams decided to go to Amsterdam to see whether he could obtain a loan from the Dutch. Franklin strongly disagreed with this decision. He did not think Americans should be wandering around Europe acting like beggars. It was not the way to win respect as a nation. Adams said he was going, no matter what Franklin thought. In Amsterdam he enrolled his two sons in a school that they both hated. Neither boy spoke Dutch, and John Quincy was dumped into a lower grade with boys half his age.[13]

Meanwhile, John got nowhere trying to persuade the Dutch to loan money to America. The Dutch were covert allies of France in the spreading war with England. Before the year 1780 ended, the British declared war on the Netherlands. Dutch leaders were not going to do anything without the approval of their far more powerful neighbor. The French ambassador to the Netherlands made it very clear that John Adams was persona non grata to Foreign Minister Vergennes.

VIII

In Braintree, Portia was enduring emotional torments of her own. Months passed without a letter from John or the boys. In June 1781 she learned from a sister of Arthur Lee, Alice Lee Shippen, that Franklin had attacked John in letters to Congress. As a partisan in her paranoid brother's camp, Mrs. Shippen made Franklin's remarks far more evil than they were in fact. James Lovell did not help matters by giving Portia "hints" of Congress discussing matters "affecting Mr. A's public character."

In a frenzy of resentment, Abigail began writing letters to a half dozen people, including Lovell, comparing Franklin to General Benedict Arnold as a traitor to true Republican virtue, as personified by John Adams. To Lovell she confessed, "When he is wounded, I blead [sic]." Soon she was concocting a totally imaginary Franklin who hated John Adams because he was so honest and had "no ambition to make a fortune with the spoil of his country or to eat the bread of idleness and dissipation." How could her husband's "zealous exertions for the welfare of his country" end in "dishonor and disgrace"?[14]

If Portia knew what was happening in the Netherlands, she would have been even more distraught. Her dearest friend had hurled himself into his effort to win a loan and diplomatic recognition from the Dutch with a frenzy of activity remarkably similar to his mother's outbursts of frantic housecleaning. Meanwhile, Charles Adams had become seriously ill with strange fevers and had lapsed into permanent homesickness. His father decided his "sensibility" was "too exquisite" for Europe, and sent him home in August 1781.

The trip would take the eleven-year-old boy five anxiety-filled months. The ship that his father chose for him, the frigate *South Carolina,* had an erratic captain, one Alexander Gillon, a Franklin-hater and friend of Ralph Izard. Gillon was supposed to transport important war matériel to America. Instead he left the cargo in Amsterdam and put into a Spanish port, where he dumped Charles and other passengers ashore and went off privateering to make some quick money. Fortunately, the Connecticut-born painter John Trumbull was one of the other passengers and looked after the boy. He and Charles finally got home aboard another American privateer in January 1782.[15]

Not long after Charles sailed, John received a letter from Benjamin Franklin. The ambassador informed him that Congress had abolished his job as sole peace negotiator with England. Adams was now one of four commissioners, including Franklin, empowered to negotiate a treaty when and if the British showed any interest in the subject. Already discouraged, Adams suffered an emotional collapse that he later described as a "nervous fever." For two months he did not write a single letter. More than once he was sure he was near death.

Then came news that rescued him from this cataclysmic gloom. An entire British army had surrendered to General George Washington and his French allies at Yorktown, Virginia.

IX

Peace and independence suddenly became possibilities. So did recognition of the United States by the Netherlands—and a loan from well-heeled Dutch bankers. The Comte de Vergennes signaled his approval of both steps to his ambassador in the Netherlands, and soon Adams was able to claim a diplomatic triumph that equaled Benjamin Franklin's. That was hardly true, but this earnest, deeply patriotic man had struggled through so many disappointments, no one had the heart to disagree with him.

In April 1782, Adams returned to Paris, where he played a major role in negotiating the treaty of peace with England. The double triumph did not make John a happy man. If anything, it exacerbated his envy of Benjamin Franklin. He filled the mail with nasty remarks about the sage. He told correspondents that Franklin's reputation was explained by "scribblers in his pay in London to trumpet his fame." He was sure that Franklin would make his grandson, William Temple Franklin, ambassador to France and himself ambassador to England—a post Adams badly wanted. He topped these wrathful thoughts with a letter to Arthur Lee, in which he predicted someone was going to propose "to name the 18th Century the Franklinian age."[16]

X

Although Abigail rejoiced in John's triumphs, she found it harder and harder to endure his absence. At one point, Portia told her dearest friend, "I am much afflicted with a disorder called the *Heartach*. Nor can any remedy be found in America." By this time, her body was succumbing to her emotional stress. "Indispositions" sent her to bed, too sick at heart to face another lonely day. Her sixteen-year-old daughter, Nabby, took over more and more of the housework.

As prices continued to soar, Abigail wondered whether she could pay the taxes on the farm. She brooded about why they were so poor, when she saw or heard about people who were living luxuriously thanks to the money that poured into Boston and other ports from privateers and smuggling. But the ultimate pain was the separation from her dearest friend. It would soon be ten years since John departed for Philadelphia in 1774—ten years of almost constant loneliness.[17]

Portia tried to console herself by telling John in several letters that "patriotism in the female sex is the most disinterested of all virtues." She made a good case for it, pointing out that women had no hope of obtaining a job or any other reward for their devotion to the country. But the lonely days and weeks and months wishing for letters that seldom came soon obliterated these attempts to find courage and patience. Not even the news that peace was about to break out did Portia much good, because her dearest friend informed her that he would probably remain in France for at least another year to help negotiate a treaty of commerce with Great Britain.

Suddenly Portia had a problem that cried out for John's presence. Their oldest child and only daughter, sixteen-year-old Nabby, was in love. A wealthy, handsome young lawyer named Royall Tyler had moved to Braintree and was boarding with Abigail's sister, Mary Cranch, and her daughters. That made the twenty-five-year-old attorney an acceptable visitor to the Adams household. However, Tyler had a reputation for being a wild man in his Harvard years. He had broken the windows of certain professors and drunk several taverns dry. He had also spent quite a lot of the fortune he had inherited from his father. In Abigail's uneasy mind, all this added up to one ominous word: dissipation.

She confided the "family problem" to her dearest friend, who predictably exploded. The man should be banned from the house, John stormed for five wordy paragraphs. But the time lapse between letters to and from Europe and their delivery enabled Tyler to charm Abigail as well as Nabby. In fact, everyone in Braintree seemed to think well of the young man, who was negotiating to buy the finest mansion in the village and was handling a wide range of cases in the courts of Boston and elsewhere. Abigail was soon telling her flummoxed dearest friend of the "esteem and kindness" her neighbors felt for Royall Tyler. John, after inquiring about him from several friends, retreated to a temporary neutrality, leaving the "family problem" to Portia.

XI

As soon as the war ended, Portia began bombarding her dearest friend with demands that she be invited to join him in Europe. John finally agreed and told Abigail to come to Paris in the spring of 1784—and bring Nabby with her. Charles, who was preparing to enter Harvard, could stay with her sister, Mary Cranch; he knew her Braintree house almost as well as his own. Their youngest son, Thomas, who was also on the Harvard track, was deposited with another Smith sister in Haverill, Massachusetts. Abigail told Royall Tyler of their planned departure. She assured him that she was not opposed to welcoming him into the family as a son-in-law. On the contrary, she told him, he had good reason to "hope."

Tyler thanked her extravagantly and accepted the separation with apparent satisfaction. So, it would seem, did Nabby, who had maintained a rather severe reserve toward her suitor—so severe at times that Abigail thought she lacked "sensibility"—something every woman was now supposed to possess. With access to her letters and diary, we know that Nabby was deeply in love with Tyler but had been badly shaken by her father's first angry reaction to him. She was even more unhappy to be told by her mother that she and Tyler were to be separated for a year.

XII

After a pleasant voyage, Abigail and Nabby arrived in England on June 21, 1784. Abigail's dearest friend was mired in more negotiations with the Dutch, but when he heard that they had survived the Atlantic, he was transformed. "I am twenty years younger than I was yesterday," he declared. A week later, John arrived in London for a reunion that capped Abigail's joy. She declared them "a happy family again." Soon they were in Paris, where John settled them in a lovely "cottage" in suburban Auteil. The house was so large, it took Abigail weeks to visit all the rooms. But it had a five-acre garden that she adored.

They began touring the city under John and John Quincy's auspices, admiring the gardens and mansions but dismayed by the prevailing stink. Nabby told one of her correspondents that the French were the dirtiest people in the world. The Adams women went to the opera and the ballet and were enchanted by the splendid interiors and the soaring music of the orchestra. Then came a shock. The ballet dancers sprang "two feet from the floor, posing themselves in the air, with their feet flying and as perfectly showing their garters and drawers." Abigail turned her face away, appalled. In a few months she was telling her sister that her disgust had "worn off" and she was now enjoying the beauty and precision of the dancers. But she lamented the sad fate of these "opera girls" who were regarded as little more than playthings for rich young men about Paris.[18]

Meanwhile, Abigail became friendly with the Marquis de Lafayette's wife, Adrienne, and several other Frenchwomen. Although she was intimidated by their gorgeous finery and intricately "frizzed" hair, she was delighted by their witty conversation, their charming manners, and their knowledge of literature. But Portia's tolerance vanished when they visited Benjamin Franklin in nearby Passy and encountered Madame Helvetius. Franklin had assured Abigail and Nabby that they were going to meet "a genuine Frenchwoman, and one of the best women in the world."

This offhand encomium left the Adams women totally unprepared for Madame when she strolled into Franklin's drawing room. Not expecting to find other guests there, she was dressed in everyday clothes, featuring a profusion of dirty muslin over a shabby blue dress. She dashed up to the seventy-eight-year-old sage crying "Helas, Frankling!" and kissed him on both cheeks, plus a smack for good measure on his wrinkled forehead. Later, she sat between Franklin and John Adams at dinner. "She carried on the chief of the conversation," Abigail reported. "Frequently locking her hand into the Doctor's, and sometimes spreading her arms upon the backs of both the gentleman's chairs, then throwing her arm carelessly around the Doctor's neck."

After dinner, Madame hurled herself on a settee, "where she showed more than her feet," Abigail reported. With Madame was her lapdog, "who was, next to the Doctor, her favorite." When the canine urinated on the floor, Madame "wiped it up with her chemise." One can almost hear Abigail's quivering indignation as she concluded, "This is one of the Doctor's most intimate friends, with whom he dines once every week, and she with him." Mrs. Adams decided she was "completely disgusted and never wish for an acquaintance with ladies of this cast." Although Madame was her "near neighbor" in Auteil, Abigail never visited her.[19]

Eventually all the Adamses had to surrender their New England simplicity, get their hair frizzed, and dress in the most shockingly splendid style. Nabby was especially stunned by the finery her father wore when he went out on diplomatic business. "To be out of fashion," Abigail wryly concluded, "was more criminal than to be seen in a state of nature"—a condition, she slyly added, "to which the Parisians were not averse."[20]

XIII

From Congress came the most satisfying news that John Adams had ever received from his fellow politicians: he was appointed America's first ambassador to England. His ailing nemesis, Franklin, was going home, removing John's often stated fear that the "Doctor," as everyone called him, would get the job. John saw his selection as amends for the cavalier way Congress had treated him during the war.

John and Abigail decided to send John Quincy home to enter Harvard. They also had to deal at long range with a warning from Abigail's sister, Elizabeth Shaw, that Charles, only fifteen, was growing very attracted to a young lady. Abigail dispatched a letter banning all such diversions. She wanted her sons to devote themselves to literature, science, and practicing virtue. It was a glimpse of how she and John had begun to trade places. Abigail was growing more severe and censorious as middle age approached, while he, having achieved some triumphs in Europe that he thought would guarantee his fame, was starting to mellow.

In London the Adamses were politely received by a few people, but most ignored them or openly abhorred them. George III was majestically polite and so was his queen, but the newspapers were full of cruel remarks. One story claimed that John's supposed inability to make a living as a lawyer explained why he had overthrown his nation's laws in a revolution.

Another story portrayed him as so awed at meeting George III, he forgot all the fine compliments he had planned to recite. But His Majesty good-naturedly forgave the stuttering American bumpkin's distress.

Abigail was soon in a permanent state of rage at the "scribblers"— especially when she learned some American newspapers were reprinting the gibes. Meanwhile, John got nowhere in his efforts to persuade the British to sign a commercial treaty with the United States, or to live up to the terms of the peace treaty he had helped negotiate in Paris.

Fortunately, so it seemed at the time, their "family problem" distracted them. Nabby had grown more and more unhappy with the few short letters she had received from Royall Tyler. Abigail's sister, Mary Cranch, annoyed that Tyler had ignored her own daughters, launched a propaganda campaign against the young man. She filled her letters to Abigail with venomous portraits of his behavior. When Nabby and Abigail departed for Europe, Tyler had reeled back to the Cranch household and cried for hours. Thereafter he vanished from Braintree, which enabled Mary to speculate that he was relapsing into his previous "dissipated" ways. Abigail passed on Aunt Cranch's dark intimations to Nabby, further fueling her unhappiness.

By this time, Colonel William Stephens Smith had become the secretary of the London legation and part of the Adams household. He was a handsome thirtyish New Yorker with a breathtaking war record. Smith had fought in almost every major battle of the Revolution, repeatedly distinguishing himself. His ultimate reward was a 1781 invitation from George Washington to join his staff. Jovial and charming, he impressed John Adams and enthralled Abigail. Nabby was more than a little distracted by the way Colonel Smith gazed at her across the dinner table. But she was still engaged to Royall Tyler and could give him no encouragement.

Colonel Smith, hinting that his heart was bruised by Nabby's apparent indifference, announced he was going to Berlin to inspect the Prussian Army. Soon after he departed, Nabby made up her mind to break her engagement to Tyler. She informed her mother, and Abigail gave her heartfelt approval. When Colonel Smith returned to London, Abigail made a point of telling him about her daughter's change of heart. The colonel did not need a map to tell him the coast was clear. His proposal was a foregone conclusion.

John and Abigail did not have a single negative thought about Colonel Smith. The "strictest scrutiny," Abigail told one of her sisters, could not find a flaw in his character or his life. Alas, strict scrutiny was precisely what the Adamses failed to give this prospective son-in-law. They never wondered how he could afford to maintain a carriage in London and spend his time with fashionable young men his own age on a legation secretary's salary. Or travel in style on the continent for six or eight weeks at a stretch. They would soon discover that Colonel Smith had a bad habit of spending a lot more money than he had in his pocket.

Ambassador Adams pulled a few strings, and the young couple obtained a marriage license in twenty-four hours. John persuaded the pro-American Bishop of St. Asaph to perform the ceremony. The newlyweds rented a fully furnished house some blocks from the Adams residence, but they came to dinner every day. John and Abigail could not resist visiting them almost as often. Soon their happiness was appreciably increased by the news that Nabby was pregnant. The colonel persuaded her to name the baby William Steuben Smith, in honor of General Friedrich von Steuben, the German-born volunteer whom Smith admired extravagantly. They vowed the next arrival would be called John Adams Smith. Abigail consoled herself by declaring the boy had "the brow of his grandpapa."[21]

XIV

In America, sons John Quincy and Charles were at Harvard. John and Abigail decided to send Thomas, too, even though he was rather young. They were afraid that they might not be able to afford a third son in the fabled college when John lost his government salary. For the moment, they were depending on John Quincy to help Charles resist the temptations that had demoralized more than one aspirant to a Harvard diploma. Abigail had an older brother, William Smith, whose inglorious career had begun with dissipation at Harvard. As a married man he had gone on drinking sprees, chased women, and accumulated awful debts. Mary Smith Cranch had a brother-in-law, Robert, who had followed a similar route to self-destruction.

John and Abigail wrote to their sons by almost every ship that sailed from London, exhorting them to study and behave. Charles had won the affection of Eliza Smith Shaw, the sister who lived in Haverhill. She predicted he would become an "engaging well-accomplished gentleman—

the friend of science, the favorite of the misses and the graces—as well as of the ladies." His younger brother, Thomas, on the other hand, had "a more martial and intrepid spirit . . . a love of business and an excellent faculty for dispatching it." Eliza thought he might have a successful career as a soldier.[22]

With their boys in the danger zone of adolescence, the Adamses' thoughts turned more and more to America. John was getting nowhere in his negotiations with the British, and Abigail found it harder and harder to deal with the "studied civility and concealed coldness" she encountered when they went to receptions at St. James's Palace. Especially humiliating was the British refusal to dispatch an ambassador to America, apparently on the assumption that the bankrupt republic would not last long enough to make it worth the trouble.

The stories that the Adamses heard from home seemed to suggest the British were right. Farmers in western Massachusetts and on the western borders of several other states revolted against high taxes in 1786, burning courthouses and beating up sheriffs. The penniless Federal government could not send a single soldier to quell the upheavals. Eventually the Massachusetts rioters had to be dispersed by gunfire from a hastily organized army from Boston and its environs. Then came news that a constitutional convention was meeting in Philadelphia to form a new government, equipped with power to deal with such crises. John decided to send his resignation to Congress. After ten years of almost total separation from his country, its politics, and its people, John Adams was coming home.

SECOND BANANA BLUES

With little to do aboard ship but brood, John Adams became convinced that he was returning home to a country that neither respected nor appreciated him. While he was in London he had written a book, *A Defence of the Constitutions of Government of the United States of America*. It stressed the importance of a balanced government, with power distributed between the legislative, judicial, and executive branches. With typical Adams bluntness, he did not hesitate to say the British government was a good working example of what he meant. The book had been assailed by some Americans who thought most if not all of the power should be given to the legislature, where the voice of the average voter would be decisive. Several critics wondered whether Adams's sojourn in the British capital had aroused a long concealed fondness for monarchy. Was he facing a future of ostracism and obloquy?

Instead, like John's nemesis, Ben Franklin, the Adamses received a splendid greeting. When their ship docked at the Long Wharf in Boston on June 17, 1788, John Hancock, the governor of Massachusetts, sent a warm note of welcome as well as his glistening coach to transport them to his mansion. Cannon boomed, church bells clanged, and the wharf was crowded with cheering people. John could see no hint of hostility among the smiling faces along the streets as they rode to Governor Hancock's opulent home.

Back in Braintree, John and Abigail were dismayed to discover that repairs and extensions to the new house they had purchased while they were in England were unfinished. The handsome furniture they had bought in London had been badly packed and was a chipped and scarred mess. The house was larger than their earlier Braintree homestead, but it still seemed small compared with the spacious quarters they had enjoyed in Paris and London. Abigail called it "a wren's house." John, ecstatic at becoming a farmer again, rushed out and bought six cows, which he presented to Abigail. She acidly pointed out they did not have a barn in which to keep them. That did not stop him from buying a herd of heifers a few weeks later.

Meanwhile there were relatives by the dozen to greet—including John's mother, still amazingly spry at seventy-nine after burying two husbands. Sons John Quincy and Charles and Thomas were among the first to embrace them. Twenty-one-year-old John Quincy had graduated from Harvard with highest honors and was reading law under a prominent attorney in Newburyport. Eighteen-year-old Charles, Abigail reported to his sister Nabby, "wins the heart as usual." Fifteen-year-old Thomas had become "the cutup of the family." The two younger boys were still at Harvard.

Nabby and her husband had sailed to New York to meet Colonel Smith's large family and settle there. She was pregnant with her second child, and Abigail decided to depart for that city as soon as possible, leaving John to figure out how to milk his six cows on his own. John's concern was more practical. He tiptoed around the subject in a long letter and finally asked Nabby what Mr. Smith planned to do for a living. Unable to disguise his own feelings as usual, he blurted out the hope that Smith would not devote himself to seeking "public employment." It was a virtual guarantee of ending up "the most unhappy of all men." He would like to see Smith become a lawyer—a profession that guaranteed a man true independence. "I had rather dig my subsistence out of the earth with my own hands than be dependent on any favour, public or private, and this has been the invariable maxim of my life," he wrote.

This was self-delusion. John Adams had now spent fourteen of the prime years of his adult life in public service, dependent on the "favour" of his supporters in Congress. He was a politician, and there was nothing wrong, and certainly nothing immoral, about a man like Colonel Smith, an authentic war hero, considering a political career. The idea that there was something low or unworthy in seeking political support from other men was John's True Whig bugaboo at work—the notion that even a smidgen of self-interest was wrong.

Nabby glumly replied that she agreed about the law as a path to personal independence, but she did not think it was a practical choice for her husband. He was too old to begin a career that required years of study and preparation. Mrs. Smith proceeded to give her father some unexpected advice. She was living with her mother-in-law in Jamaica, Long Island, not far from New York City, and was picking up lots of political vibrations from her in-laws. "The general voice" that she was hearing in New York agreed that George Washington was certain to be the nation's first president. But the second-highest honor, the vice presidency, was by no means decided. Many people had told Nabby the post belonged to John Adams. "I confess I wish it, and that you may accept it," she wrote.[1]

To Braintree came corpulent, affable General Henry Knox, another soldier who had decided to devote himself to public life. He spoke as a representative of General Washington's former aide, Alexander Hamilton, who had become the leader of the country's first political party, the Federalists. They had been the backers of the Constitution in the struggle to win its ratification. Their opponents were called "Anti-Federalists" at the moment and were widely scorned for failing to recognize the need for a strong central government. Knox reported that Colonel Hamilton thought John Adams deserved to be vice president and wanted to know how he felt about the office. Adams replied that he was not in any way, shape, or form seeking the job. But if it was offered to him, he intimated that he would accept it.

That meeting made John Adams vice president. Only much later did he learn that Hamilton had considered a half dozen other candidates but learned Adams had the backing of New England Federalists. Hamilton had to accept him or create a breach in the party. That was not Hamilton's only worry. Under the new constitution, each state chose electors who cast the decisive votes for the presidency. But the drafters of the constitution had carelessly decided to let the candidates for president and vice president run on the same ballot. Whoever got the most votes would win. What if one or two electors, for reasons unknown, did not vote for Washington? If they and everyone else voted for Adams, he would become president. That was unthinkable as far as Hamilton was concerned.

Hamilton wrote letters to the leaders of several states, asking them to make sure their electors dropped three or four votes for Adams. His goal was modest—to have Adams come in second by perhaps a dozen votes. But in his hurry, Hamilton forgot that there were several other candidates on the ballot. These men, too, attracted electoral votes for vice president. Early in March the final tally reached the Adamses in Braintree. Washington had received all sixty-nine electoral votes and was elected president unanimously. John Adams was vice president—with thirty-four votes.

For a while, the "scurvy manner" in which he was chosen made John consider resigning. He declaimed to one correspondent that it was "an indelible stain on our country, countrymen and constitution." Only fear that his resignation might endanger the fragile new federal system, which depended on support from all parts of the nation, persuaded him to accept the election.

John's journey to New York to take the oath of office was satisfyingly rich in receptions and plaudits in various cities along his route. He left Abigail behind to run the farm until he located a suitable house in which they could live. When the new government convened, Adams became the president of the Senate. He solemnly informed the senators that he needed their advice about what to do when and if President Washington addressed their august body. While the two men were in the Senate, were they equal in power and authority? How should he address the president, and how should Washington address him?

Although the Constitution specified that the chief executive would be called "the president of the United States," Adams insisted on forming a committee that recommended, with his backing, "His Highness, the President of the United States, and Protector of the Rights of the Same." In letters and formal addresses, he thought Washington should be called "His Majesty." He said the vice president deserved the same title.

Adams was oblivious to the large political fact that the Senate and the House of Representatives had many members who belonged to the Anti-Federalist party and who feared the new government was going to transmute into something very close to monarchy if given too much power. Everything the vice president said seemed to confirm these fears.

Honest John became the butt of jokes for his titular extremism. Congressmen and senators began calling each other "Your Highness" with grins on their faces. Senator Ralph Izard, whose acid tongue had left stains on Benjamin Franklin's reputation in Paris, won the ridicule prize by nicknaming Adams "His Rotundity"—a label that stuck. Meanwhile the House of Representatives, under the leadership of James Madison, voted overwhelmingly to call General Washington "The President of the United States." Defiant to the bitter end, Vice President Adams could only watch the Senate agree, after mocking and finally consigning his magnificent but absurd titles to oblivion.

A dismayed and disconsolate John Adams wrote to Abigail, begging her to come to New York as soon as possible. If she did not have enough money available, she should borrow it from a friend. "If you cannot borrow enough, you must sell horses, oxen, sheep, cowes [*sic*], anything at any rate rather than not come on. If no one will take the place [the farm] leave it to the birds of the air and the beasts of the field. . . . It has been a great dammage [*sic*] that you did not come with me."[2]

II

Abigail was soon on her way. With her came Charles, who had graduated from Harvard—and become a worry. He had succumbed to "being spoilt by the . . . caresses of his acquaintance[s]," as Abigail put it. Charles loved being everyone's favorite companion, and his good looks and genial temperament gave him a head start over most men his age. Young women admired his style on the dance floor and gravitated to him. He soon developed a fondness for liquor and at one point led a campus rebellion. Another report, although fragmentary, seems to connect him to running naked, either solo or with a group, across Harvard Yard. John Quincy, who observed him for a year, came away fearing the worst. "Charles does not like to be censured," he said. This sensitivity soon made him almost morbidly averse to letters from his father or mother, exhorting him to behave.[3]

The Adamses' youngest son, Thomas, had another set of problems. His brothers, above all John Quincy, had attractive personal and intellectual gifts. Thomas was shy and often melancholy. He was the only child who never went to Europe—which may explain his surly refusal to write letters to his parents when he was old enough to do so. Abigail's sisters thought it was a mistake to send Thomas to Harvard at the age of fifteen. John and Abigail paid no attention to them—or to John Quincy's warning that Thomas was "too young to be left so much to himself." Abigail compounded this error by writing the boy strident letters, scolding him for failing to study and running wild in various ways. She had no evidence for these accusations; she simply assumed on the basis of most freshmen's conduct that Thomas was guilty.

In fact, Thomas was studying far into the night and angrily accused his mother of slandering him. Abigail apologized—but in the four-month gap in sending and receiving letters between Britain and America, Thomas had lots of time to brood about the way his parents treated him. He was probably not cheered by Abigail's apology—she added to it a lecture on virtue. Nothing less than perfection should be his goal in conduct and studies, the already stretched student was told.[4]

John Quincy was not such an obvious worry to his parents. Studious almost beyond belief, he was fluent in French, Latin, and Greek. But he, too, felt the pressure of their high expectations, especially after his graduation. He had also inherited his father's youthful interest in the opposite sex. He told a female cousin that he found women "irresistible" and fell in and out of love several times a month. On the other hand, as an Adams, he disapproved of this predilection. At Harvard he gave a speech before the Phi Beta Kappa society in which he condemned marriages based on passion.

Although John Quincy could not bring himself to admit it, he had little enthusiasm for spending three years in Newburyport becoming a lawyer. He told his aunt, Mary Cranch, that it was a place where "he cared for nobody and nobody cared for him." Slipping into typical Adams gloom, he told his diary: "I am good for nothing and cannot even carry myself forward in the world." Before long this gifted young man was calling himself a "cypher" and begging God to "take me from this world before I curse the day of my birth." His mother, oblivious to psychological explanations, diagnosed his problem as an acid stomach.[5]

III

In New York, Vice President Adams rented a large, attractive country house, Richmond Hill, a mile outside the 1789 city limits. (Today the site is in Greenwich Village.) The house had a lovely garden and a superb view of the Hudson River and the New Jersey shore. When Abigail arrived, she was delighted to find that Nabby and her husband had already moved in with their two boys. Colonel Smith was still in search of a way to make a living. John Adams was growing more and more disillusioned with him, and so was Nabby. But Abigail remained a captive of Smith's roguish charms and she adored the children, especially the older boy, who bore a strong resemblance to Grandpa John.

Meanwhile, John was being harassed via the mails by dozens of people who sought jobs in the new federal government. One of the most disturbing came from their old friend Mercy Otis Warren, who asked John to help her husband, James. In the heady days of 1776 he had been John's favorite correspondent. But the decade of separation had left them semi-strangers. James Warren had refused to support the Constitution, thanks largely to a feud with John Hancock, and drifted into political isolation. Mercy Warren denounced the ungrateful citizens of Massachusetts for their mistreatment of him.

Vice President Adams was not even slightly sympathetic. He curtly told Mercy that he had no patronage to dispense and if he had any, neither his own children nor close friends like the Warrens would get any of it. That would be a violation of his principles. Mrs. Warren retreated into aggrieved silence. In years to come, she would exact exquisite revenge.

Friends and relatives continued to bombard Adams with pleas for help. He kept saying no, no, no. But at home, John found himself forced to surrender his true whig principles to domestic pressure. Colonel William Smith wanted an appointment as U.S. marshal for the district of New York. Nabby and Abigail added their pleas, and the vice president asked President Washington to make the appointment. Washington did so without the slightest hesitation.

On another front, John decided to make his son Charles a lawyer. He dispatched him to Alexander Hamilton's office with a note, asking the New Yorker to take him under his guidance. Hamilton was about to become secretary of the treasury in President Washington's cabinet. Although it was evident to him and others that Charles had little or no enthusiasm for the profession, Hamilton arranged for him to study with another experienced New York attorney.

Although everyone had rejected his grandiose titles, John was determined to display a lifestyle worthy of the vice presidency. Each morning he rode from Richmond Hill in a handsome coach, often accompanied by Charles. Presiding in the Senate, John wore a powdered wig and the expensive clothes he had bought while serving as ambassador in London. Critics began calling him "The Duke of Braintree." A Boston writer assailed him in a ferociously satiric poem that warned him not to "sully your fame" by "daubing patriot" with a "lacker'd name."[6]

Abigail was deeply upset by this and similar assaults and did her best to defend John in numerous letters. She was inclined to agree with her husband's preference for titles. In some letters she playfully referred to herself in the third person as "Her Ladyship." At the same time she was anxious to retain her republican humility. She asked Mary Cranch to warn her if she (Abigail) showed any sign of treating her friends and acquaintances with arrogance or condescension. Abigail found it especially galling that no one criticized President Washington for his lifestyle. He had "powdered lackies" waiting at his door to announce visitors—and this scribbler accused John of aristocratic faults!

Typically, John Adams met his critics head on by writing another book, *Discourses on Davila,* which ran in newspapers in weekly installments. Among many topics, the book included a ferocious attack on the idea of equality. John took special aim at the French Revolution's frequent proclamations on this subject, dismissing their faith in human perfectibility as a fable. Mankind was not going to improve anytime soon. The only way to achieve social happiness was to maintain a balanced government that took into consideration humanity's inequalities and limitations.

The reaction to these ideas was so negative that Adams was forced to abandon newspaper publication. John and Abigail were especially dismayed when Secretary of State Thomas Jefferson, with whom they had been friendly in France, referred to the book as a collection of "political heresies."

IV

The Adamses entertained often and lavishly at Richmond Hill and sometimes dined with the Washingtons at their house on Cherry Street. Abigail's fondness for Martha Washington grew pronounced. She was "dignified and feminine, not even the tincture of hauteur about her," she told her sister, Mary Cranch. Abigail ruefully added that Martha, although somewhat plump, had "a much better figure" than she did.

Later in the year Abigail took Nabby and Charles to one of Martha's weekly receptions, where she reported the president displayed "grace, dignity and ease" in chatting with his guests. She thought he was far better company than King George III at similar receptions in his London palace. Abigail was invariably invited to sit beside Martha Washington. In the shifting conversations, other ladies occasionally occupied this place of honor. The moment President Washington noticed that Abigail had been displaced, he would lead her back to the coveted position and explain that it belonged to Mrs. Adams. This small gesture of concern pleased Abigail enormously.

Much as she liked Richmond Hill, Abigail found it very expensive to maintain, especially in the winter, when the fireplaces devoured cord after cord of costly wood. She regularly gave dinner parties for as many as twenty-four people. Counting her relatives and servants, there were eighteen people in the house to be fed daily. John Adams's salary was only $5,000 a year—perhaps $100,000 in modern money—and the farm at Braintree was producing little or no income. Abigail's favorite relatives, her sister Mary Cranch and her husband, Richard, were getting old and frequently ran short of cash. Abigail loaned them modest sums without hesitation.

V

John and Abigail were dismayed by Congress's decision to move the national capital to Philadelphia for ten years and thereafter settle themselves in Washington, D.C. The Adamses had to leave Nabby and William Smith in New York, along with Charles Adams, who was still studying to become a lawyer. Their son Thomas had joined them after graduating from Harvard, and John decided to take him along as his secretary. The duties of the job were minimal, and his parents began debating how he should make his living. In the end, the only solution that satisfied them was the law. John would find a Philadelphia lawyer under whom he might study.

Thomas began the three-year slog to the legal profession with as much good humor as he could muster. Like his brothers, he was dependent on his parents' checkbook, and he hesitated to strike out on his own for fear of disappointing them. John Quincy summed up the prevailing psychology when he wrote that he feared he would make John and Abigail "lament as ineffectual the pains they have taken to render me worthy of them."

The connection between their father's fame and fear of failure burdened all three sons. Thomas soon found himself struggling with what he called "the blue devils"—depression.[7]

VI

John Quincy Adams was not much happier. He had completed his three years of study at Newburyport and had opened an office in the front room of an Adams-owned house on Court Street in Boston. His clients were few and he made a botch of his first case, discovering that he had no ability to speak extemporaneously before a jury. He was also passionately in love with Mary Frazier, a beautiful young woman he had met in Newburyport.

Throughout the first six months of 1790, John Quincy saw Mary constantly. He refrained from mentioning her to his mother and father. But he confessed his passion to his brother Thomas and a few other friends. Mary inspired him to write poems to her beauty, several of which were published in local newspapers, exciting a dream of becoming a writer in John Quincy's troubled head. "All my hopes of future happiness in this life center in the possession of that girl," he told his former Harvard roommate, James Bridge.

But time was running out for the lovers. Although his law office was still virtually bare of clients, John Quincy told his sister Nabby that he was thinking of marrying Mary. Nabby apparently told Thomas, who told Charles—and Abigail. Charles, in typical younger-brother style, mocked Mary's charms: "Nothing so like perfection in human shape, [has] appeared since the world began."[8]

Abigail Adams was not amused. Nor was her mind changed by warm letters from her sisters, Eliza Shaw and Mary Cranch, urging her to bless the match. Abigail unleashed a barrage of letters on her oldest son, telling him that his romance was unacceptable and must be abandoned without even momentary hesitation on his part. John Quincy promised to obey, but he found it emotionally impossible. In desperation, he tried to persuade Mary Frazier to agree to an informal engagement. But her family intervened, warning her that a woman who remained linked to a man in that way soon endangered her reputation and hopes of marrying anyone else.

After months of agonizing, John informed his mother she could stop worrying about Mary Frazier. He was "perfectly free" and there was no need to fret about any future entanglement that might "give you pain." But to Eliza Shaw, he poured out his bitterness. He predicted that he would never be able to love anyone again. This tender-hearted woman wrote him a wise, consoling letter that he kept among his papers for the rest of his life.

VII

In spite of his personal woes, John Quincy retained a lively interest in politics—hardly surprising after his heady exposure to the subject at the age of eleven. Writing under pen names such as Publicola, Columbus, Marcellus, and Barnveld, he argued with skill and fervor for a strong central government and attacked, often ferociously, the anti-federalists who were coalescing into a political party led by Thomas Jefferson. John Quincy sided emphatically with Washington's commitment to a strong presidency and backed him when he issued a proclamation of neutrality in the war between England and France. His parents were delighted by his forays. The vice president said there was more "mother wit" in these essays than he had heard in the Senate "in a whole week."[9]

On June 3, 1794, a letter from the vice president informed John Quincy that he had been named American ambassador to the Netherlands at a salary of $4,500 a year. Along with the glorious news came a pious claim that John Adams had had nothing to do with the appointment. President Washington was rewarding John Quincy for his vigorous support of the Federalist Party in the newspapers—and was aware, thanks to his youthful years in Europe, that he had the background and experience to handle the job. Unmentioned were the vice president's several conversations with Edmund Randolph, who had succeeded Thomas Jefferson as secretary of state, in which John had not too subtly urged his son's appointment.

John Quincy's first reaction was nausea. He took to his bed for the better part or two weeks, apparently wrestling with the inescapable fact that he had to accept this offer—and simultaneously loathing the idea that he was once more obeying the parents who had destroyed his hope of happiness with Mary Frazier. But he finally rose and packed his trunk for the trip to Philadelphia to get his instructions for the job. He told one of his friends that his father "was more gratified than myself at my appointment." But he admitted the bright side of it to another friend— he was "my own man again."[10]

VIII

Not a little of John and Abigail's determination to manage the careers of John Quincy and their younger sons may have flowed from the continued woes of Abigail Adams Smith. Four months after Nabby gave birth to a daughter—her third child in four years—Colonel Smith quit his job as U.S. marshal and headed for England—only a few days before Christmas. Abigail found his timing almost as hard to take as his decision to leave the government payroll. Smith told Nabby and his numerous friends that he was going to make a fortune from collecting debts owed to his merchant father in England.

Nabby was left alone once more, with barely enough money to feed her family. Abigail could do nothing but write frantic letters from Philadelphia. The Adamses' vice-presidential expenses were so high in the city of brotherly love that they had little or no cash to send Nabby. Then came news of a woebegone letter from Colonel Smith, telling a New York friend from whom he had already borrowed money that he could not leave England unless the friend sent him more cash to pay his debts.

Once more, salvation came from George Washington. Whether John Adams intervened or Smith's friends appealed to the president for help remains unclear. At any rate, the vice president urgently informed the colonel that he had been appointed supervisor of revenue for New York State. His chief responsibility would be the collection of money from the federal liquor tax. The job paid $800 a year and a percentage of the money he took in. John urged him to return to America without delay. Abigail, underscoring their joint anxiety, wrote another letter, calling the salary "handsome" and all but begging him to accept the job "as soon as possible."[11]

This gift of the political gods soon proved to be no more than a temporary solution to the fortunes of Nabby and her wayward spouse. A year later, Smith abandoned his handsome salary and took his entire family to England, this time certain he would make an immense fortune. He was deep in a speculative bubble that would entrance him and many other Americans for the rest of the decade. All these gamblers were certain that mountains of money could be made selling millions of acres of American land to gullible European investors. Robert Morris of Philadelphia led the way, making immense sums on his first two speculations.

The defects of land speculation soon became apparent. The money was never paid in full. The seller was obligated to obtain a clear title to the land from resident Indians and other local claimants, including state governments. The seller had to survey the tract and divide it into salable parcels. Then he and the purchasers had to hope some settlers would show up, because the buyers left themselves numerous loopholes to escape from the deal if the golden promises turned to dross. This started to happen because of the appearance of that frequent historical intruder, the unexpected. War broke out between England and Revolutionary France early in 1793, indefinitely postponing large-scale immigration to America. Colonel Smith got into this game just as the land bubble was beginning to deflate.

Operating on little but nerve and faith in his luck, Smith returned to New York exuding affluence. He had reportedly bought five townships in northern New York and had sold thousands of acres to eager British investors. In New York, he bought twenty-three acres of land along the East River and began building a mansion that he dubbed Mount Vernon. On paper, it was to be a replica of President Washington's home. But his creditors were closing in, and the colonel never managed to complete more than the frame before he was forced to sell it to stay out of debtors' prison.

Meanwhile, Smith inveigled Charles Adams into his schemes. To John and Abigail's dismay, Charles had married Smith's sister Sally in 1795. The Adamses thought one Smith in the family was more than enough trouble for a lifetime. They had written numerous letters to Charles warning him against a premature marriage, but he had mastered the technique of ignoring their advice.

For a while, Charles seemed to be prospering as a lawyer. When John Adams visited him in December 1795, he was living in a house with a fine view of the East River. Friends told the vice president that his son had "twice or thrice the employment he ever had before." John was delighted to find him in his office, conversing with three new clients. He was equally pleased to learn that many people considered Charles a wit. Not long after he married Sally Smith, a man asked him if "the fever" was spreading in New York. Charles replied, "Do you mean the yellow fever or the Smith fever?" In the large Smith family, marriage had become so frequent it was being described as an epidemic.[12]

John Lawrence, the attorney under whom Charles studied law, was a speculator who also infected Charles with dreams of a quick killing. All he needed was cash—and thanks to John Quincy, he suddenly had a supply. John Quincy had asked Charles to invest a handsome sum John and Abigail had bestowed on him when he became a diplomat. Charles started putting his brother's money into speculative ventures, no doubt thinking of how much he (Charles) would make if the gambles succeeded. But the time when speculators could make a quick fortune was gone. Cheerful Charles was sowing seeds that would humiliate and destroy him.

IX

In spite of Nabby's woes and the wider worries over the turmoil the French Revolution was stirring in America, John and Abigail remained devoted partners. But they had little or no enthusiasm for participating in the political world of Philadelphia. The mounting feud between Thomas Jefferson and Alexander Hamilton was creating two political parties that attacked each other with rabidly partisan venom. Abigail disliked the French styles that prevailed in the Quaker city. She thought they were much too revealing and encouraged the sexual license that had become associated with France's revolution. She had no difficulty persuading John that they could save a great deal of money if she retreated to Massachusetts and he pursued a bachelor life in Philadelphia while Congress was in session.

Around this time they decided to give their Massachusetts home a name—Peacefield. They may have been motivated by the town of Braintree's decision to upgrade its name to Quincy in 1792. But Peacefield was also a revealing indication of their feelings about acrimonious Philadelphia. A pleasant dividend of Abigail's decision to flee the political fray was the letters they began writing to each other with much of their old warmth and candor. Both now began their reports with "My Dearest Friend."

In one of his first letters, John regaled Abigail with the way Governor John Hancock revealed his petty envy and hunger for popularity. "I would not entertain you with this political tittle-tattle," he added, "if I had anything of more importance to say. One thing of more importance to me, but no news to you, is that I am yours with unabated esteem and affection forever."

Abigail told him of the pleasure of spending Thanksgiving Day with their family—mostly Smiths and John's mother. She was alarmed by John's reports of the way the country was dividing into two political parties. He took some consolation from the way he was unanimously elected for another term as vice president by the states who backed the Federalists—and just as unanimously opposed by the states controlled by the anti-Federalists. He was especially dismayed by "the blind spirit of party" that had seized the soul of his friend Thomas Jefferson. Somewhat ruefully, John noted that the vice presidency was regarded as an office that played no part in the raging partisan quarrels. "Poor me . . . I am left out of the question and pray I ever may."[13]

Abigail disagreed with John that he was becoming—or had already become—a political nonentity. Although she took a dark view of the turbulent political scene—"The halcyon days of America are past, I fully believe"—she told John that in spite of the "limited office you hold" he had a "weight of character" thanks to his "former exertions and services" that was bound to exert a "benign influence" on the partisan quarrels. In fact, she was happy to note from the resolutely cheerful tone of his letters that "the only fault of your political character—which had always given me uneasiness was wearing away"—his "irritability." It had sometimes thrown him off his guard and revealed "that a man is not always a hero."[14]

No wonder John wrote in reply, "One day spent at home would afford me more inward delight and comfort than a week or a winter in this place." Abigail's letters, he assured her, "give me more entertainment than all the speeches I hear." This frequent reiteration of his boredom as vice president makes it evident that the job had been another bad career choice for this complex man. John should have accepted election as a senator from Massachusetts, where his oratorical skills and political insights could have been influential in the political struggle that raged throughout these years. He might have developed a following that would have added weight to his presidential ambitions.

Instead John and Abigail convinced themselves that he was better off remaining aloof. The implications of this embrace of political purity would become painfully visible when John Adams became the second president of the volatile American republic.

PARTY OF TWO

As the presidential election of 1796 approached, John Adams vacillated between hungering for the honor and dreading the abuse he would receive in the vicious political atmosphere of the times. "I am weary of the game," he told Abigail in February. "Yet I don't know whether I can live without it."[1] Both Adamses seemed to think the presidency was John's by right of seniority. A lot of people disagreed. Thomas Jefferson accepted the Republican Party's nomination, with Aaron Burr as his vice president. John acquiesced to Thomas Pinckney of South Carolina as his vice president on the Federalist ticket. Pinckney's claim to fame was a treaty he had negotiated with Spain in 1795, which opened the Mississippi to western farmers eager to sell their surpluses to a hungry world.

Adams was elected president by three electoral votes. If New England had not stood solidly behind its native son, he would have lost. Thomas Jefferson was second and became vice president. Pinckney ran third. John immediately saw winning by three votes as a humiliation, which it was to some extent. His mood was not improved when he learned, just before the electoral votes were counted, that Alexander Hamilton had again intrigued behind the scenes. This time he had tried to make Pinckney the president by persuading some electors not to vote for John. The Federalist leader had decided that Adams would be difficult to deal with.

The new president's response to Hamilton's treachery was strange. It was common knowledge that the cabinet Adams inherited from Washington were all followers of Hamilton. Nevertheless, Adams kept them on the job. Numerous historians have described this decision as bizarre. But Adams, without a political following, had few friends and no allies with whom to replace them. Later, he claimed that he feared the public would react violently if he had swept out the entire Washington cabinet. As for his personal relationship with Hamilton, Adams vowed to "maintain the same conduct toward him that I always did—that is, to keep him at a distance."[2]

Abigail had a darker view of Hamilton. She told John to "beware of that spare Cassius." The words from Shakespeare's *Julius Caesar* occurred to her almost every time she saw that "cock sparrow." She had "read his heart in his wicked eyes many a time. The very devil is in them . . . or I have no skill in physiognomy."[3] In a letter to a friend, she portrayed the Federalist leader as a dangerous man who was pursuing a "Machiavellian policy." He wanted a weak president like Thomas Pinckney—someone that a "Master Hand could work" behind the scenes.[4]

But Abigail did not try to persuade John to dismiss the Hamilton-controlled cabinet. She did not arrive in Philadelphia for almost two months after his inauguration. As early as March 13, a little more than a week after he became president, John told her, "I must go to you or you must come to me. I cannot live without you." In another letter he all but wailed: "I never wanted your advice and assistance more in my life." He warned her, "You and I are entering on a new scene, which will be the most difficult and least agreeable of any in our lives."[5]

Abigail still delayed her departure from Quincy. Her excuse was the lingering, ultimately fatal illness of John's aged mother and the tragic death of a young niece. Neither of these small, sad dramas required her presence. There were numerous relatives in and near Quincy who could have cared for both women. The real reason for Abigail's delays was her lack of enthusiasm for the ordeal she and John faced in Philadelphia. At one point she told John of a dream in which she seemed to be in a battle and cannon were firing huge black balls—all aimed at her.[6]

The challenge that haunted Adams's presidency was post-revolutionary France. With the Reign of Terror over and its Jacobin leaders executed, the country was in the hands of a Directory, a five-man group who were as hostile to America as their blood-soaked forerunners. Soon after Adams took office, they ordered their navy and privateers to seize American ships at will. If it was not a declaration of war, it was the next best (or worst) thing. Adams responded by calling a special session of Congress for May 15, 1797. Behind the scenes, he rushed a letter to Mount Vernon, offering to resign so George Washington could take charge of the nation once more. It was a painful glimpse of John's lack of self-confidence in the shadow of Washington's fame.

Abigail finally arrived in Philadelphia on May 10, after an exhausting trip on muddy roads and over rain-swollen rivers. In a touching scene, she told her sister Mary Cranch how John met her about twenty-five miles outside the city and she "took my seat by his side" in his carriage. They stopped for dinner at Bristol, giving themselves most of the day to talk things over. Here, Abigail probably confided to John her hope that their friend, Vice President Thomas Jefferson, might be an ally in spite of their political differences. She had already expressed this opinion in a letter from Quincy.[7]

Six days later, John went before Congress and told the legislators he was prepared to negotiate America's differences with France but was determined to preserve America's neutrality in the ongoing war between the two major powers. To do that, it was time to start creating a strong navy to protect the nation's merchant ships from French depredations. The Jeffersonians' reaction swiftly destroyed Abigail's hopes for help from the vice president. Republican newspapers and orators in Congress dismissed John's speech as "a presidential war-whoop." They defended France's policies and called for a virtual repudiation of George Washington's declaration of neutrality. Jefferson never said a word on Adams's behalf.

Abigail would have been even more disappointed if she had known that while the Republicans were publicly excoriating President Adams, Vice President Jefferson was secretly telling the French chargé d'affaires in Philadelphia that there was no need to take him seriously. Jefferson advised the French to pretend to negotiate and wait for the American people to assert their friendship for France. President by only three electoral votes, Adams had no popular support.[8]

II

Along with this public uproar, private woes tormented the Adamses. Nabby's marriage to the erratic William Smith showed no sign of improvement. When they returned from England, the colonel dumped her and their offspring in a tiny house in Westchester County while he roamed northern New York looking for bargains in real estate on which to speculate. Abigail visited her daughter on her way to Philadelphia and was appalled by her unhappiness. She had four young children to raise and did not have a neighbor within twenty miles with whom she could converse.

Abigail was almost as worried by news from John Quincy that he had fallen in love with Louisa Catherine Johnson, the seventeen-year-old daughter of a wealthy Maryland-born merchant living in London. Although John Quincy was twenty-eight, and a diplomat so successful that outgoing president George Washington had made a point of praising him, Abigail still wrote to him as if he were a teenager. She worried about Louisa's foreign birth and wealthy background. Would John Quincy be able to support this young woman in her accustomed style? The ambassador struggled to reassure his mother in long, elaborate letters.

Charles Adams was a far worse worry. With a wife and two children to support, he was often pressed for money and declined to seek help from his parents, after so long and so stubbornly ignoring their advice. He was an easy target for his brother-in-law William Smith and his smooth-talking relatives. They lured him into one of their speculative schemes, and when the bubble burst, Charles lost almost every cent of the money John Quincy had given him. It was a humiliation that Charles was psychologically incapable of enduring. He reached for the liquor bottle, and soon he was on his way to poverty and despair.

III

In spite of these personal sorrows, First Lady Abigail Adams did not neglect her duties. She received as many as sixty callers a day at the presidential mansion on Philadelphia's Market Street and often entertained forty at dinner. But her enthusiasm for her job was constantly undermined by the city's opposition newspapers. Worst of all the "scribblers," as Abigail called them, was Benjamin Franklin Bache, Ben Franklin's grandson, who was editor of the *Philadelphia Aurora*. He regularly accused President Adams of being in his dotage. He called John "the advocate of a kingly government and of a titled nobility . . . to keep down the swinish multitude."

When John sent a three-man delegation to France to try to resolve the crisis, Bache called it an attempt to deceive the American people. In his speech to Congress, Bache raged, the president never said a word about the way the British navy arrogantly boarded American ships on the high seas and seized sailors that they claimed were deserters from their ranks. Adams acted as if he governed a nation of "ourang outangs" instead of intelligent men. Someone should tell "His Serene Highness" that he had been elected president by three votes.[9]

By the time Abigail finished one of these Bache screeds, she was seething with outrage. It did not help to know that Bache was Ben Franklin's grandson; it only reignited Abigail's (and John's) enmity for Franklin at a time when they needed to stay calm and clearheaded. Even more unhelpful was the way Bache constantly claimed that Adams was a secret monarchist because of his reputed insistence on resounding titles for President Washington and for himself. Did Abigail blame herself for not accompanying John to New York in 1789 until his frantic letters pleading for her presence reached her in Quincy?

When President Adams appointed John Quincy minister to Berlin, capital of Prussia, the billingsgate from Bache and other editors got worse. They claimed John Quincy's salary was ten thousand dollars a year (it was actually $4,500) and made it sound as if the president were sending him the money as a bonus. Was there any limit to the malice of these scribblers? Abigail wondered.

IV

In February 1798, the Adamses reeled under another blow. A group of fashionable Philadelphians announced that they planned to stage a ball and dinner to celebrate George Washington's birthday. These celebrations had taken place every year while Washington was president. Though John and Abigail had never approved of them—they claimed to admire Washington but did not think he should be feted as a king or a Roman emperor—they kept this opinion to themselves. A celebration when John Adams was president was another matter. A celebration to which the president and his wife were invited, as if they were just another pair of citizens? The Adamses reacted with outrage. John's concealed envy of Washington exploded and ignited similar emotions in Abigail.

In a fiery letter to Mary Cranch, Abigail wrote, "I do not know when my feelings of contempt have been more called forth." She poured most of her spleen on the Philadelphians who were staging the ball, calling them "a strange set of people" lacking "the least feeling of real genuine politeness." How could they dare to invite the president of the United States to appear as a "secondary character"? It would hold him up before all the nations of the world "in that [secondary] light." It was "ludicrous beyond compare."[10]

President Adams wrote "DECLINED" on the face of the invitation and added a snarling denunciation of the affair. Abigail warmly approved—and then expressed horror when this communication appeared in the *Aurora* with sneers and gloats of delight from editor Benjamin Franklin Bache. The Federalists were dismayed, and they began quarreling among themselves when some party stalwarts cancelled their acceptances to the celebration.

Vice President Jefferson was delighted by the uproar, remarking in a letter to James Madison that the "birthnight" had split the Federalist Party. Abigail remained grimly convinced that she and John had done the right thing. She told her sister Mary Cranch that the Americans now knew they had a president "who would not prostrate their dignity and character, neither to foreign nations, nor the American people." This was, to put it mildly, inflating a molehill into an imaginary political mountain. More and more, it became apparent that John and Abigail were almost totally lacking in political savvy. They were a party of two.[11]

V

A few weeks later, a startling turn in the crisis with France upended everything. The three envoys that President Adams had sent to Paris were rejected by the French government, after they refused to pay a bribe of $250,000 to Foreign Minister Charles Maurice de Talleyrand and agree to a loan of $10 million as a precondition to "peace" talks. The envoys responded with "No, not a sixpence," which was soon transformed by Federalist newspapers into "millions for defense but not one cent for tribute." President Adams sent the correspondence with the French negotiators, disguised as X, Y, and Z to Congress in response to the Republicans' demand to see the documents. When the Jeffersonians read the stunning evidence of French arrogance and greed, they were, Abigail gleefully reported, "struck dumb and opend not their mouths."

The desperate Republicans attempted to keep the correspondence secret. The Federalist-controlled Senate voted to publish fifty thousand copies and distribute them throughout the country. Americans from Boston to Savannah were infuriated and called for war. Suddenly, President Adams was the most popular man in America. People cheered him when he appeared in public. At one point, eleven hundred young Philadelphians marched in a long column to the presidential mansion to present him with pledges of their support.

One of the most emphatic war-wishers was Abigail. In a letter to her daughter Nabby, she pointed out that America was already at war on the high seas and had been for months. She was eager to see a formal declaration of war by Congress, "the sooner the better." At present the country was suffering "the miseries and misfortunes" of war without striking back. A declaration of war would also enable the president to do something about the lying newspapermen and the French agents that Talleyrand reportedly had swarming throughout the nation. To her nephew William Shaw, Abigail declared that France was plotting to subjugate America. The French were spreading their amoral principles, their atheism, and their "depravity of manners" in every part of the United States. There was only one answer to this insidious attack: "Let every citizen become a soldier and determine as formerly on Liberty or Death!"[12]

With Abigail saying such things at the dinner table, it was hardly surprising that President Adams also wanted a declaration of war. But Congress was too divided to produce one. They compromised by creating a twelve-ship navy and a "provisional" army of ten thousand men. Adams asked George Washington to take command of the army, and he reluctantly agreed. But he informed the president that he had no intention of exercising actual command. That large task called for a younger man, and Washington chose Alexander Hamilton as his chief deputy.

Here was a new humiliation that tormented both Adamses. For weeks the president struggled to evade appointing Hamilton. In a raging memo to Secretary of the Treasury Oliver Wolcott, Adams called it "the most difficult action to justify" that he had been forced to take in his "whole life."[13] But Washington made it clear that he would resign if he did not get Hamilton, and the president was forced to capitulate.

With Abigail's warm approval, the president attempted to name William Stephens Smith as a brigadier general in the new army. When

Washington saw his name on the list, he exploded: "What in the name of military prudence could have induced [this] appointment?" An enraged President Adams told Secretary of State Timothy Pickering that his son-in-law was "far, very far superior to Hamilton" as a soldier. Pickering saw to it that the Senate vetoed Smith. It was not a difficult task. The colonel's reputation as a bankrupt speculator was well and widely known. An overwrought Abigail cried in a letter that "secret springs" were at work. But Smith remained defunct—another painful political humiliation.[14]

VI

While John Adams was fighting a losing struggle over control of the army, the Federalists in Congress decided to silence the newspapers that were slandering and insulting them and the president. Abigail Adams was an enthusiastic backer of these "Alien and Sedition Acts," as they were soon called. She told Mary Cranch she wanted to "punish the stirrer up of sedition, and the writer and printer of base and unfounded calumny." Such laws would add immeasurably "to the peace and harmony of our country."[15]

The acts were actually four in number, dealing with different aspects of the situation that seemed to be threatening the United States. One bill revised the naturalization laws, requiring a fourteen-year wait to become a citizen. Two other acts empowered the president to deport "aliens"—noncitizens he deemed dangerous to the nation's peace and safety. The chief target of these laws were the 25,000 refugees from the French Revolution in America. Some came from France, but a far larger number had fled from French West Indian islands, which had undergone their own revolutions. The law also made immigrants from Ireland, England, and other nations liable to expulsion if they criticized the government.

Abigail approved of these alien laws. She thought a "more careful and attentive watch ought to be kept over foreigners." But it was the Sedition Act that brought rejoicing to her lips. The act called for fines and imprisonment for anyone who wrote, printed, or uttered "any false, scandalous and malicious" statements against the government of the United States, either house of Congress, or the president "with intent to defame" these public servants, or bring them "into contempt or disrepute." Here was a law that would protect Portia's dearest friend from slanderers like Bache.

To no one's surprise, John Adams signed the bills into law. He saw them as "war measures"—temporary weapons that would enable the government to retain control of the country. The bills stated that the special powers they authorized would expire in 1801, unless Congress renewed them.

The Republicans, thrown on the defensive by the XYZ revelations, were desperate for an issue. They seized on the Alien and Sedition Acts as a heaven-sent weapon. Thomas Jefferson and his ally James Madison drafted resolutions that were passed by the legislatures of Kentucky and Virginia denouncing the acts. Jefferson's Kentucky resolutions declared a state had the right to nullify acts of Congress that violated the Constitution. Federalists saw these declarations as little short of a call to revolution.

Up and down the country, Republican orators fulminated and newspapers published hysterical denunciations of the acts. They were condemned as a first step toward setting up a Federalist dictatorship, with a president and senate elected for life. The vituperation flung at John Adams reached hitherto unimaginable heights. In July 1798, a bewildered, exhausted Abigail reeled back to Quincy with her politically battered husband—and collapsed.

VII

Abigail's symptoms were similar to those that had laid her husband low when he took to his bed in Amsterdam in 1781, demoralized by his losing struggle with Benjamin Franklin's fame. Rheumatic-like pains roamed Portia's body, diarrhea contracted her intestines, and insomnia wracked her nights. Most of the time she was close to a helpless invalid, too weak to rise from her bed. Today's doctors would probably call her condition a nervous breakdown. John hovered over her day and night for the next four months, letting the federal government operate without him. The thought of losing Abigail left him virtually unable to function. When letters warned him against giving Hamilton unchecked control of the army, he responded that Abigail's "dangerous sickness" had left him too depressed and agitated to do anything about it.[16]

In November, the president returned to Philadelphia without his dearest friend. Abigail was tormented by guilt. She felt she had "quitted" her "post" at John's side. But she was mentally and physically incapable of facing more political warfare. She followed the ongoing turmoil from a soothing distance, relying on letters from John and from his secretary, their nephew William Shaw.

During these months alone in his mansion, the president began to rethink the international situation. From several sources, including his diplomat son, John Quincy, he learned that the French Directory and its spokesman, Foreign Minister Talleyrand, were now saying they did not want a war with the United States. They realized it would inevitably convert the Americans into allies of the English. At home, American voters were not enthusiastic about the taxes the Federalists levied to pay for the new navy and army. One of these measures was a stamp tax, not much different from the one that had helped launch the revolution against George III. Then there was the matter of the man Adams now unreservedly hated, Alexander Hamilton, parading around Philadelphia in the uniform of a major general. Wouldn't it be delicious if his army suddenly became superfluous?

The president decided to dispatch another envoy to see whether an understanding could be reached with the Directory. To the astonishment of everyone in the Federalist party, on February 18, 1799, Adams sent a note by messenger to Vice President Jefferson and asked him to read it to the Senate. The staggered solons learned that Adams was appointing William Vans Murray, the young ambassador to the Netherlands, as minister plenipotentiary to France to see if peace could be achieved. Politicians all, the Federalists in Congress saw Adams destroying the issue that had won them unparalleled popularity.

A delegation of congressional leaders called on the president to discuss the decision with him. According to Secretary of State Pickering, Adams flew into a "violent passion" and rebuked the visitors for questioning his judgment. When the Federalists threatened to reject Murray as a nominee, Adams told them that if they did, he would resign and hand the presidency over to Jefferson. Senator Theodore Sedgwick of Massachusetts lost his temper and called Adams's decision "the wild and irregular starts of a vain jealous and half frantic mind."[17]

Throughout the country, a large percentage of the Federalist Party shared Sedgwick's reaction. One New York Federalist leader said the Murray appointment had given "almost universal disgust."[18] The Federalists thought Adams was risking another series of runarounds and humiliations by submitting so eagerly to reports of France's changed mood.

Instead of staying in Philadelphia and combating this reaction, when Congress adjourned, President Adams departed for Quincy and the company of the other member of his party of two. Not even an armed revolt against federal taxes in Pennsylvania, led by a German American named Fries, delayed him. He issued a proclamation declaring the rebels guilty of treason, ordered the army to seize them, and left for Quincy. He stayed there for the next seven months—the longest absence of any president from the seat of government in America's history.

The other member of his party of two professed total admiration for the president's peace initiative. In a fervent letter, Abigail told John his decision was a "master stroke" that had "electrified the country." She dismissed the Federalist senators and congressmen who questioned John as "dupes of intrigue." The president had exercised the power given him by the Constitution. "Time will discover who is right and who is wrong," she wrote.

Portia particularly enjoyed hearing that several people had declared that if "Mrs. Adams" had been in Philadelphia, she would have stopped the president from making such an awful mistake. "That ought to gratify your vanity enough to cure you," John wrote. When he reached home and added intimate details of the uproar, Abigail became even more enthusiastic.[19]

Her support was not enough to prevent John from slipping into a black depression as he pondered his wounded presidency. One hot July day, three old friends, led by General Henry Knox, rode out from Boston to see him. John sat in the parlor reading a newspaper while they tried to converse with him. He did not offer them so much as a sip of cold water before they stalked out, wondering if the president was more than a little crazy. A few days later, a group of young men from Boston and some officers from the newly commissioned frigate USS *Constitution* paid him an unannounced visit. Adams gave them a snarling, raging lecture about their bad manners, while Abigail watched, appalled. He was almost as irritable with the house servants and farmhands.

Even Abigail was the target of John's barbed remarks. But she forgave him and gently coaxed him out of his gloom. She persuaded him to attend the Harvard Commencement and a Fourth of July celebration in Boston. She monitored his mail, withholding letters that warned him of chicanery in his cabinet and elsewhere because, in his present condition, they would upset him to no purpose. Her loving concern slowly restored the president's emotional balance. The visible evidence of how desperately he needed her help inspired Abigail to banish her own nervous tremors. Soon, John was taking an interest in the political scene again. When the tax rebel Fries was captured and sentenced to death for treason, Adams demanded to see all the papers related to the trial and verdict, and decided to pardon him.[20]

VIII

Not until November 1799 did President Adams respond to worried letters urging him to take charge of the government again. The Hamiltonians in his cabinet were doing everything in their power to sabotage his peace initiative. Before he left Philadelphia, Adams had agreed to appoint two older Federalist politicians to bolster the youthful Murray in the negotiations. Secretary of State Pickering repeatedly delayed their departure for Europe. Adams grimly resolved that he would see to it that these envoys sailed as soon as possible.

En route to Philadelphia, John stopped at Nabby's house in Westchester County. He found his daughter and her children surprisingly contented. The president had managed to wangle William Stephens Smith a colonel's commission in Hamilton's army. For once there was some money in the family exchequer. But there were unexpected guests—Charles Adams's wife, Sally, and her two children. With tears on her cheeks, Sally told the president that Charles had become a hopeless alcoholic. His law practice had collapsed and he had vanished into New York's back streets, running up ruinous bills in taverns and consorting with prostitutes.

Frantic with grief and rage, John turned to the only person who could share his anguish. He told Abigail how much he "pitied . . . grieved . . . mourned" for Sally and her children. Charles was "a madman possessed of the devil . . . I renounce him." The president made no attempt to see his son. He had urgent business awaiting him in Philadelphia. He resumed his journey, "loaded with sorrow," begging Abigail to "write me every day.[21]

The nation's capital was in the grip of another yellow fever epidemic. The government had moved across the Delaware to the capital of New Jersey, Trenton. There, Adams confronted his cabinet, who told him the latest news from France: the Directory had been overthrown by a young French general, Napoleon Bonaparte. Secretary of State Pickering argued that there was no point in sending a peace mission now. Adams was unconvinced and no one was able to change his mind, not even Major General Hamilton, who rushed to Trenton to add his arguments to the contretemps. The president ordered the peace commissioners to sail for France as soon as possible. A disgusted General Hamilton fired off a deploring letter to General Washington, obviously hoping to bring his influence into play.

IX

Stunning news from Mount Vernon distracted everyone: George Washington was dead. President Adams expressed his genuine grief in his message to Congress. "I feel myself alone, bereaved of my last brother," he wrote. Black bunting shrouded the door of the presidential mansion and the entrance to Congress Hall. Abigail told her sister Mary Cranch that "no man ever lived who was more deservedly beloved and respected."[22]

On December 26, John and Abigail joined a host of distinguished mourners at Philadelphia's Christ Church to hear Congressman Henry Lee of Virginia extol Washington as "first in war, first in peace and first in the hearts of his countrymen." The Episcopal service lasted almost five hours, and the Adamses followed it with a presidential dinner for thirty guests. Over the next weeks, more and more extravagant eulogies of the dead hero appeared in the newspapers and Abigail began to grow impatient with their rhetoric. When a New England clergyman called Washington "Liberty's spotless high priest," and hoped President Adams had the ability to become "Columbia's second savior," she called it "a mad rant of bombast." It was time for someone to declare that "no one man" was or ever could be the country's savior. These idolaters did not seem to realize that by exalting one character, they "degrade that of their country."[23]

Without the fear of Washington's disapproval, President Adams and Abigail became outspokenly critical of the Hamilton loyalists in the president's cabinet. Abigail was especially hostile to Secretary of State Timothy Pickering. His manners, she told one correspondent, were "forbidding," his temper was "sour," and his resentments were "implacable." He fancied himself having the power to "dictate every measure," thanks to Hamilton's backing. The party of two concluded that this situation was intolerable. But the president did nothing about it for months.[24]

The weakest member of the cabinet was Secretary of War James McHenry, an affable former Revolutionary War aide to Washington who strove to be agreeable to everyone. On the night of May 5, 1800, President Adams summoned him from a dinner party to discuss a routine matter. He suddenly asked what McHenry knew about General Hamilton's activities in New York, where Adams had heard he was constantly criticizing the administration. McHenry claimed to know nothing about Hamilton's hostile words or actions. The president exploded into vituperative rage: "You are subservient to him. It was you who biased General Washington's mind . . . and induced him to place Hamilton on the list of major generals!"

Stamping up and down his study, Adams called Hamilton "the greatest intriguant in the world, a man devoid of moral principle—a bastard and . . . a foreigner." He would rather be a vice president under Jefferson than be indebted "to such a being as Hamilton for the presidency." Moreover, McHenry was a total failure as secretary of war. The army was a mess! The soldiers lacked decent uniforms! "You cannot, sir, remain any longer in office!" Adams bellowed.

The overwhelmed McHenry said he would resign at once. Regaining a semblance of self-control, Adams apologized. He admitted McHenry was a man of integrity. McHenry returned to his office and wrote a vivid account of his dismissal, which he sent to Hamilton. The secretary of war was soon describing the scene to other people, telling them that he thought Adams was "actually insane."[25]

A week later, the president fired Secretary of State Pickering. Abigail reported the dismissals in a letter to a cousin: "You will learn that great changes have taken place in the cabinet—some will mourn, some will rejoice, some will blame, others will confuse, all this was foreseen." In fact the firings were foreseen by no one. The president had given nobody a hint of what he was thinking of doing—except his fellow member of their party of two.[26]

X

By this time the presidential election of 1800 had begun. President Adams was seeking reelection; Thomas Jefferson was his opponent. The Republicans' chief hit man was Scottish journalist James Thomson Callender. In

a blazing pamphlet, *The Prospect Before Us,* "the wretch," as Abigail called
Callender, described John Adams as "that strange compound of ignorance and ferocity, of deceit and weakness." He said the voters' choice lay
"between Adams, war and beggary, and Jefferson, peace and competency."
As usual, Abigail read every word of these clotted pages of invective, shud-
dering with each blow.

The party of two returned to Quincy for the summer, and the newspapers
told them of even more damage inflicted on John's hopes for reelection. In
a series of trials, hot-tempered Judge Samuel Chase of the Supreme Court
jailed numerous editors for criticizing, among other things, the president's
supposed "thirst for ridiculous pomp, foolish adulation and selfish avarice." Chase inflicted heavy fines on those found guilty, along with jail
sentences. One of the prime victims was Callender, whom Judge Chase
excoriated with special savagery. The Republicans denounced these trials
as a violation of the first amendment, and many people agreed with them.
Abigail became so distressed that she wondered whether American elections ought to be less frequent. She began to think an excess of democracy
would destabilize the country.

The party of two's only hope was the peace mission to France. But no
news of this divisive venture emerged from the vast Atlantic Ocean. For
reasons that remain obscure, the three envoys did not begin negotiating
until March 1800. Meanwhile more bad news reached John and Abigail.
The president had disbanded Hamilton's provisional army, turning their
political differences into a personal vendetta. The former general was
passing the word that "Mr. Adams must be sacrificed." The party's vice
presidential nominee, Charles Cotesworth Pinckney of South Carolina
(older brother of Thomas Pinckney), had to be elected president if the
party were to survive.

To guarantee this outcome, General Hamilton decided to write a *A Letter from Alexander Hamilton Concerning The Public Conduct and Character of John Adams Esq, president of the United States.* Fifty-four pages long,
it was as thick with abuse as Callender's Republican screed. Hamilton
denounced Adams's "ungovernable temper" and "paroxysms of anger."
At least as bad was the president's "disgusting egotism" and his "bitter
animosity" toward his own cabinet. These "great intrinsic defects of character" made him a menace to stability and order.

Hamilton originally planned to circulate this interminable missive only to a select group of Federalist leaders to persuade them to choose Pinckney over Adams. But several Republican operatives obtained copies and published excerpts in the *Aurora* and other newspapers. Blinded by his rage at Adams, Hamilton decided to publish the entire letter as a pamphlet, which soon circulated throughout the nation. The general's intemperate blast split the Federalist Party and all but guaranteed Jefferson's election. A heartbroken Abigail told her sister Mary Cranch that Hamilton had defeated himself—and John.[27]

XI

Heartbreak of another kind awaited the party of two as they departed from Quincy for the final months of John's presidency. John left first, and on his journey through New York he again refused to see his son Charles. He had declared that he renounced him, and he grimly kept his word. Abigail could not be so hardhearted. "My journey is a mountain before me but I must climb it," she told her son Thomas, who was now practicing law in Philadelphia. She found Charles unmistakably dying. His wife, Sally, was with him in a furnished room paid for by a generous friend. He was bloated and incoherent, in the final stages of alcoholism. His doctor told her there was no hope.[28]

Bearing this terrible burden, Abigail struggled on to Philadelphia, where she was consoled by her son Thomas. From there she journeyed to the new capital of the country, Washington, D.C., and joined her husband in the huge unfinished mansion the government had constructed for the president. Abigail was fascinated by the even more unfinished capital city, which she described as "romantic but wild." John gave his final State of the Union address to Congress, and Abigail tried to cope with heating the gigantic house, which was being called the president's "palace." The term White House was more than a decade in the future.

In December the results from the last states to vote trickled into Washington. (There was, as yet, no single election day.) The contest was surprisingly close. But by December 16, it became apparent that Jefferson had won by eight electoral votes. By that time, crushing personal news had reached John and Abigail from New York: Charles was dead. Abigail wrote a touching letter to his widow, Sally, recalling how lovable Charles had been as a boy. She assured her that the president mourned for his lost son, "as he has for a long time."[29]

As for the lost election, Abigail confessed that her first reaction was relief. "I shall be happier at Quincy," she told Thomas Adams. But to a cousin she admitted, "I lose my sleep often and I find my spirits flag. My mind and heart have been severely tried."[30] She worried about how the president would react to this abrupt loss of power and prestige. She feared that returning to his farm, "a world so limited and circumscribed," would plunge him into permanent gloom.[31]

But Portia's dearest friend found consolation in the peace treaty with France, the chief accomplishment of his administration. A copy of the treaty did not reach Washington, D.C., until after the election was lost. The terms were so unsatisfactory that the Senate at first refused to ratify it. The envoys had ignored instructions to seek millions of dollars to repay American merchants and ship owners for the vessels the French had seized in the Quasi-War. But the agreement ended the shooting war, enabling John to claim a victory over his enemies in his own party, if not over the Jeffersonian Republicans. He found special satisfaction in thinking of the treaty as a triumph over Alexander Hamilton. The president vowed he would go home and write his autobiography, answering the ex-general's slanders. Above all, he looked forward to years and years in the company of his beloved Portia, the only person in the world who appreciated him.

REMEMBERING SOME OTHER LADIES

John Adams never completed his autobiography. But he enjoyed seventeen happy years of Abigail's companionship. They remained partners in mind and heart, deeply involved in the lives of their children, grandchildren, nieces, and nephews. At various times, many of the grandchildren lived with them. As the years advanced, Abigail's health became fragile. But she refused to slow down. Her "uncontrollable attachment to the superintendence of every part of her household," John told John Quincy, worried him. "She must always be writing to you and all her grandchildren . . . she takes upon herself the duties of her granddaughter . . . maids, husband." He might have added that she was also managing the life and career of her youngest son, Thomas. Acknowledging that he had no hope of restraining her, John could only grumble, "I say . . . she must, because she will."[1]

During these years, several other women became part of John Adams's life. One drove him almost berserk, another broke his heart, and a third brought him love and consolation when he needed them most. Mercy Otis Warren was the disrupter of his peace. This remarkable woman, the sister of James Otis, the early agitator against British rule, was an old friend who attracted John's attention with her trenchant prose style even before the Revolution. He praised her satires and hard-hitting criticism of loyalists and conservatives. We have seen how their friendship collapsed when John refused to help her husband, James Warren, get a federal appointment in 1789.

In 1805, Mercy Warren published what she and many others considered her masterpiece—three formidable volumes, titled *History of the Rise, Progress and Termination of the American Revolution, Interspersed with Biographical, Political and Moral Observations.* President Thomas Jefferson bought copies and urged his friends to do likewise. Inevitably, John and Abigail read these long and remarkably well-researched books—and were dismayed and enraged. Although Mrs. Warren praised John's domestic life for its "morality, decency and religion," she made it clear that she took a dim view of John Adams as a revolutionary leader. She contended that he was a failure as a diplomat—he was "ridiculed by the fashionable and polite society of France" because he was deficient in the *"je ne sais quoi* so necessary in polite society."

In England, merciless Mercy claimed Adams had been "corrupted" by his close contact with monarchy and came home enamored of aristocracy, replete with "all the insignia of arbitrary sway." A "large proportion of his countrymen" thought he had "forgotten the principles of the American Revolution." As president he was a virtual betrayer of the Revolution, leader of an anti-republican administration. Even worse were her comments on John's character: he was driven by "pride of talents and much ambition." Too often "his passions and prejudices were . . . too strong for his sagacity and judgment."[2]

An enraged Adams fired off ten blazing letters to Mrs. Warren, condemning her habit of presenting him "in an odious light . . . to lessen and degrade" him. Where did she get this disposition to "wink him out of sight?" he virtually bellowed. Why was he deficient in *je ne sais quoi* but not Franklin, Jefferson, and a host of others? How dare she call him "corrupted"? He challenged her to produce a single fact justifying this insult, "from my cradle to this hour!" He reminded her that her brother, James Otis, had predicted in the 1760s that "John Adams would one day be the greatest man in America!"

Mrs. Warren was not even slightly intimidated by John's extended tantrum. She replied that she was unable to understand "the rambling manner in which your angry and undigested letters are written." She had set out to do him "complete justice" and could not understand why John did not agree with her statements. She could only conclude that he thought "his fame had not been sufficiently attended to." As even more vituperative letters arrived on her desk, she accused him of "meanness as well as malignancy"—he was trying to "blast a work" that had won the praise of "many of the best judges of literary merit." It came down to this —her book was "an inadequate panegyric of your life and character."[3] John's criticisms struck her as "the ravings of a maniac." She closed her ripostes by declaring: "As an old friend, I pity you. As an Historian, I forgive you."[4]

Poor Abigail was trapped in the middle of this savage exchange. She sided with John but probably regretted his uncontrollable rage, as she had "mourned" the mess he had made of firing Secretary of War McHenry. Several months later, she was astounded to receive a letter from Mercy in which her old friend told her that disagreements on subjects that involve "the great bustle of the world" ought not to interfere with true friendship. Mercy even thought it was possible to denounce someone's politics and retain feelings of love and admiration.

Abigail did not answer this letter, nor did she write a letter of sympathy when Mercy's husband died the following year. She had been hoping Mercy would reread John's letters and "acknowledge her errors." When no such acknowledgment came, Abigail could not pretend she had any friendly feelings for her. The politics of the party of two were still alive and well, seven years after John had left the presidency.[5]

But Abigail had too many shining memories of her friendship with Mercy Warren to allow rancor to obliterate them. As a lifelong advocate for the equality of women, it pained her that John had acquired such a low opinion of a woman who had demonstrated her ability to write and discuss history and philosophy as intelligently as any man of her time. Abigail gradually resumed corresponding with Mercy and in 1812, visited her in Plymouth, where hours of conversation began a reconciliation between "the two ancient friends," as Abigail called them. In a moment of courageous candor, Abigail said Mercy and John "were both in the wrong." She tried to explain that John no longer harbored any "personal animosity" toward Mercy but he was not ready to be entirely reconciled. Time would be the best healer of the "reservation" that the ex-president still nursed about Mercy's book.

Mercy sent Abigail a ring as a memento of their "former amities," and toward the end of 1812 Abigail sent Mercy a lock of her hair and a snippet of John's sparse gray locks "at his request." A year later some newspaper critics attacked Mercy's play, *The Group,* which she had written during the Revolution. They maintained that it was too complex and historically insightful to have been written by a woman. John Adams leaped to Mercy's defense in a vigorous letter, declaring he could confirm she had written the drama. In brusque Adams style, he described Mercy as the most gifted woman of her generation, a rare combination of intellect and artist. A delighted Mercy invited him to visit her in Plymouth. On the sidelines, Abigail smiled contentedly. The reconciliation was complete.[6]

II

Another woman who loomed large in John's old age was his daughter, Nabby. She had long been a presence in his life, of course. He had writhed and fumed countless times when he heard about how badly her husband, Colonel William Smith, was treating her. As Nabby's unhappiness intensified, her father became more and more meaningful to her, while her relationship to her mother deteriorated. Abigail's repeated attempts to persuade her to leave Smith may have had something to do with Nabby's inability—or refusal—to write to or talk meaningfully with her mother. She may also have blamed her for the way Abigail had interfered in her romance with Royall Tyler, and then all but thrust her into the arms of Colonel Smith in London.

It was one of those tragedies of good intentions. Abigail had frequently complained about Nabby's silent ways, even in her girlhood. As eager for her to achieve perfection in all things as she was for Nabby's brothers, Abigail did not realize that a feeling of inferiority can be the result of too much exhortation and scrutiny. In Nabby's case the tendency may have been intensified by the feeling that she could never equal her mother. In letters to her Cranch cousins, Nabby often expressed delight when she was permitted to visit friends and relatives for a week or two and escape from Abigail's critical eyes.

As the dimensions of her unhappy marriage became starkly apparent, Nabby's silences grew more extensive. She ignored Abigail's attempts to encourage or advise her. Only to her father was Nabby able to open her anguished heart—but Abigail began intercepting her letters and rebuking Nabby for sending them. As the politics of the vice presidency and presidency grew more turbulent, Abigail feared John would be driven over the edge by the details of his daughter's suffering. This policy only deepened the rift between Abigail and Nabby; soon the mother was complaining that her daughter was "unapproachable."[7]

In the final year of his presidency, John appointed Colonel Smith surveyor of the port of New York, one of the most lucrative jobs in the federal government. For a while, Nabby's problems seemed solved. Smith had more than enough money. But he spent it as fast as he made it, and continued to associate with hard-drinking men who passed their days gambling on fast horses and cards and talking endlessly about how to make a quick fortune. Still an Adams to the bone, Nabby despised Smith's cronies. When her brother John Quincy visited her in New York, she invited him into a remote room in her house and played chess while her husband partied with his raucous friends.

Frustrated by the lack of speculative opportunities in the United States after the bubbles of the 1790s had burst, Colonel Smith began conspiring with South American revolutionaries to fund an invasion of Venezuela and the overthrow of the Spanish government there. It would be the beginning of a revolutionary tide that would sweep through South America and make its leaders immensely rich. Such enterprises were difficult to keep secret; the talkative Smith did not bother to try. Instead, he boldly sent Francisco Miranda, the leader of the rebels, to President Jefferson to ask his help.

The president declined to participate, but Smith and Miranda went ahead anyway. When the American ship *Leander* sailed with a vanguard of revolutionaries aboard, the president accused Smith of making war on a country with whom the United States was at peace and arrested him. Meanwhile the *Leander* was captured by the Spanish navy. Smith was not even moderately repentant; he claimed the charges against him were a Jeffersonian plot to remove him from his job.

Once more Abigail implored Nabby to leave her husband. Instead, she moved into a cottage on the prison grounds and lived with him while he argued in court that he had been acting on Jefferson's orders. The president and Secretary of State James Madison declined to appear at the trial, and the judge freed Smith for lack of evidence. Smith soon discovered freedom was a mixed blessing. Numerous people whom he had conned into financing the Miranda expedition wanted their money back. Virtually penniless, barely able to pay the rent for a crumbling house on a New York back street, the embittered Nabby subsided into total silence for almost a year, while Smith's behavior toward her grew more and more hostile.

Early in 1807, Nabby's anguish exploded in a searing letter to her father, telling in detail Smith's awful abuse. Once more, Abigail intercepted the letter and rebuked Nabby for threatening John's emotional equilibrium. This was around the time that the ex-president was engaged in his violent controversy with Mercy Otis Warren. Tormented by his creditors, Colonel Smith abandoned Nabby and her daughter, Caroline (the only child still living at home), and fled west. John Quincy Adams brought the two women to Boston, where Nabby chose to live with him and his wife rather than endure Abigail's attempts to mother her at Peacefield.

A year later, Colonel Smith appeared in Boston and abruptly ordered Nabby to pack her trunk. He had found refuge in Chenango, New York, with his relatives, and it was her duty to join him. To her parents' horror, Nabby consented and vanished into northern New York state, where she did not write to anyone in the family for the next three years.

Early in 1811, Abigail received a stunning letter: Nabby feared a "hardness" in her right breast might be cancer. She described her symptoms, and Abigail rushed to confer with the best doctors in Boston. All agreed that Nabby could not get decent care in northern New York, a virtual wilderness in 1811. After several delays she decided to come to Peacefield, where the Boston physicians seemed to think her condition could be cured by hemlock pills. Nabby, meanwhile, was reading about her illness and discovered a treatise on breast cancer by Dr. Benjamin Rush of Philadelphia, an old and close friend of her parents. She wrote to him, describing her symptoms, and he recommended immediate surgery.

On October 8, 1811, four surgeons, led by Dr. John Collins Warren, amputated Nabby's cancerous right breast. She remained conscious during the twenty-five minute operation, never making a sound. The doctors marveled at her courage. For Abigail and John, the operation was a nightmare that left them barely able to speak or think. John said he felt as if he were living the Book of Job. Abigail said she felt as if she had survived a session in the biblical fiery furnace.[8]

Accentuating the sense of nightmare was the almost simultaneous illnesses and deaths of Abigail's sister Mary Cranch and her husband, as well as the heartbreaking decline of Sally Smith Adams, Charles's widow, who was dying of consumption in another room in Peacefield. The ex-president was also a patient. He had rushed outdoors to see a comet passing overhead and fallen over a stake, inflicting an ugly gash on his leg. Compounding the family's misery, their son Thomas, whom Abigail had persuaded to return to Boston to practice law, had been thrown by his horse, and his doctors feared he might be crippled for life. Indomitable as always, Abigail struggled to nurse everyone. But Nabby was her deepest anguish. Three months after the surgery, she could barely sit up in bed and was unable to use her right arm. When her husband took her back to Chenango six months later, the arm was still in a sling. Almost frantically, Abigail told John Quincy and other correspondents that the doctors had assured her that Nabby's recovery would be complete.

In northern New York, Colonel Smith amazed everyone by running for Congress as a Federalist and winning. He departed for Washington, leaving Nabby with little or no money, suffering bouts of excruciating pain that Smith offhandedly diagnosed as a familiar Adams illness, rheumatism. Little more than a year later, Smith's sister, Nancy, told Abigail that Nabby was being consumed by cancer. A contrite Colonel Smith mournfully informed Abigail that Nabby wanted "to die in her father's house." The word "father's" was not accidentally chosen.

In agonizing pain, Nabby endured three hundred miles in a jolting carriage over the primitive roads of northern New York to reach Peacefield on July 26, 1813. Abigail soon realized that she had nothing to offer her stoic daughter. She fled the dying woman's sickroom and John Adams, revealing wellsprings of pity and love that his crusty exterior usually concealed, took charge of nursing Nabby for the next two weeks. At times the pain was so intense that she lay doubled up. Only opium gave her occasional relief. No one knows what the father and daughter said to each other during these horrendous hours. Silent tearful embraces were probably more meaningful than words.

On the morning of August 15, Nabby asked her father to bring her a hymn book. He helped her open it to her favorite song, "Longing for Heaven." She invited her mother, her husband, and her son John and daughter Caroline to join her and her father in singing it:

Oh could I soar to worlds above
That blessed state of peace and love

How gladly would I mount and fly
On Angels wings to joys on high.[9]

Nabby died peacefully in her father's arms a few hours later. In a letter to Thomas Jefferson, with whom he had begun a correspondence, John Adams called her "a monument to suffering and patience." Jefferson wrote Abigail a sympathetic letter and she responded with a fragment of a poem:

Grief has changed me since you saw me last
And careful hours, with time's deformed hand
Hath written strange defections o'er my face.[10]

III

Besides Abigail and Nabby, the woman who played the largest role in John Adams's old age was Louisa Catherine Johnson, John Quincy's wife. Her first appearance in the family could not have been more unpromising. Abigail had been nurturing a dislike for her ever since her son became engaged to this "heiress," ignoring his mother's repeated warnings that she would make him miserable. The prediction acquired unexpected weight when Louisa's merchant father went bankrupt and was unable to deliver the handsome dowry he had promised. John Quincy's New England conscience—and Louisa's delicate dark-haired beauty—persuaded him to marry her anyway.

The marriage was fraught with strains from the start. To the bridegroom's financial disappointment Louisa added a highly independent spirit that constantly put her at odds with her short-tempered husband about everything from wearing makeup to raising children. John Quincy's decades of listening to and reading Abigail's lectures on virtue and accomplishment and whom he should not marry—with a special emphasis on Mary Frazier—had left him determined not to take a woman's advice no matter how good it was. He even went so far as to tell Louisa that women should have "nothing to do with politics"—a startling statement from Abigail's favorite son.[11]

Abigail was enraged when John Quincy did not even bother to tell her the date of his wedding. "My son, you will see as I do by the papers, is married," she acidly informed her younger sister, Elizabeth Shaw Peabody.[12] When John Quincy and Louisa returned to America in 1801 and she met her mother-in-law for the first time, everything seemed to go wrong. Louisa had an atrocious cold and was exhausted by the journey from Washington, D.C. Abigail studied her slim figure, listened to her croaking cough, and concluded that her stay on the planet would be of "short duration." This did not distress her nearly as much as the way John Quincy's anxiety about his wife had added "a weight to his brow." Also in the mix was Abigail's displeasure that the couple had named their first son George Washington Adams instead of John Adams. Abigail was sure this was Louisa's doing.

Peacefield was soon crowded with Adams relatives, who came to ogle the exotic English creature whom John Quincy had found in his travels. In the small-town world of Quincy, Louisa was more than a new bride; she was a curiosity. Abigail had propagandized everyone in the family about her moneyed background. Too sick to enjoy the special dishes Abigail prepared for her, Louisa was soon labeled "proud."

Louisa was equally unimpressed with her in-laws. Their nasal Boston accents, their old-fashioned clothes, and their strange hairstyles bewildered her. "Had I stepped into Noah's Ark I do not think I could have been more astonished," she recalled years later. Only one person seemed to receive her without suspicions or reservations: John Adams. She sensed the "old gentleman" was the only person in the family to whom "I was literally and without knowing it [more than] a fine lady."[13]

This rapport would grow stronger over the next twenty years, as John Quincy's star rose in the American political firmament. Louisa played a crucial role in this ascension. She was more politically astute than Abigail, never hesitating to disagree with John Quincy, who was prone to the Adams habit of ignoring public opinion. When he became a U.S. senator, Louisa relentlessly criticized his carelessness about his appearance, even enlisting Abigail's help to reform his tailoring, as well as his eating habits.

IV

During these decades, Abigail remained the center of John's emotional life. He struggled in vain to persuade her to slow down as her health became increasingly fragile. In October 1818, she took to her bed with an unidentified malady. A visit from their family doctor revealed she was suffering from typhoid fever, which happened to be epidemic in and around Quincy at the time. When the bad news reached Louisa in Washington, where John Quincy was secretary of state in the cabinet of President James Monroe, she wrote to John, expressing her own concern as well as her husband's. John Quincy was too incapacitated by a psychic clash between love and guilt to write a word.

The doctors struggled in vain to save Abigail. To conserve her dwindling strength, they even forbade her to speak or be spoken to. But the primitive medicine of the day could do nothing to help her. She died on Wednesday, October 28, 1818, at the age of seventy-four. Her eighty-three-year-old husband was the only person in the sickroom who remained calm. "I wish I could lay down beside her and die too," he said.[14]

Although John's health remained remarkably good, and grandnieces and nephews and grandchildren and friends visited him, Peacefield seemed empty without Abigail. John told one friend it had become a "region of sorrow." Into this void stepped Louisa Johnson Adams. She began sending John daily excerpts from her diary, filling him in on things political and social in Washington, D.C. John called the letters "a reviving cordial" and exulted over the way she was admitting him "into the characters of statesmen, politicians, philosophers, orators, poets, courtiers, convivialists, dancers, dandies and above all ladies of whom I should know nothing without your kind assistance."[15]

Next, Louisa enlisted him in a project to translate Plato from the original Greek into English. John told her the idea stirred "feelings of curiosity, astonishment and, excuse me, risibility." He could not imagine or conceive "a subject more to my taste." When she sent him a translation of Plato's discussion of the Athenian general Alcibiades, John was all but overwhelmed. He wondered how it was possible "that a gay lady of Washington, amid all the ceremonies, frivolities and gravities of a court and a legislature, can write so many and so excellent letters to me . . . and at the same time translate Plato's *Dialogues*?"[16]

Month after month for the next six years, Louisa enchanted John with her diary excerpts and letters full of witty and wry observations on Washington politics. Sometimes she cautioned him that her reports were opinionated—she made no secret of her antagonism toward slavery during the uproar that swirled around extending the peculiar institution into the territories—a dispute that foreshadowed the Civil War. John told her not to worry about letting her own opinions speak. "I am myself too much under the influence of prejudices to have ever reproached you seriously for yours," he wrote—something he would never have admitted to another human being, even Abigail.

Often, John reminded Louisa that she was at the vital center of the political world while he was "in a part of the world where nothing happens but morning noon and night, new moons and full moons, spring, summer winter." Without her he would "vegetate in solitude." This fulsome appreciation grew more and more important to Louisa as she and John Quincy began to debate whether he should seek the presidency. John Adams's approval gave her the self-confidence to urge her humorless workaholic husband to put aside his tormenting Adams conscience and demoralizing self-doubts and reach for the job. Louisa told him how they could do it. He would continue to excel as secretary of state, and she would put to political use her talents as a hostess.

Soon Mrs. Adams's dinners and balls were the talk of Washington, and John Quincy was on his way to the White House. In Quincy, John Adams heartily approved of this partnership, which reminded him of the one that had sustained him throughout his life. "Two such industrious honey bees as John Quincy Adams and his wife were never connected before," he declared.[17] It is not hard to imagine how much these words delighted Louisa. She was being told that she was the equal—and even the superior—of her formidable mother-in-law.

V

In 1820, when John Adams was eighty-five years old, another woman reappeared in his life after an absence of almost sixty years. The former Hannah Quincy was now a widow, almost as old as John. She persuaded her relative Josiah Quincy to take her to Peacefield for a visit with her old flame. When she walked into the Adams's study unannounced, John gazed at her in astonishment. His face came aglow and he said, "What! Madam, shall we not go walk in Cupid's Grove together?" This reference to the lovers' lane where they had strolled so long ago pleased Hannah immensely. With a smile that was a reminder of her coquettish youth, she replied. "Ah sir, it would not be the first time we have walked there!"[18]

Josiah Quincy was awed by "this flash of old sentiment." He was a frequent visitor to Peacefield and he found it totally unexpected. He was discovering that in the very old, the fires of youth are banked by time but by no means forgotten. Suddenly he glimpsed John Adams's whole complex life, from the yearning bachelor to the tormented president to the ancient sage now confronting him.

VI

None of these other women ever replaced Abigail Smith in John Adams's mind and heart. To the end of his life he constantly spoke of her "equal fortitude and firmness of character." Thinking of the sorrows they had shared, he was equally praiseful of her "resignation to the will of heaven." Above all, she had never "by a word or look" tried to dissuade John from "running all hazards" for the salvation of the American nation he had been among the first to envision and then helped to create. She was a unique combination of courage and forbearance, patience and pride. Without her, stumpy, potbellied, perpetually insecure John Adams would never have become a founding father.

When John died in the late afternoon of July 4, 1826, at the age of ninety, he inadvertently bore witness to this indubitable truth. Among his last words was a choked whisper: "Thomas Jefferson survives." (In fact, Jefferson had died several hours earlier.) Honest John seemed to be uttering a final sigh of regret for the way those tall Virginians had outmatched him again in the contest for fame. Here, as so often before, he needed his Portia to tell him he was a great man in spite of his flaws.[19]

The future Martha Washington was twenty-six and still married to Daniel Custis when this portrait was painted in 1757. That same year Custis died and Martha became the wealthiest widow in Virginia.

Washington-Custis-Lee Collection, Washington and Lee University, Lexington Virginia

General Washington wore this miniature of Martha by Charles Willson Peale beneath his shirt throughout the Revolution. She wore a Peale miniature of him around her neck.

Courtesy of the Mount Vernon Ladies' Association

This regal portrait of Martha was painted in 1878 for the White House. The head is based on a life portrait by Gilbert Stuart. The painting still hangs in the mansion's East Room.

White House Historical Association

Sally Cary Fairfax's coquettish personality is visible in this portrait. It was painted around the time she first entranced sixteen-year-old George Washington while he was visiting his older half brother Lawrence at Mount Vernon.

Virginia Historical Society

This only known portrait of Deborah Franklin was painted when she and Ben had become wealthy. It does not take much effort to imagine her brow furrowing and angry words exploding from her lips.

Emmet Collection, Miriam and Ira D. Wallach Division of Art, Prints and Photographs, The New York Public Library, Astor, Lenox and Tilden Foundations

This solemn portrait of Madame Helvetius suggests her aristocratic dignity but it lacks the wit and effervescence of the woman to whom seventy-six-year-old Ben Franklin proposed marriage.

Yale University Library

Sally Franklin Bache inherited her father's winning personality and leadership gifts. She raised thousands of dollars to buy shirts for the ragged soldiers in the American army.

Reproduced from the Collections of the Library of Congress

This is the Abigail Smith Adams who won John Adams's turbulent heart and was soon telling him to "remember the ladies" in the new government he was helping to found in Philadelphia.

Massachusetts Historical Society

There is more than a hint of sadness and resignation in this Gilbert Stuart portrait of Abigail Adams. It was painted in 1800 when she was fifty-six. Her husband's controversial presidency had been a harrowing experience for her.

Reproduced from the Collections of the Library of Congress

Abigail Adams Smith, known to her family and friends as Nabby, married erratic speculator William Stephens Smith and became a constant source of sorrow and concern to her parents.

Print Collection, Miriam and Ira D. Wallach Division of Art, Prints and Photographs, The New York Public Library, Astor, Lenox and Tilden Foundations

Louisa Catherine Adams, John Quincy Adams's wife, filled the void in John Adams's life after his wife's death. With John's warm approval, Louisa managed John Quincy's rise to the presidency.

Reproduced from the Collections of the Library of Congress

Born to wealth, Elizabeth
Schuyler Hamilton was
the "good girl" in contrast
to her two high-spirited
sisters. Her tentative smile
conveys an impression
of insecurity, even
melancholy. She could
not compete with her
husband's first love, fame.

*Emmet Collection, Miriam
and Ira D. Wallach
Division of Art, Prints
and Photographs, The
New York Public Library,
Astor, Lenox and Tilden
Foundations*

After her husband's
tragic death, Elizabeth
Hamilton became a
decisive, independent
woman. She pioneered
care for orphans in
New York City. In her
nineties, she played a key
role in the building of
Washington's monument
in the nation's capital.

*Collection of the New-
York Historical Society*

This sole surviving portrait of Angelica Schuyler Church only hints at the woman who mesmerized men from Paris and London to New York.

The Belvidere Trust

After witnessing her father's agony when her mother died in 1781, Martha Jefferson Randolph devoted the rest of her life to Thomas Jefferson. She also managed to raise eleven children and deal with an embittered, jealous husband.

Thomas Jefferson Foundation/Monticello

Gifted painter and musician Maria Cosway entranced forty-four-year-old ambassador Thomas Jefferson for two exhilarating weeks in 1786 in Paris. Then came a fateful attempt to vault a hedge and a shattered wrist that ended the idyll.

Reproduced from the Collections of the Library of Congress

Gilbert Stuart's portrait of Dolley Madison brings to life the ebullient hostess and loving wife who invented the office of First Lady and won women a meaningful role in American politics.

White House Historical Association

BOOK FOUR

Alexander Hamilton

BASTARD SON AND WARY LOVER

Marital turmoil seems to have been endemic in Alexander Hamilton's family. His maternal grandmother, Mary Uppington Fawcett, separated from her dour, aging husband, Dr. John Fawcett, and moved from the West Indian island of Nevis to neighboring St. Kitts with her only surviving child, Rachel. Mary must have been extremely unhappy; she gave up her rights to Dr. Fawcett's considerable estate in return for fifty-three pounds and four shillings in annual support. Rachel Fawcett matured into a beautiful and spirited woman and proceeded to replicate her mother's history—and then some.

In 1745, Rachel and her mother moved to the Danish-owned island of St. Croix, where at age sixteen Rachel married twenty-eight-year-old John Michael Lavien, a merchant with a murky background and a fondness for splendid clothes. Lavien seemed rich. Apparently he thought Rachel, who had inherited some St. Croix property in her father's will, was also rich. Both assumptions were wrong.

Like many other West Indian merchants, Lavien's fortunes fluctuated as violently as the weather and the market for sugar, the island's main crop. While living on a plantation named Contentment, he and Rachel had a son, Peter. As Lavien's debts mounted, they descended to a far less genteel house called Beeston Hill, and husband and wife began quarreling. Soon Rachel was expressing her discontent in a highly visible way. She began sleeping with other men. Backed by eyewitness testimony, the outraged Lavien had her arrested for being "twice guilty of adultery" and thrown into jail in the fort that guarded the harbor of the island's main port, Christiansted.[1]

Hoping she had learned her lesson after several weeks in a narrow cell, Lavien petitioned the judge to grant Rachel's release. She decamped to St. Kitts with her mother, abandoning Lavien and four-year-old Peter. On St. Kitts, Rachel began living with James Hamilton, the fourth son of Alexander Hamilton, laird of an estate called the Grange in Ayrshire, Scotland. These Hamiltons were a lesser branch of the dukes of Hamilton, a proud name in Scottish history. James was a handsome, charming loser who virtually specialized in going bankrupt.

Rachel and James Hamilton soon moved to nearby Nevis, the tiny island of her birth, and for the next fourteen or fifteen years lived a precarious existence punctuated by Hamilton's various attempts to make money as a merchant, all of which ended in disaster. In keeping with the relaxed sexual mores of the West Indies, there is no record of a formal marriage. Rachel gave birth to two sons, James and Alexander. Alexander entered this world either on January 11, 1755 or 1757. Scholars have been arguing about the two dates for decades. Hamilton himself has been no help, giving various birth dates during his tumultuous life. Recent biographers have inclined toward 1755, making Alexander not quite as youthful a prodigy as others have claimed.[2]

In 1759, John Lavien decided to remarry and obtained a divorce from Rachel in the St. Croix courts. It described her as "having shown herself to be shameless, rude and ungodly . . . and given herself up to whoring with everyone." Lavien termed Alexander and his brother "whore-children" and persuaded the court to rule that they could never inherit his property. Rachel ignored the summons to defend herself and her sons, if it was ever delivered to Nevis. The court permanently severed Lavien from Rachel and forbade her to marry again.

Rachel and her sons followed James Hamilton to Dutch-owned St. Eustatius in the course of his business peregrinations. In 1765 they returned to the scene of Rachel's early disgrace, St. Croix, to collect a debt due James's employer on Nevis. Exactly what happened next remains undocumented. But it seems likely that Rachel emulated her mother and told James Hamilton that she no longer had any desire to see him in her bedroom—or anywhere else in the house.

Some biographers have described the breakup as a desertion on James Hamilton's part. But all his life, Alexander spoke of his father with great sympathy, describing him as a man whose early failures never gave him a chance. In a letter to his brother at the close of the Revolution, Hamilton asked, "What has become of our dear father? . . . My heart bleeds at the recollection of his misfortunes and embarrassments." Equally significant is the dolorous fact that in the tens of thousands of words Hamilton wrote in his hyperactive life, he never mentioned his mother with affection. On the contrary, when he was about to marry, he felt compelled to ask his fiancée if she would share "every kind of fortune with him." He attributed his anxiety to his experience with "a female heart" who declined to tolerate a husband's failure.[3]

It seems probable that Rachel, still beautiful at thirty-six, decided she could prosper on bustling St. Croix without an albatross like James Hamilton around her neck. He meekly departed to spend the rest of his life drifting through the islands, eking out a living as a clerk or some other equally menial job in the counting houses or on the sugar plantations. When he left, Alexander was eight years old. He never saw his father again.

II

Rachel opened a small store in Christiansted and joined the local Anglican church. She apparently tried to regain a modicum of respectability in the teeth of the divorce decree that Lavien had procured against her in 1759. She got the goods for her store from two American merchants, David Beckman and Nicholas Cruger, who had offices just down the street. Rachel had also inherited nine slaves from her mother. She rented them out to various households on the island, augmenting her business income.

Three years later, Rachel and Alexander were stricken with one of those nameless fevers so rampant in the islands—possibly typhoid. After a week of agony, Rachel died. Alexander and his brother James inherited her modest estate, including the nine slaves—until John Lavien appeared with a lawyer and declared that his son, Peter, was Rachel's only legal heir, and the "bastard children" of her whoring days had no right to a cent. The St. Croix court found his argument legally irresistible. The half brother, Peter, twenty-three, soon arrived from South Carolina and claimed the estate.

What did the thirteen-year-old Alexander Hamilton think and feel, to hear his mother once more branded a whore and himself a bastard? It could only have further complicated his conflicted feelings about Rachel and about marriage in general. In this crisis, there is no mention of James Hamilton. No one seems to have considered summoning him to collect his sons. Bankruptcy and repeated failure had eliminated him as an option.

Fortunately, the two orphaned boys had relatives on St. Croix. Rachel's aunt, Ann, had married into the well-to-do Lytton family. Her son appeared in court on behalf of James and Alexander and managed to salvage part of the estate by claiming that Rachel had given two of the slaves to the boys. The Lyttons were in the process of going broke and had no money to give the Hamiltons. But at least they let them know there was someone in the world who cared about them.

Another friend was merchant Thomas Stevens, who invited Alexander into his home—and conspicuously ignored his brother James. Not a few biographers have wondered whether Stevens was Hamilton's real father— a complication that might be an added explanation for the way Rachel's marriage to James Hamilton expired. The older brother was apprenticed to an aging carpenter, and Nicholas Cruger took Alexander into his counting house as a clerk. The Crugers had a network of stores throughout the islands, in New York, and in Bristol, England.

III

Alexander Hamilton seems to have demonstrated an aptitude for buying and selling from the start. But he had ambitions beyond the Cruger counting house and the island of St. Croix. At the age of fourteen he wrote a letter that summed up much of his past and future life. He sent it to Thomas Stevens's son, Edward, who was a year older. The two young men looked so much alike that many people wondered if they were brothers. "Ned," Hamilton wrote, "My ambition is so prevalent that I contemn the groveling condition of a clerk or the like to which my fortune, etc condemns me, and would willingly risk my life, though not my character, to exalt my station. I'm confident, Ned, that my youth excludes me from any hopes of immediate preferment, nor do I desire it, but I mean to prepare the way for futurity."

Alexander admitted he had a propensity for building "castles in the air"

but noted that "such schemes" were sometimes successful "when their pro-
jector is constant." He closed with a fervent hope for an event that would
accelerate his rise to fame and fortune: "I wish there was a war."[4]

At sixteen, Hamilton was put in charge of the St. Croix office while
Cruger was away on a voyage to New York. The self-confident teenager
bought and sold everything from slaves to flour to mules and managed to
make a steady profit, even though the flour was moldy and the mules half
dead when they arrived from the American mainland.

On the side, he wrote poetry that he published in *The Royal Danish
American Gazette,* the local English-language paper. He saw it more as a
way to improve his chances of moving up in the world than an attempt
to launch a literary career. One poem was particularly noteworthy in the
light of his mother's reputation. It described how he found a lovely young
woman sleeping beside a brook and stole a kiss from her. She awoke and
there were more kisses, until

> *a rosy red o'er spread her face*
> *and brightened all her charms.*

Instead of instant bliss, however, the poet took his beloved to church
where "hymen join'd our hands." The happy rhymer ended his love story
with a brief sermon:

> *Ye swains behold my bliss complete;*
> *No longer then your own delay*
> *Believe me love is doubly sweet*
> *In wedlock's holy bands.*

Another poem revealed the sexual sophistication of this ambitious teenager:

> *Coelia's an artful little slut;*
> *Be fond, she'll kiss, et cetera—but*
> *She must have all her will;*
> *For, do but rub her 'gainst the grain*
> *Behold a storm, blow winds and rain*
> *Go bid the waves be still.*

So, stroking puss's velvet paws
How well the jade conceals her claws
And purrs; but if at last
You hap to squeeze her somewhat hard
She spits—her back up—prenez garde;
Good faith, she has you fast.[5]

IV

On August 31, 1772, St. Croix was struck by a ferocious hurricane. It killed over thirty people and leveled most of the buildings on the island, causing an estimated one million pounds in damage. Hamilton wrote an account of the storm in a letter to his father that demonstrated the young man had a gift for vivid language:

It seemed as if a total dissolution of nature was taking place. The roaring of the sea and wind, fiery meteors flying about it in the air, the prodigious glare of almost perpetual lightning, the crash of the falling houses and the ear-piercing shrieks of the distressed, were sufficient to strike astonishment into angels.

The letter continued for several more pages, in which Hamilton reflected on the impermanence of life and ended with a plea for help to heal the island's grievous wounds. He even threw in a lecture to the white masters of St. Croix who "revel in affluence," urging them to bestow some of their "superfluity" on the afflicted, a majority of whom were the island's twenty thousand black slaves. Their ramshackle cabins undoubtedly vanished in the first five minutes of the storm.[6]

Hamilton showed a copy of his letter to the Reverend Hugh Knox, a Presbyterian minister from the College of New Jersey who had recently begun preaching on St. Croix. Knox thought it was good enough to warrant publication in *The Royal Danish American Gazette*—and he praised the young author extravagantly to everyone he met. An idealistic man, Knox had already noticed Hamilton and pitied his harsh fate. On the strength of the letter, he launched a campaign to send the young man to America to get an education.

Armed with introductions from Knox and Cruger and the encouragement of his grandaunt, Ann Lytton, Hamilton arrived in New York in
June 1773. For a year he lived across the harbor in Elizabethtown, New
Jersey, attending a school run by a talented teacher, Francis Barber. His
recommendation from Knox gave him entrée to the homes of two leading
Presbyterians, Elias Boudinot and William Livingston. Hamilton's boy-
ish good looks and effervescent personality charmed both men. They set
about turning Hamilton into a Presbyterian and a student at the College
of New Jersey in Princeton.

Hamilton lived for a while in the Livingston home, Liberty Hall, in
Elizabethtown, and hugely enjoyed the company of their four beautiful
daughters. Gertrude Atherton, whose novel *The Conqueror* launched a
Hamilton revival in 1902, was convinced that twenty-two-year-old Kitty
Livingston introduced the eighteen-year-old Hamilton to "the fascination
of her sex"—a notion that seems almost ludicrous in the light of Alexander's island upbringing in Rachel Lavien's ménage. He was attracted
to Kitty for another reason. Her father was one of the leading political
controversialists of the day, constantly attacking the Anglican church's
attempts to create a clerical establishment in America and fiercely defending American rights. Kitty was equally fascinated by politics and presented
a very different creature to Hamilton's attention—the thinking woman.

In their correspondence, Hamilton admitted that he liked the way she
was "content with being a mere mortal" and required "no other incense
than is justly due you." He therefore resolved to talk to her "like one (in)
his sober senses." In spite of "amorous transports" whenever he thought of
her, he would pay her "all the rational tribute applicable to a fine girl."[7]

For the time being, Hamilton was more preoccupied with getting into
college. Pleasing his Presbyterian mentors, he applied to the College of
New Jersey in Princeton. According to President John Witherspoon, his
request to dash through the curriculum as fast as possible was rejected by
the board of trustees. Since no trustee ever said or did anything without
the imperious Witherspoon's approval, this was a polite way of saying no.
Hamilton switched to King's College, the future Columbia, an Anglican
school in New York City, which had given him permission to set up his
own course of study.

To disappoint men who had labored so hard to make him a Presbyterian and to select a school they despised for its British sympathies shows

a remarkable willfulness in one so young. It was a pattern that Hamilton would replicate all his life—first charming fathers, then repudiating them. He was his mother's son in his determination to have his own way. At a deeper level, unconscious anger was probably at work.

V

At King's College, Hamilton worked his youthful magic on the president, ruddy-faced Myles Cooper, charming him as he had charmed Boudinot and Livingston. He was soon a regular guest at Cooper's table, at which Anglo-American politics was discussed with a vigor that more than equaled the Livingstons' intensity. But here the talk was on the other side of the quarrel. The genial Cooper was a king's man to the core. Hamilton displayed his Presbyterian influences by stoutly resisting his arguments. Worse, he wrote a pamphlet demolishing the royalist reasoning of Cooper's friend and predecessor, Dr. Samuel Johnson, who had written a tract warning the rebellious Americans of retribution if they went too far in their defiance of the mother country.

Cooper stirred a filial response in the young West Indian. It may have had something to do with the president's lavish hospitality. At Cooper's table, sherry, port, and Madeira flowed freely. In the idealized spirit of academia, he tolerated Hamilton's contrary opinions. A year later, when the quarrel between England and America had exploded into bloodshed at Lexington and Concord, Hamilton demonstrated his filial feelings in a dramatic way.

America was aflame with anger and not a little of it was directed at men like Myles Cooper, who had been outspoken supporters of George III. On the night of May 10, 1775, a mob of four hundred torch-carrying self-styled Sons of Liberty showed up at the gates of King's College, determined to tar and feather and perhaps kill the Tory president. Hamilton and his roommate, Robert Troup, met them at the door of the main building, which stood at what is now Barclay and Church streets.

Leaping up on the stoop, Hamilton began haranguing the mob. The terrified Cooper, looking out the window in his nightshirt, screamed: "Don't listen to him. He's a crazy man!" As Hamilton's words reached him above the hubbub, Cooper realized the young man was trying to save him. He was urging the crowd not to disgrace the struggle for American liberty by mobbing a defenseless man.

Leaving Troup on the steps to continue his exhortation, Hamilton dashed upstairs and urged Cooper to dress and flee. They went out the back door and over a fence as the mob brushed aside Troup and stormed upstairs in search of their prey. Hamilton and Cooper scurried downtown along the bank of the Hudson and eventually made their way to the home of Cooper's friend Peter Stuyvesant. The next day, Cooper slipped aboard a British ship and retreated to London.[8]

War between England and America grew inevitable. A month after Cooper's flight, the Battle of Bunker Hill made the situation clear to everyone. Hamilton joined a military company with his college classmates and began drilling regularly under the guidance of a former British officer, Edward Fleming. By March 1776, the West Indian youth was a captain in command of an artillery company. He served his guns throughout the disastrous early battles of 1776, following George Washington and the collapsing Continental Army in their retreat across New Jersey to the far side of the Delaware. When Washington made his Christmas night strike at the exposed royal garrison in Trenton, Hamilton and his guns played a key role in the attack. His cannon swept the main street of the town with grape shot, breaking up every attempt by the enemy to form a battle line.

With the Revolution's spirit revived, the Continentals retreated to Morristown. There, perhaps at the suggestion of William Livingston, who had become governor of New Jersey, Washington invited Captain Hamilton to become his aide. It was the beginning of a relationship with an ultimate father that would last for the rest of Hamilton's life. The job meant an instant promotion from captain to lieutenant colonel, and a chance to shine at the very top of the American hierarchy. The combination of hurricane and war was sweeping Hamilton up the high road to fame.

VI

Becoming a member of Washington's official family also meant unremitting toil. "I give in to no kind of amusement myself," the American commander wrote. "So those around me can have none." All day every day, aides slaved at their desks, drafting letters for Washington's enormous correspondence. Hamilton's gift for language swiftly made him the champion in this department. As time went on, the General would simply tell Hamilton the overall direction of a reply and Hamilton would draft it for his signature. Sometimes Washington copied Hamilton's drafts in his own hand, word for word. Slowly, steadily, he became Washington's ghostwriter—and to some extent, his alter ego.

Washington gave Hamilton amazing responsibilities for such a young man. He dealt with congressmen and with Washington's potential military rivals, such as General Horatio Gates. He operated as an "intelligencer"— running several spies inside British-held New York. He wrote essays denouncing war profiteers in and out of Congress and flayed another rival general, Charles Lee, who had tried to undermine Washington. For a pen name, Hamilton chose Publius, after a first-century B.C. Latin writer who had been born a slave in distant Syria and brought to Rome, where his literary gifts won him the favor of his master and his eventual freedom. The choice suggests Hamilton had not forgotten he was an outsider in this American world.

In spite of Washington's stern description of an aide's life, the younger members of his military "family" did not work twenty-four hours a day. Most of them were in their twenties, and wherever they camped, they lost no time discovering the prettiest young women in the neighborhood. Hamilton was apparently a champion in this department. Martha Washington, observing him in action with the experienced eye of a former Virginia belle, nicknamed the headquarters tomcat "Hamilton" as an oblique compliment to his charms.

In Morristown, for a while Hamilton wooed Cornelia Lott, daughter of a wealthy farmer, then shifted his attention to someone named Polly, whose last name has escaped the records. He also continued to pay compliments in letters and in person to Kitty Livingston and her sister Susannah, who visited the encampment to nurse sick soldiers. Married officers imitated Washington and brought their wives to winter quarters—and their grown daughters if they had any. Vivacious young ladies were often guests at dinner, where Hamilton displayed a gift for making flattering toasts to their beauty.

In 1779, Hamilton began thinking of his future. The war was far from won, but with France as an ally, it was hard to believe that the Americans would lose if they "stayed in the game," as Washington sometimes put it. When Hamilton's closest friend among the aides, John Laurens, departed for a visit to his native South Carolina, "Ham" dashed off a frolicsome letter, asking him to find a wife for him:

Such a wife as I want will, I know, be difficult to be found, but if you succeed, it will be the stronger proof of your zeal and dexterity. Take her description— she must be young, handsome (I lay most stress upon a good shape), sensible (a little learning will do), well-bred (but she must have an aversion to the word ton [fashionable]), chaste, and tender (I am an enthusiast in my notions of fidelity and fondness), of some good nature, a great deal of generosity (she must neither love money nor scolding, for I dislike equally a termagant and an economist). In politics I am indifferent what side she may be of. I think I have arguments that will easily convert her to mine. As to religion a moderate stock will satisfy me. She must believe in God and hate a saint.

Next he added words that some people have found crass, others poignant:

As to fortune, the larger stock of that the better. You know my temper and circumstances and will, therefore, pay special attention to this article in the treaty. Though I run no risk of going to Purgatory for my avarice, yet as money is an essential ingredient to happiness in this world—as I have not much of my own and as I am very little calculated to get more either by my address or industry; it must needs be, that my wife, if I get one, bring at least a sufficiency to administer to her own extravagancies.

One can see more than a little evidence that the young Hamilton was trying to avoid some of the mistakes that had ruined his mother's two marriages. His modesty about his own ability to earn money also suggests that he carried within him wounding memories of his father's ineptitude in this department. After all these revelations, Hamilton recoiled from any serious commitment and assured Laurens he was only joking:

I am ready to ask myself what could have put it into my head to hazard this jeu de folie. Do I want a wife? No—I have plagues enough without desiring to add to the number that greatest of all [Hamilton's underlining].[9]

VII

Within a few months of writing this anti-romantic epistle, the swaggering soldier fell in love with a young woman who satisfied many of the specifications he had laid down. Elizabeth Schuyler came to the Morristown winter camp with her father, Congressman (and Major General) Philip Schuyler, and her sister Margarita. Schuyler was part of a committee that Congress sent to improve communications between the national legislature and the army. He was one of the richest men in America, owner of great swaths of land along the Mohawk and Hudson rivers.

Was it a love match? Fellow aides, such as Tench Tilghman, seemed to think so. "Hamilton," Tilghman wrote a mutual friend, "is a gone man." By February 1780, Hamilton was writing to Margarita Schuyler that "by some odd contrivance or other your sister has found out the secret of interesting me in everything that concerns her." In fact, he had become so absorbed, he was forced to insist on the basis of military necessity that she be "immediately removed from our neighborhood" so he could concentrate on winning the war.

To his friend Laurens, he described Betsey in terms that revived the themes of his wife-search letter: "She is a good natured girl who I am sure will never play the termagant; though not a genius she has good sense enough to be agreeable, and though not a beauty, she has fine black eyes— is rather handsome and has every other requisite of the exterior to make a lover happy . . . Believe me, I am a lover in earnest."[10]

He was soon writing letters to Betsey that were little short of impassioned: "You engross my thoughts too entirely . . . You not only employ my mind all day, but you intrude on my sleep. I meet you in every dream." In another letter he praised "the sweet softness and delicacy of your mind and manners, the elevation of your sentiments, the real goodness of your heart."[11]

But the lover was not above a lecture on self-improvement to further perfect his beloved. He urged Betsey to devote more of her leisure to reading. He wanted her to excel her sex in "distinguished qualities" as well as amiable ones. This urgent request betrayed a slight uneasiness about Betsey in Hamilton's mind. It is perhaps worth noting that his friend Tilghman referred to her somewhat sarcastically in one of his letters as "the little saint."

In the Schuyler family, Betsey was the good girl who obeyed her domineering father in all things—while her two beautiful sisters, the older Angelica and the younger Margarita (called Peggy), made a practice of defying him. Angelica had already eloped with an Englishman, John Barker Church, and Margarita would later follow her example, scampering into the night with a rich American, Stephen Van Renssalaer. Hamilton's choice of the less beautiful, more docile sister again reveals his desire to steer a course that would avoid the disasters of his father's marriage to his tempestuous mother.

Hamilton's French friend the Marquis de Fleury saw the marriage as something less—or perhaps more—than a love match. He wrote Hamilton congratulating him on his engagement "for many reasons." The first reason was the undoubted fact that henceforth Hamilton would have the backing of the Schuyler family's "influence." Second, it should put him in "a very easy situation, and happiness is not to be found without a large estate." Third, the Marquis hoped to be Hamilton's brother-in-law. He was pursuing Margarita Schuyler.

This European, even cynical view of the marriage is belied by other evidence. After the engagement was announced, Hamilton's anxiety about money and marital happiness surfaced almost uncontrollably in his letters. No matter how docile Betsey was, he could not forget that she was the daughter of a rich man. "Do you soberly relish the pleasure of being a poor man's wife?" he asked. He urged her to consider "the dark side" of their coming marriage and worried that they might be playing a comedy of deluded lovers. He wondered whether Betsey should "correct the mistake before we begin to act the tragedy of the unhappy couple."[12]

Betsey reassured Hamilton and remained his fiancée. He agreed to curb his impatience and wait until December 1780 for the wedding. Hamilton had pressed for an immediate ceremony in Morristown, but the Schuylers wanted to see Elizabeth married in their Albany mansion, The Pastures, to make up for the pain they had suffered over Angelica's elopement. As usual, Betsey succumbed to her parents' wishes, and Hamilton had little choice but to go along.

The illegitimate boy who had resented being a "groveling clerk" overcame his fears and married the rich man's daughter on December 14, 1780. A month later, Hamilton wrote a letter to Margarita Schuyler that again reveals he considered marriage a perilous venture. Betsey had written a note to Margarita telling her how happy she was, ending with: "Get married, I charge you." To it, Hamilton added a postscript that Betsey may not have seen.

He advised Margarita not to let Betsey make her "marriage mad." Marriage was a "very good thing when their stars unite two people who are fit for each other." But it was a "dog of a life when two dissonant tempers meet, and 'tis ten to one but this is the case." He advised Margarita to wait for "a man of sense, not ugly enough to get pointed at—with some good-nature—a few grains of feeling—a little taste—a little imagination—and above all a good deal of decision to keep you in order, for that I foresee will be no easy task."[13]

VIII

Rachel Lavien's banishment of James Hamilton left his son with another problem: he found it difficult to deal with substitute fathers—including George Washington. Living and working intimately with the general, Hamilton proved the adage that no man is a hero to his valet. Exhausted by the seemingly endless war, Washington was often irritable and impatient. Exacerbating Hamilton's feelings was his hunger for fame. The General had permitted other aides to escape his military family and return to leading troops. But he refused to part with Hamilton—a tribute to his talents, to be sure. Hamilton discounted the compliment. In the shadow of the Great Man—a phrase that Hamilton began to use derisively in private letters—an aide would always be a cipher.

On February 16, 1781, at Washington's headquarters in New Windsor, north of New York, Hamilton dashed downstairs with an important letter that fellow aide Tench Tilghman was to rush to the Commissary Department. On his way upstairs, he met the Marquis de Lafayette, and they paused on the landing to discuss another piece of business. At the head of the stairs he found a fuming Washington, who said, "Colonel Hamilton, you have kept me waiting these ten minutes. I must tell you, sir, that you treat me with disrespect."

"I am not conscious of it, sir," Hamilton replied. "But since you have thought it necessary to tell me so, we part."

"Very well, sir," Washington said. "If it be your choice."

Hamilton retreated to his room. Within ten minutes, Tench Tilghman knocked on the door bearing an apology from Washington. He said the General wanted "to heal a difference which could not have happened but in a moment of passion." He added that Washington retained full confidence in Hamilton's abilities and integrity.

Hamilton stonily declined to accept Washington's apology or his offer to meet with him and apologize in person. He was determined to leave the General's "family." He would stay only until Washington found a replacement. It would be up to the commander in chief to continue dealing with him as if nothing had happened.

"Thus we stand," Hamilton told his father-in-law, Philip Schuyler, having described in explicit detail how the breach occurred. He added that he had always disliked the "personal dependance" [sic] of an aide-de-camp and had accepted the job on the crest of patriotic enthusiasm in 1777 and "an idea of the General's character which experience soon taught me to be unfounded." For the past three years, Hamilton declared, "I have felt no friendship for him and professed none."

His rage building, Hamilton continued: "The truth is our dispositions are the opposite of each other & the pride of my temper would not suffer me to profess what I did not feel . . . You are too good a judge of human nature not to be sensible how this conduct in me must have operated on a man to whom all the world is offering incense."[14]

This portrait of Washington, reaching out to Hamilton as a father figure and sensing but not understanding his rebuff—and resenting it—is all too clear. Also painfully clear is the pleasure Hamilton took in inflicting this discomfort on the Great Man. Hamilton's memories of his failed father made him as wary of being a surrogate son as he was of becoming a husband. Rachel Fawcett Lavien and James Hamilton had inflicted grievous wounds on the soul of this enormously gifted young man.

THE WOMAN IN THE MIDDLE

The only portrait of Elizabeth Schuyler Hamilton painted during her husband's lifetime is rich in psychological interest. The artist was Ralph Earl, a hard-drinking New Englander who was no genius but had a knack for creating a good likeness. He had studied under the far more famous American painter, Benjamin West, in London. Earl saw Betsey as an attractive women, with deep-set dark eyes under thick brows. She is stylishly dressed and coiffed. But her tentative smile conveys an impression of insecurity, even melancholy. There is not a trace of the energy and self-confidence that emanates from almost every portrait of her husband.

Betsey's portrait was painted in 1787. By that time, Alexander Hamilton had closed his military career by persuading George Washington to give him command of a light infantry regiment at the siege of Yorktown, where he led a charge that made him a popular hero. After the war he emerged as one of New York's most successful lawyers and a political thinker of formidable stature. He played a crucial role in persuading Americans to junk their unworkable first constitution, the Articles of Confederation, and create a new national charter.

Riding this political whirlwind absorbed most of Hamilton's waking hours. When he was not talking politics with thinkers such as James Madison of Virginia or economics with merchant kings such as Robert Morris of Pennsylvania, he was persuading a reluctant George Washington to become the nation's first president. Meanwhile he was churning out brilliant essays defending the Constitution and urging its ratification, and waging an uphill fight against popular Governor George Clinton to win New York's approval of the disputed charter. Elizabeth Hamilton slowly realized that she was destined to play a minor role in the central drama of her husband's life—his pursuit of fame.

An additional reason for the melancholy and uncertainty that stained Betsey's sensitive face was the presence of another woman in her life—and in her husband's life: Angelica Schuyler Church. Her portrait reveals a dramatically different personality from shy, submissive Betsey. Angelica had an angular face with mocking dark eyes and a sensuous mouth that seems poised to make a witty—or a seductive—quip. Sophistication, worldliness, and even a certain recklessness emanate from Angelica. She had charmed bluebloods in Paris and London and New York. Gouverneur Morris, Gotham's best-known rake, had been so dazzled that he claimed Angelica could have whatever part of his body she preferred. But there seemed to be only one man who aroused this temptress's deepest feelings. It was not her fat, stolid, monosyllabic British husband, who had made a fortune during the Revolution supplying the American and French armies. The man was her brother-in-law, Alexander Hamilton.

Angelica made no secret of her passion when she wrote to Betsey. In 1784, when she was in Europe with her husband, she mentioned Hamilton's name no less than fourteen times in a single letter. In another letter she rather abruptly told Betsey to stop sending her fruit, which was spoiled by the time it reached her. What she wanted was newspapers—especially "those that contain your husband's writings." She ordered Betsey to tell Hamilton how much "I envy you the fame of so clever a husband, one who writes so well; God bless him and may he long continue to be the friend and brother of your affectionate . . . Angelica."[1]

Angelica's husband bought a palatial country estate near London and became part of the circle around the Prince of Wales, a social whirl in which the men gambled themselves into near bankruptcy and drank themselves into insensibility and the idea of marital fidelity drew guffaws. Almost everyone imitated the prince and had a mistress. Whether Church was one of these sexual adventurers remains uncertain. He seems to have had only two passions, business and gambling. As early as 1785, Angelica was telling Hamilton that "he no longer hears me." Ignoring her advice, Church bought himself a seat in Parliament, where his inadequacy as a speaker left him a nonentity. In one letter, Angelica lamented to Hamilton that he lacked "your eloquence." Clearly, she was an unhappy woman.

By 1787, Angelica was writing letters to Hamilton that contained a subtle code of endearment. She played games with commas, inserting them where they gave special meaning to certain words. "Indeed my dear, Sir," she wrote. Hamilton was deep in composing *The Federalist Papers,* the brilliant commentary that persuaded thousands of voters to support the new constitution. But he instantly noticed Angelica's misplaced comma and was stirred: "There was a most critical comma in your last letter. It is in my interest that it should have been designed; but I presume it was accidental." Still, he had a high opinion of her *"discernment."*(He underlined the word.) That tempted him to hope that she would read his reply in a certain mood that would enable her to "divine that in which I write it."

Angelica had thanked him for strewing so many "roses" (compliments) in her path in a previous letter. Hamilton told her that these pretty phrases had become "only a feeble image" of what he now wanted to convey to her. Whereupon he closed this confession of love with a misplaced comma of his own: "Adieu, ma chere, soeur. A. Hamilton."[2]

In 1789, as the new federal government began to operate, Angelica persuaded—or perhaps browbeat—her husband into financing a trip to New York. She came alone—a statement in itself. Hamilton, who functioned as John Barker Church's American business representative, handled the money he sent with her. Mrs. Church stayed five months, from May to November. One biographer has found the financial records that Hamilton kept for her in his office ledger very revealing. For an unknown reason, the cash was divided into "Monies paid to Yourself" and "For You." Certain sums were destined for a "last" landlady—not the one with whom Angelica stayed and whom John Barker Church paid. Was this "last" landlady in charge of another set of rooms, where the lovers secretly met?

It may not be entirely coincidental that during these months, Hamilton's political enemies launched a savage attack on him. He had infuriated a great many politicians who had opposed New York's ratification of the new constitution. Hamilton replied in equally ferocious style, under the pseudonym "H.G." One attacker, writing under the pen name William Tell, replied by taking the low road: "Your private character is still worse than your public one, and it will yet be exposed by your own works, for

you will not be bound by the most solemn of all obligations! *******."
Insiders had no trouble reading the seven asterisks: *wedlock*.[3]

When Angelica was summoned back to England by her husband, her
letter revealed an anguish that transcends the pain of parting from a mere
brother-in-law and favorite sister. *"Me voilà très cher bien en mer and le
pauvre coeur bien effligé de vous avoir quitte."* [Here I am, my precious darling, all at sea and my poor heart so distressed at leaving you.] In a burst of
defiance, she added: "I have almost vowed not to spend [more than] three
weeks in England."

Angelica shifted her attention to Betsey—a pattern that became habit-
ual in her letters to Hamilton. Her sister was ill, and Angelica felt guilty
that she would not be able to stay and care for her. She vowed to return
soon, and urged Hamilton "to remember this yourself, my dearest brother,
and let neither politics nor ambition drive your Angelica from your affections."

Another letter went from her ship as it headed into the wintry Atlantic.
Angelica wondered how she could be content when she was parting with
"my best and most valuable invaluable friends." (Again there is the hint of
a coded message. Were some of her friends valuable and one invaluable?)
She begged Hamilton to remember that he had told her she was "as dear
to you as a sister." She hoped he would "keep your word" and "never forget the promise of friendship you have vowed." Then came the inevitable
"A thousand embraces to my dear Betsey." Angelica hoped she would not
have "so bad a night as the last."

Ailing Betsey had not accompanied Hamilton to the ship to say goodbye
to Angelica. Back home, Hamilton reported to Angelica that he found
Betsey "in bitter distress" but "much recovered from the agony in which
she had been." He "composed her" by assuring her that Angelica would
survive the voyage. Then he and his oldest son, Philip, and Revolutionary
War hero Baron Friedrich von Steuben, whom Angelica had apparently
charmed, strolled to the Battery at the tip of Manhattan Island, where
"with aching hearts and anxious eyes" they watched her ship depart. "We
gazed, we sighed, we wept; and returned home," Hamilton wrote, "to
give scope to our sorrows and mingle without restraint our tears and our
regrets."[4]

This can be read as a perfectly honest and deeply affectionate tribute
to a remarkable sister-in-law. It can also be read as evidence that Betsey

had by this time become aware of Hamilton's passion for Angelica, and thinking about it had made her ill. The imputation is supported by Hamilton's enemies' gibe about wedlock. But it is disputed by a tender letter that Betsey wrote to Angelica, which Hamilton enclosed with his own impassioned epistle.

Betsey was too overwhelmed by grief at losing "My very dear beloved Angelica" to write more than a few lines. She begged her to tell Mr. Church that he must bring her back to America as soon as possible for the sake of her (Betsey's) happiness—and the happiness of "my Hamilton who has for you all the affection of a fond brother."[5] The emotions expressed on both sides seem to go far beyond the *amitié amoureuse* of a Benjamin Franklin with Madame Brillon. These people were in their thirties, in the full vigor of love and desire. Put another way—did Hamilton's head control the obvious longings of his heart? Perhaps. Perhaps not.

II

Hamilton's program to rescue the newly constituted United States of America from national bankruptcy and economic stagnation stirred enmity in Thomas Jefferson and many other people, who took seriously if not literally the claim in the Declaration of Independence that all men were created equal. Hamilton's proposals, which were aimed at persuading—and rewarding—the rich to invest their capital in the United States seemed unwise and often unjust to these men—a betrayal of the Revolution's ideals. These feelings were exacerbated by the turmoil of the French Revolution and its denunciations of aristocrats and kings.

Hamilton defended his ideas in the newspapers, claiming his policies were the only alternative to the anarchy and mob rule that were sweeping France. Like his political clashes with Governor Clinton in New York, the contest with Jefferson soon grew personal. Supporting the French Revolution left a man open to the accusation that he was an atheist. From that gibe, it was only a few steps to viewing him as a libertine, preaching amorality and sexual license to everyone. The widower Jefferson was a natural target for such allegations.

In the midst of this imbroglio, Hamilton did not forget Angelica. On January 7, 1790, while he was working twenty hours a day to prepare his financial program for submission to Congress, he found time to tell her he was "very busy and very anxious." But he could not let the London packet sail without a letter telling her that "no degree of occupation can make me forget you."[6] Angelica would remain beyond Hamilton's reach in London and Paris for the next six years, the zenith of Hamilton's career. But the intensity of his attraction to her remained a spiritual force in his life, slowly eroding his feelings for Betsey. There was a side of Hamilton's personality that wanted more drama, more ecstasy, than Betsey could inspire with her devotion and fidelity. Sexuality became intermingled with his political triumphs and his growing fame—a phenomenon that would be repeated by more than one American politician in future decades.

III

The first years of Hamilton's service as George Washington's secretary of the treasury were a series of political victories over his critics. In the no-holds-barred partisan style of the day, Jefferson inspired—and sometimes hired—newspapermen who began printing violent attacks on Hamilton and his policies. The treasury secretary was portrayed as a would-be despot, ready to assert his power in every imaginable way—in Congress, the courts, the voting booth, and the bedroom. The Jeffersonians saw him as a truly dangerous man who had to be destroyed before he corrupted America. Hamilton's West Indian birth, with its overtones of illegitimacy, added spice to the accusations.

Hamilton accepted the rules of this rough game and responded in kind. Writing as Catullus, he assailed Jefferson's character. There would soon come a time, he predicted, "when the visor of stoicism is plucked from the brow of the epicurean, when the plain garb of Quaker simplicity is stripped from the concealed voluptuary." Allies joined him in castigating Jefferson's concealed hunger for power—and his even more secret sensuality. Two centuries later, the "you're another" quality of the accusations are obvious and almost tiresome. But Americans of the 1790s, experiencing their first immersion in partisan politics, were enthralled.

In the early stages of this brawl, Hamilton prevailed. He won Congress's approval to create the Bank of the United States, which assumed the $40 million Revolutionary War debt and sold its shares to eager, mostly wealthy, investors. Soon there was an embryo stock market flourishing on Wall Street, another British innovation that Jefferson and his friends deplored. As we have seen, in return for tolerating this financial program, the Jeffersonians demanded that Hamilton agree to move the capital to Washington, D.C. While the federal city was under construction, Philadelphia would be the temporary capital. Like John and Abigail Adams, Hamilton had no enthusiasm for this transfer. New York was his hometown and New York State the base of his political power. But he made the personal sacrifice to rescue his financial program. He was convinced that the nation's salvation depended on it.

Thus, Hamilton and Betsey and their four children found themselves in the City of Brotherly Love in 1791. As the summer began, a letter arrived from Betsey's father confessing that he and Mrs. Schuyler were worried about their daughter and their grandchildren. Philadelphia was infamous for its outbreaks of yellow fever during the warm months. Betsey and the children were soon on their way to Albany.

Not long after they departed, Hamilton was visited by an attractive brunette who introduced herself as Maria Reynolds and blurted out a tearful story of neglect and abuse. She was a New Yorker, she said, a relative of the Livingstons, married to a man named James Reynolds, who had been in the Continental Army's commissary department during the Revolution. Her husband had left her for another woman, and she hoped that Hamilton, as a fellow New Yorker, would give her enough money to get back to the Empire State and sympathetic relatives.

This, it should be emphasized, is Hamilton's version of the story. Hamilton said he did not have much cash in his house (an oddity in itself) but would be glad to lend her the money later in the day. Maria gave him her address, 134 South Fourth Street, only a block away from Hamilton's Third Street home, and departed. That evening, Hamilton put a thirty-dollar bank bill in his pocket and strolled to the Reynolds house. It was as expensive and comfortably furnished as the one Hamilton was renting—a fact that should have aroused his suspicions. A servant led him upstairs to Maria Reynolds's bedroom, where she greeted him with effusive gratitude and accepted the money. They talked for a while and, as Hamilton later described it, he became aware that "other than pecuniary consolation would be acceptable."

Unquestionably, if Hamilton's version is true, this ranks as one of the easiest seductions on record. But his story raises far more questions than it answers. It seems unlikely that this was the first time Hamilton had met Maria Reynolds—unless he was ready to go to bed with almost any woman who extended an invitation. It seems more than likely that she and her husband, who was a stock speculator, had managed to inject themselves into some part of Hamilton's social life in New York or Philadelphia, where she had attracted his amorous attention.

Reynolds was a petty player, but he was not a complete nobody. During the Revolution, he did business with William Duer, later Hamilton's right-hand man at the Treasury. In 1789, Robert Troup, Hamilton's college roommate and still a close friend, had given Reynolds a letter of recommendation for a job in the Treasury department. On the record is also a Hamilton meeting with Reynolds in New York, when the speculator played the political ally and reported a Philadelphia smear campaign that portrayed Hamilton as a secret agent working to put one of George III's sons on an American throne.

Among the Jeffersonians, Reynolds was a name that was synonymous with Hamilton's supposedly corrupt influence. In 1790 the speculator had traveled to North Carolina and Virginia and bought up promissory notes given to former Continental Army soldiers in lieu of cash during the Revolution. He had conned these patriots into selling him their government paper for next to nothing, then hustled back to New York and made a killing when Hamilton announced his plan to pay the government's debts at par—face value.

Reynolds had obtained lists of the ex-soldiers' names from a crooked clerk in Hamilton's Treasury office. This unsavory deed turned James Madison into a Hamilton opponent. It is hard to believe that this story, which swirled through Congress and became a key reason for Jefferson's and Madison's hostility to Hamilton, did not reach the secretary of the treasury's ears.[7]

Did pity play a part in Hamilton's seduction? Did Maria remind him of his mother's travails with an angry husband and several lovers? Why didn't the secretary of the treasury consider that he, the most powerful man in Philadelphia, except for President George Washington, might be the target of a plot to ruin his reputation? Was he so deep in his romance with fame that he considered himself invulnerable? Was Maria Reynolds a kind of damaged version of Angelica Church—a woman who made no secret of her attraction to men and her awareness that men desired her? That may be the best answer to Hamilton's readiness to offer Maria "consolation"—a significant choice of words.

Somehow she must have represented a version of erotic love that fit Hamilton's new vision of himself as a man of power, with the freedom to seek the kind of pleasure that his wife could not give him. This "grand passion," as one biographer has called it, would be invoked by thousands of lovers in the dawning age of romanticism to justify their scorn for moral restraint. The ancient Greeks called it *eros*; they saw it as a force that virtually annihilated the self while simultaneously expanding it to mystical dimensions. But in Hamilton's case, the energy—and not a little of the ecstasy—flowed from the dynamo of fame.[8]

Another connection to Angelica may also have been on Hamilton's mind. Jefferson's newly formed Republican Party had humiliated him in New York by defeating his father-in-law, Philip Schuyler, in his bid for reelection to the Senate. The winner was a youthful politician who was emerging as a Hamilton rival: Aaron Burr. This failure may have damaged Hamilton's relationship to both Betsey and Angelica—especially the latter. A rebuke from Angelica—even the thought of one—would have made Hamilton susceptible to Maria, who was appealing to him as a man of power, pleading for rescue from her abusive husband.

Throughout that humid Philadelphia summer of 1791, Hamilton repeatedly visited Maria in her bedroom—or invited her to join him in the bed he shared with the absent Betsey. In his later account of his infatuation, Hamilton confessed that most of his "frequent meetings" with Maria were "at my own house." In all these encounters, Maria proved herself a consummate actress. She portrayed herself as a once naïve woman who had married the scoundrel Reynolds at the age of fifteen, had a child by him, and found herself trapped in an ongoing nightmare. He turned her into a prostitute who handed over all the money she received from her clients so he could gamble in the stock market. This victimized version of Maria's marriage—she actually seems to have been closer to a partner in crime—only increased Hamilton's passion, without arousing his suspicion.

IV

In the summer of 1791, Hamilton was riding the crest of the huge wave of public enthusiasm he had created by persuading Congress to approve his plan for overhauling America's public credit. On June 20, when shares of the Bank of the United States went on sale, the first offerings in New

York and other cities had sold out in a matter of minutes. Secretary Hamilton had attempted to check speculation in the bank's stock by making it expensive to buy. A $400 share required $100 down, the rest to be paid in four semiannual installments. (In modern money, this was roughly $6,000 a share with $1,500 down.) But Congress, already demonstrating an eagerness to please as many people as possible, reduced the opening payment to $25. For this amount, the purchaser got a certificate, soon nicknamed "scrip," which entitled him to buy the full share at par.[9]

In less than an hour, the $8,000,000 first issue was oversubscribed by $1,600,000. In five weeks, the value of the scrip soared from $25 to $325. The low opening price enabled almost everyone to get into the game; "scrippomania" swept the nation. Newspapers began printing daily stock quotations. An agitated James Madison told Thomas Jefferson that no one in New York talked about anything but "stock jobbing." Congressman Henry Lee reported that everyone he knew was investing in the bank. Hamilton, who could have made a fortune by buying shares for himself at far-below-market prices, remained financially pure. But this veritable explosion of fame redoubled his ardor for Maria Reynolds.[10]

Interesting evidence of the psychological connection between Maria and Angelica is visible in the abrupt fall-off of Hamilton's letters to his sister-in-law and the no-longer-rapturous expressions of affection in them. In one letter he almost offhandedly remarked that Betsey approved of his loving Angelica "as well as herself"—not the kind of extravagance she had come to expect. Soon he was writing, "I think as kindly as ever of my dear sister in law." Betsey did not fare much better during these months. In several letters, after a perfunctory profession of love, Hamilton assured his wife there was no need to hurry back from Albany. When she complained that his letters were infrequent, he vowed that he had been writing her almost every day and wondered if sinister conspirators were intercepting his letters.[11]

When Betsey returned from Albany with the four children, Maria's visits to the Hamilton home ceased. By this time the treasury secretary had rented a house on Market Street, only a few doors from President Washington's residence. But Hamilton continued to visit Maria throughout the fall of 1791, while he was deep in another battle with Congress about his proposal to create a major manufacturing metropolis, in part financed by the government, on the banks of the Passaic River in Paterson, New Jersey. It was part of his long-range program to make America an industrial rival to Great Britain. This vision of America as an economic powerhouse clashed violently with the Jefferson-Madison vision of a nation of virtuous small farmers, deepening their conviction that Hamilton was a menace.

V

Hamilton wrote two versions of his affair with Maria. In the first, which was never published, he described her as "play-acting" her love for him. In his second draft, he maintained that she had fallen in love with him and this made it extremely difficult to break off their relationship. He blamed his belief in her "real fondness" for him on his vanity. The two drafts probably describe the evolution of their relationship. As a man of the West Indian world with a mother who had slept with many men and may have discussed them in his earshot, he could have had no illusions about a woman's ability to fake affection.

As the affair continued, Hamilton had to justify it by convincing himself that he had won Maria's love. This illusion brought pity into play. Rather than break off the affair abruptly, he told himself a "gradual discontinuance" would be best. This self-deception enabled him to continue the affair for another year. Maria contributed to the illusion of heartbreak, Hamilton later ruefully recalled, by producing "all the appearances of violent attachment and of agonizing distress" at the merest hint that the affair would have to end.[12]

In the late fall of 1791, reality came crashing into Hamilton's imaginary romance. James Reynolds returned to his wife and professed shock and outrage to discover that she had been seduced. In an interview with Hamilton, he hinted that only a job in the Treasury Department would satisfy him. The secretary put him off with evasions, but in a few days Reynolds was back demanding money. The desperate Hamilton agreed to pay him $1,000 (the equivalent of 20,000 modern dollars) in two installments. This only led to more requests for cash, sometimes described as loans, which Hamilton often had to borrow from friends to pay. Meanwhile, the infatuated secretary continued to visit Maria, with Reynolds's tacit consent.

Reynolds had another reason for bearding Hamilton. He was in danger of doing jail time for fraud. He and a fellow speculator named Jacob Clingman had made $400 claiming that Clingman was the heir of a soldier on Reynolds's list to whom the government owed money. Unfortunately, the soldier turned out to be alive and the comptroller of the treasury, Oliver Wolcott Jr., urged the state of Pennsylvania to arrest both crooks. Like most people, Clingman dreaded the prospect of going to prison; a stay in the vile jails of the eighteenth century was often a death sentence. Clingman expressed skepticism when Reynolds assured him that Hamilton would protect them. One day Clingman waited in the street while Reynolds entered Hamilton's house and came out with a supply of dollars.

Clingman professed shock at this revelation. He was a political ally of the Jeffersonians and saw Reynolds's access to the secretary of the treasury as possible proof that Hamilton was using him as a front man to speculate in the market. Clingman turned to a Federalist congressman for whom he had clerked, Frederick A. C. Muhlenberg, briefly the speaker of the House of Representatives. A former minister whose brother had served with distinction as a general in the Revolution, Muhlenberg was a political moderate. He felt sorry for Clingman, who claimed the fraud charge was a misunderstanding; he thought the soldier had died and his claim against the government was legitimately for sale.

Muhlenberg was far more concerned by Clingman's claim that Reynolds had some sort of illicit connection with Hamilton. The congressman formed an impromptu committee to investigate it. He invited Virginia Senator James Monroe, a passionate ally of Jefferson, and Congressman Abraham Venable, also a Virginia Republican, to join him. They interviewed both Reynolds and his wife, and obtained from them unsigned but seemingly incriminating letters that Hamilton had written to them. Reynolds told them he could prove "the misconduct . . . of a person high in office."[13] Maria assured them that her husband could "tell them something that would make the heads of [government] departments tremble."[14]

Deepening their suspicions was the way Hamilton dealt with the fraud charge against Clingman and Reynolds. When they agreed to return the money and surrender Reynolds's leaked list of veterans, Comptroller Wolcott dismissed the charge. Whereupon James Reynolds disappeared. Clingman maintained that Hamilton had paid the speculator a great deal of money to depart for some distant destination, perhaps the West Indies.

This was explosive stuff, especially in the overheated atmosphere of December 1792. Jefferson had recently written a scorching letter to President Washington, accusing Hamilton of plotting "to undermine and demolish the republic." He had, the secretary of state claimed, "corrupted the legislative branch by dealing out Treasury secrets to his friends."[15]

At first, Muhlenberg was inclined to send the committee's findings to the president. But after much discussion, they decided to confront Hamilton first. His initial response was furious indignation, but he swiftly calmed down when they showed him the incriminating letters in their possession. The secretary of the treasury said he would explain everything if they came to his house that evening.

There, a mournful Hamilton admitted he had written the letters. But they had nothing to do with illegal speculation. Reynolds was blackmailing him for his affair with Maria. The senator and two congressmen expressed surprise at this confession. Hamilton went into painful detail about the way Reynolds had tolerated and even encouraged his later visits to Maria. Within minutes, the embarrassed politicians were saying they had heard more than enough to convince them that the situation had nothing to do with corruption in the Treasury Department. But Hamilton insisted on sharing other intimate letters between him and Maria. By the time he finished, his accusers were all but begging his forgiveness for their invasion of his privacy. As gentlemen, they all agreed that Hamilton's adultery would remain a secret they would share with no one. The treasury secretary said he was deeply grateful for the way they had handled the humiliating incident.

Congressmen Muhlenberg and Venable departed expressing deep sympathy for Hamilton. Senator Monroe, Hamilton noticed, was "more cold." Monroe could barely conceal his disappointment that Hamilton had apparently escaped unscathed. He wrote a full report of the confrontation for Jefferson and Madison and during the next several weeks met several more times with Jacob Clingman, who reiterated his belief that Hamilton was guilty of far more than infidelity. Clingman made a point of telling Monroe that Mrs. Reynolds had burst into tears when she learned that Hamilton had been exonerated.

When Hamilton had time to think about the confrontation, he began to wonder whether Senator Monroe and his friends would keep their word. He asked Monroe to give him copies of the letters the committee had shown him. They would be important if he ever had to defend himself against fresh accusations. Monroe agreed, and asked the clerk of the House of Representatives, John Beckley, to make the copies. Beckley was a major behind-the-scenes political operator. The request was tantamount to sharing the secret with the entire Jeffersonian Republican Party. Beckley made copies for his own files, almost certainly hoping that that they would prove useful at some later political moment.

VI

While this drama was playing on a separate stage, Betsey Hamilton thought she had returned from Albany to a loving husband. In one sense of that term, she was correct. She welcomed an apparently amorous Hamilton into her bed and was soon pregnant with her fifth child. In the busy social world of Philadelphia, she proudly accompanied her husband to receptions at the president's mansion and to dinners and balls at the even more splendid mansion where Anne and William Bingham entertained the city's elite. Anne Bingham was the equal, perhaps even the superior, of Angelica Schuyler Church when it came to high style and witty conversation. Betsey never tried to compete with her; she was more than content to be Mrs. Alexander Hamilton. She let her husband manage the repartee and the flirtatious remarks.

At home, Betsey supervised the children's education. Each morning at breakfast, the boys and their sister Angelica read aloud a chapter from the Bible. Betsey ran the house and servants with a masterful hand. Secretary of War James McHenry once teased Hamilton about how well Betsey supervised his household. He claimed she "had as much merit as your treasurer as you have as treasurer of the wealth of the United States."[16]

Betsey had excellent taste. She decorated their various homes with Louis XVI–style chairs and portraits by the best painters. Her dresses were stylish and beautifully cut. Her model was Martha Washington. Betsey, too, felt that the demands of public life were distractions from the real source of happiness, her home and children. She agreed wholeheartedly with Martha's comment that public occasions were mostly "a waste of time." But Betsey later admitted that as a younger woman, she often enjoyed the parties and balls to which her famous husband escorted her. "I mingled more in the gaieties of the day," she said.[17]

Once she accepted her secondary role in Hamilton's all-consuming pursuit of fame, the only thing that troubled Betsey was her husband's lack of religious faith. He was not hostile to religion; as a young man he had been devout. His college roommate, Robert Troup, recalled that Hamilton got down on his knees and prayed each night and morning. He had been more than agreeable when Betsey insisted on having their first three children christened at Trinity Church in New York. But he rarely accompanied her and the children to services on Sunday, and when he went, he never received communion. He may have been imitating George Washington, who also declined the sacrament when he went to church with Martha. It was a common practice among thoughtful men in this era. Many of them had been influenced by the renowned English scientist Joseph Priestley's 1782 book, *The Corruptions of Christianity,* which called for a rational scientific approach to the Bible. Betsey never sat in judgment on Priestley's book or its readers. Instead, in her quiet, steadfast way, she let Hamilton know more than once that she hoped he would someday regain his youthful faith in a Christian God.

VII

For the next three years Hamilton was engulfed day and night by the political battles of the decade. Again and again, Jeffersonian Republicans in Congress called for investigations of the Treasury Department that required immense labors by Hamilton and his staff. They submitted thousands of pages of documents for scrutiny. Each time, the investigating committee found not a scintilla of corruption to back up their vitriolic shouts and murmurs about Hamilton's evil ways. But in the back of his mind throughout these years, the memory of Maria Reynolds and the meeting with the congressional investigating committee in 1791 loomed as a demoralizing threat. Some historians speculate that Maria was the reason why Hamilton never tried to push John Adams aside and run for president.[18]

In his last years as secretary of the treasury, Hamilton was harassed not only by endless political attacks; Betsey continued to have children, and he was running short of money. The first time he tried to resign to return to the private practice of law, Washington grew so angry that the two men almost had a quarrel as potentially disastrous as their clash when Hamilton had been the general's restless aide. Not until January 1795 did the president agree to let him go. But Hamilton continued to advise Washington by letter, and was constantly consulted by his successor at the Treasury, Oliver

Wolcott Jr., and by the new secretary of state, Timothy Pickering. As we have seen, President John Adams kept these men as well as Hamilton's old friend, Secretary of War James McHenry, in his cabinet. Hamilton's political power remained immense. He was still the leader of the Federalist Party, in charge of selecting candidates and orchestrating foreign and domestic policies.

The Jeffersonian Republicans were well aware of Hamilton's central role. After John Adams became president and his detestation of Hamilton became virtually public knowledge, the Jeffersonians made their move. Their hatchet man, John Beckley, exhumed the Maria Reynolds papers from his files and gave them to the most scurrilous journalist in America, James Thomson Callender, abuser of John Adams and every other Federalist politician of note, including George Washington.

Callender was of course delighted with the Reynolds story. Maria had by this time obtained a divorce from her vanished husband, with the help of New York Senator Aaron Burr. She was living in Maryland with Jacob Clingman. Part of Callender's motive, he claimed, was Hamilton's pretensions to being "a master of morality"—a reference to his attacks on Jefferson as a secret sensualist. The journalist gleefully declared he would now reveal letters confessing that this "father of a family" had an "illicit correspondence with another man's wife." Whereupon Callender published the incriminating letters Hamilton had given Senator Monroe and his fellow investigators in 1791.

Callender and the Jeffersonians were not satisfied with exposing Hamilton's adultery. "So much correspondence could not refer exclusively to wenching," the newsman declaimed. "No man of common sense will believe that it did . . . Reynolds and his wife affirm that it respected certain speculations." This was the heart of the matter as far as the Jeffersonians were concerned. Callender's investigation led him to conclude that "there appears no evidence [of a liaison] but the word of the Secretary."[19] He accused Hamilton of forging Maria's misspelled love letters to him to conceal the real crime, James Reynolds's speculations in the market with the money Hamilton surreptitiously "loaned" him.

What to do about this? Various friends urged Hamilton to ignore Callender. No one else had bothered to answer his abuse. Most Federalists were indignant and on Hamilton's side. A close friend warned that opening a debate only played into Callender's hands. He lived on controversy.

But two of the documents that Callender had published wounded Hamilton in a way he could not tolerate. One was the memorandum the Frederick Muhlenberg committee members had signed after their meeting with Hamilton, saying "we left him under an impression our suspicions were removed." There was more than a faint implication of deception in those words. Even more inflammatory was a memorandum of Senator Monroe's interview with Jacob Clingman, in which the con man claimed that Maria denied an affair with Hamilton and insisted the story was fabricated to conceal his speculations with her husband. These documents inflicted potentially fatal damage on Hamilton's reputation as the creator of the nation's financial system, the essence of his hope for fame.[20]

At first, Hamilton tried to obtain from the senator and the two congressmen a statement that they had fully believed his explanation of the Reynolds imbroglio. Muhlenberg and Venable promptly complied. But James Monroe was not inclined to do any favors for Hamilton. He was dealing with grievous political wounds of his own. President Washington, in an attempt to better relations with France, had sent Monroe to Paris as America's ambassador. Monroe had performed unsatisfactorily, often siding with the French against his own government. Washington had recalled him—and Monroe had no doubt that Hamilton had persuaded him to do it.

Monroe told Hamilton he would confirm the first statement ("Our suspicions were removed."), but he curtly declined to say he disbelieved Clingman. An enraged Hamilton accused Monroe of breaking his solemn promise and leaking the documents. Vitriolic letters flew back and forth, and the two men came close to fighting a duel. Only the intervention of Senator Aaron Burr, who said he believed Hamilton's version, averted an exchange of gunfire.

Still infuriated, Hamilton decided to do what he had repeatedly done with his critics during his years as secretary of the treasury: silence them with an avalanche of facts. Retreating to Philadelphia in July 1797, he began working sixteen hours a day on a refutation of Callender's slurs. The result was a mini-book, ninety-five pages long; about a third was a detailed history of his obsession with Maria; the rest was letters and affidavits confirming the truth of what he had said and done.

In its wild-eyed way, the pamphlet was a masterful example of Hamilton's thrust-and-parry style. He opened his case with a crucial question:

if he had been a crook, out to make a fortune, why would he choose such a penny-ante operator as James Reynolds as a confederate? He claimed the whole story was the result of a "conspiracy of vice against virtue." The virtue was on the side of the Federalist Party, the vice on the side of the Jeffersonian Republicans, whom Hamilton called "Jacobins"—the French radicals who had guillotined twenty thousand mostly innocent people during their infamous reign of terror.

Seldom, if ever in the history of politics had any man been pursued with such rancor and venom for so little cause, Hamilton declared. But he was sustained by his "proud consciousness of innocence." He was, of course, referring to his financial integrity—the bulwark of his fame.[21]

A genuine confession of the heart—if not the head—lay semiconcealed beneath these bellows of defiance: "This confession is not made without a blush . . . I can never cease to condemn myself for the pang which it may inflict in a busom eminently entitled to all my gratitude, fidelity and love."

These touching words went unnoticed by most people. Hamilton's friends were almost unanimously appalled by his interminable narrative. Jefferson, Madison, and their fellow Republicans were delighted. Madison called it "a curious specimen of the ingenious folly of its author." Jefferson maintained that the narrative "strengthened rather than weakened the suspicions that Hamilton was guilty of the speculations." James Thomson Callender mocked the author with practiced savagery. He said Hamilton was whining that he was grossly charged with being a speculator "whereas I am only an adulterer." He again accused Hamilton of forging the Reynolds letters, pointing out inconsistencies in Maria's supposed misspellings and her unusual vocabulary. Some historians have been inclined to agree with him.[22]

One reader of the pamphlet had a very different reaction. Within a week of its publication, a package containing a silver wine cooler arrived at the Hamilton home on Cedar Street in New York City. The gift was from George Washington. The ex-president said he was sending it "not for any intrinsic value the thing possesses, but as a token of my sincere regard and friendship for you and as a remembrance of me . . . I pray you to present my best wishes, in which Mrs. Washington joins me, to Mrs. Hamilton and the family, and that you would be persuaded that with every sentiment of the highest regard, I remain your sincere friend and affectionate honorable servant."[23]

These astonishing words—every one of them carefully chosen—were Washington's way of saying that he was still a believer in Hamilton's integrity and his patriotism. There are biblical overtones here—the American father is forgiving his prodigal son.

VIII

The ostensible reason for Washington's gift was the birth of another Hamilton son. But the baby went unmentioned in the ex-president's note—underscoring his deeper reason for the present. For Hamilton, Elizabeth's pregnancy had been an additional complication in these months of political and emotional upheaval. In 1794, she had suffered a miscarriage when Hamilton rode off to western Pennsylvania to crush a tax rebellion that many people feared would escalate into a civil war. Hamilton was concerned that his Maria Reynolds confession might trigger another possibly fatal miscarriage. He seems to have delayed the publication of his booklet until his wife gave birth to a sturdy boy on August 4, 1797.

Another complication in Hamilton's life was the presence of Angelica Schuyler Church. Angelica had persuaded her husband to return to America with her. Church was immensely wealthy by now. They bought a magnificent house in New York and began entertaining with royal panache. Angelica's stylish gowns and dripping diamonds reminded more than one New York Jeffersonian Republican of British arrogance. When the Reynolds confession exploded, Angelica was fiercely loyal to Hamilton.

Once the baby was born, Hamilton told Betsey about the forthcoming confession. She decided to retreat with the infant to her parents' home in Albany. This was a graphic statement of how deeply Hamilton had wounded her. Angelica, perhaps sensing that she was partly responsible for the affair with Maria, did her utmost to repair the damage. She wrote her sister a long letter, describing how depressed Hamilton was when he returned from escorting Betsey to the Hudson River sloop to Albany. "You were the subject of his conversation the rest of the evening," she wrote. She begged Eliza (the name that gradually replaced the youthful "Betsey") to "tranquilize your kind and good heart" and accept her ordeal as almost inevitable for a woman married to a man who achieved fame. If she had married less "*near the sun* [her italics]," Eliza would never have experienced "the pride, the pleasure, the nameless satisfactions."[24]

In her girlhood home, Elizabeth tried to ignore the shrieks of the Republican press when the confession appeared. In the *Aurora,* Benjamin Franklin Bache cried: "He acknowledges . . . that he violated the sacred sanctuary of his own house, by taking an unprincipled woman . . . to his bed." New York editor James Cheetham expressed shock that Hamilton had "rambled for 18 months in this scene of pollution and squandered about $1,200 to conceal the intrigue from his loving spouse." Another Republican paper found humor in Mrs. Reynolds's "violent attack" on "the virtue of the immaculate secretary . . . in Mr. Hamilton's own house."[25]

On her knees, Eliza Hamilton sought the sort of help that was beyond the reach of her worldly older sister. She struggled to forgive her sinner husband, as Jesus had forgiven the woman taken in adultery. It was not easy. Only a few weeks earlier, in July 1797, Eliza had come across a copy of the *Aurora* in which Bache had mockingly asked how she could tolerate Maria Reynolds as her husband's mistress. Eliza had handed the newspaper to John Barker Church with a gesture of contemptuous dismissal. Church had told Hamilton that the accusation "made not the least impression on her." She assumed the story was concocted by a conspiracy of scoundrels. Now she had to somehow accept the harsh truth.[26]

Hamilton did his utmost to testify to his continuing love for her by becoming extraordinarily attentive to their children. He hoped Eliza would see this as a kind of penance—as well as a way of saying that their wounded love still shared this form of devotion. About a month after the Reynolds pamphlet was published, their oldest son, fifteen-year-old Philip, contracted typhoid fever. Hamilton left him in the care of a physician he did not know well, Dr. David Hosack, and departed to argue a legal case in Hartford, Connecticut.

Philip grew alarmingly ill, and Hosack sent a messenger galloping to Hartford urging Hamilton to return immediately. As Philip's pulse fluttered and his eyes grew vacant with impending death, his mother was so violently upset that Hosack decided on extreme measures. He immersed the dying boy in a tub of hot water full of Peruvian bark and literally restored him to life. Hamilton arrived later in the crisis-thick night and thanked Hosack with tears streaming down his cheeks.

For the next several days, Hamilton became Philip's nurse. Dr. Hosack later wrote that he administered "every dose of medicine or cup of nourishment that was required."[27] But this attempt to make amends would soon be overshadowed by a new opportunity to pursue that irresistible prize, fame.

LOVE'S SECRET TRIUMPH

*O*ver the next year, the Hamiltons struggled to restore their damaged marriage. The Jeffersonian Republican press was no help. Robert Troup told a mutual friend, "For this twelvemonth past this poor man . . . has been violently and infamously abused by the democratical party. His ill-judged pamphlet has done him incomparable injury."[1]

For a while he remained a power behind the scenes thanks to his influence over President Adams's cabinet. But Hamilton repeatedly told Eliza that he had lost his enthusiasm for public life. "In proportion as I discover the worthlessness of other pursuits, the value of my Eliza and of domestic happiness rises in my estimation," he told her. When Elizabeth made another visit to Albany, her absence made him realize "how necessary you are to me." He looked in vain for "that satisfaction which you alone can bestow." When one of New York's senators resigned and Federalist governor John Jay offered to appoint Hamilton in his place, he ostentatiously declined.

In 1798, George Washington offered Hamilton command of the army that Congress had created to confront the threat of a French attack. Hamilton accepted instantly. It was a temptation he could not resist. It turned Washington's forgiveness for the Maria Reynolds affair into a public political redemption. Hamilton again became a contender for the loftiest ranks of fame.

The appointment soon became the most time-consuming and controversial public office Hamilton ever held. Eliza Hamilton must have wondered about her husband's apostrophes to the joys of domestic happiness while he traveled continually to Philadelphia and other cities to confer with fellow generals and stimulate recruiting. Meanwhile, President John Adams's hatred and the opposition of the Jeffersonian Republicans transformed what seemed the realization of a lifelong dream into a nightmare. Adams avoided a war with France and soon dissolved General Hamilton's army. Hamilton retaliated with his savage attack on Adams, virtually guaranteeing Thomas Jefferson's election as president. Almost simultaneously, George Washington died, leaving Hamilton without the huge advantage of his support. In a single stunning year, Hamilton went from being the commander of the American Army and the leader of the ruling Federalist Party into an ex-general and a defeated, discredited politician.

Throughout this roller-coaster ride on fame's treacherous trajectory, Hamilton remained conscious that he was a man with a marital debt to pay, even though he seemed to be in danger of defaulting on it. His letters to Eliza were a mixture of affection and apology for the hours he was spending away from her and the children. He frequently reiterated his devotion. "Indeed, my dear Eliza . . . your virtues more and more endear you to me and experience more and more convinces me that true happiness is only to be found in the bosom of one's own family." He wrote these words in 1801, when the Hamiltons had been married twenty years. In this perspective, the words acquire a somewhat artificial quality. Hamilton is virtually confessing that for most of these twenty years, Eliza's virtues had not endeared her to him. Worse, it carelessly repeats the same sentiments he had written in the immediate aftermath of the Reynolds confession. The letter underscores the sad truth that Eliza was never Hamilton's dearest friend at the level of sharing achieved by the Washingtons or John and Abigail Adams.

That same year, 1801, the year of Jefferson's elevation to the presidency, brought excruciating personal grief. Hamilton's oldest son, Philip, now nineteen, got into a nasty quarrel with one of Thomas Jefferson's triumphant supporters, a twenty-seven-year-old lawyer named George Eacker. Philip took exception to a Fourth of July speech Eacker had made, in which he portrayed President Jefferson as the rescuer of the Constitution against a coup d'etat by power-hungry General Hamilton. Philip and a friend named Richard Price climbed into Eacker's box at the Park Theater and called him some extremely unpleasant names. Both intruders were probably drunk; around this time, Robert Troup, still one of Hamilton's close friends, referred to Philip as a "sad rake."[2]

The confrontation led Eacker to call the two intruders "damned rascals." This was a term no would-be gentleman of the era could tolerate. Philip and his friend Price promptly challenged Eacker to a duel. The lawyer exchanged four shots with Price, without spilling a drop of blood. Then it was Philip's turn. Hamilton advised his son to fire in the air, a tactic called a *delope*. This alternative would enable Philip to escape the odium of killing Eacker. The *delope* was also a show of superior courage and often a statement of contempt for an opponent. Hamilton did not think he was putting his son in mortal danger. Only about one man in five was killed in a duel—and Eacker had already demonstrated he was a poor shot.

The two young men met in Paulus Hook, on the New Jersey shore of the Hudson River. Though dueling had been outlawed in New York, it was still legal in the Garden State. Philip, obedient to his father's advice, waited for Eacker to fire. But Eacker also waited, obviously hoping that the quarrel could be settled with an apology. Finally, to draw Eacker's fire, Philip leveled his pistol but did not fire. Eacker's bullet ripped through Philip's body and lodged in his arm. The young man died in agony twenty-four hours later. Beside him on the deathbed lay his weeping father and mother, frantically clutching him in their arms.

At the funeral, General Hamilton could barely stand erect. "Never did I see a man so completely overwhelmed with grief," said Robert Troup. Not long after the funeral, Philip's beautiful younger sister, Angelica, had a mental breakdown. She sank into a miasma of fear and confusion that made her incapable of a normal life. On the harp and piano, she played over and over again songs that Philip had loved.

Watching Eliza struggling for the strength to accept this double tragedy and continue to be a mother to their seven surviving children, Hamilton began to wish he still retained his youthful religious faith. It would give him a bond with his wife that was deeper than the dutiful love that seems to have been the best he could offer her.

II

Angelica Church was still very much in the picture, tempting Hamilton to reach for the consolation of the ecstatic eros that had led him into Maria Reynolds's arms. Angelica's admiration and devotion remained intense in spite of Hamilton's fall from power. She made no secret of her hope that he could and would somehow transform his present humiliation into an even more spectacular return to fame's summit.

In 1804, a worried Robert Troup began telling friends that Hamilton and Angelica had resumed their once torrid affair. Conclusive evidence has never been found, but Hamilton was obviously spending a great deal of time with her. Since 1798, Hamilton saw Elizabeth and the children only on weekends. They were living at The Grange, a handsome mansion he had built on Harlem Heights. He spent his weeknights at their New York City home on Cedar Street. Angelica was living only a few blocks away in the Church mansion, ignored by her dour husband, who spent almost all his nights gambling for high stakes at various card games.

Angelica—and Hamilton—listened to their mutual friend, Gouverneur Morris, who had accepted the appointment to the U.S. Senate that Hamilton had rejected. Morris's loathing for Jefferson and his low-or-no-tax minimal style of government was pervasive. The disgusted senator repeatedly predicted the country's imminent collapse. Not a few of Hamilton's Federalist friends in New England were saying similar things. They feared that President Jefferson's 1803 purchase of the Louisiana Territory from France was going to give the Virginian the power to create more states that would owe their allegiance to him. Rather than tolerate becoming a minority, the Yankees were proposing to secede from the union.

In 1804, the American political scene underwent a dramatic upheaval. President Jefferson and his vice president, Aaron Burr, quarreled, and Burr tried to launch a third political party, composed of moderates from both sides. He called it "a union of honest men." As a first step, Burr announced he would run for the governorship of New York. He intimated that if he won, he would not be averse to joining the New England secession movement.

Hamilton detested and feared Burr even more than he loathed Jefferson. He considered Burr a demagogue and thought his third party was a sham. He was especially troubled by Burr's tacit support of the secession movement, which Hamilton deplored. Hamilton tried to rally New York's Federalists behind another gubernatorial candidate. They ignored him, underscoring how totally political power had slipped from his grasp. Hamilton was so disgusted by this rejection, he announced in the *New York Evening Post* that he would never again accept public office in either the state or federal governments. But he added a very significant exception: he would be willing to serve "in a civil or foreign war."

President Jefferson urged his followers in New York, led by Hamilton's old enemy Governor George Clinton, to destroy Aaron Burr. In one of the most viciously partisan campaigns on record, they smeared Burr as an embezzler of his legal clients, a womanizer who had destroyed dozens of marriages, and an abuser of the men he commanded during the Revolution. Burr lost badly, leaving him as powerless as Hamilton.

Deeply depressed, Burr challenged Hamilton to a duel for the harsh things he had said during his attempt to rally Federalists against the vice president. Hovering over the challenge was the knowledge both men possessed of New England's plan to secede. If the Yankees acted on it, there would be a civil war. From it would emerge a victorious general who would restore the union by force of arms. If that man possessed political wisdom as well as military skills, he would achieve transcendent fame as the rescuer of the republic.

Alexander Hamilton wanted to be that man. So did Aaron Burr. Both knew that if Hamilton evaded Burr's challenge by apologizing, he would never be acceptable to the nation's soldiers. This irrepressible hunger for ultimate fame was why Hamilton refused to apologize to Burr for his largely ignored remarks and made a duel inevitable.

III

On July 4, 1804, with the time and place for the duel not yet chosen, Hamilton wrote a letter to his wife. It reveals a man struggling with terrific guilt. It also tells us how far Hamilton had traveled in his attempt to retrieve his lost Christian faith:

This letter, my dear Eliza, will not be delivered to you unless I shall first have terminated my earthly career; to begin, as I humbly hope from redeeming grace and divine mercy, a happy immortality.

If it had been possible for me to have avoided the interview, my love for you and my precious children would have been alone a decisive motive. But it was not possible without sacrifices which would have made me unworthy of your esteem. I need not tell you of the pangs I feel, from the idea of quitting you and exposing you to the anguish which I know you would feel. Nor could I dwell on the topic lest it would unman me.

The consolations of religion, my beloved, can alone support you; and these you have a right to enjoy. Fly to the bosom of your God and be comforted. With my last idea I shall cherish the sweet hope of meeting you in a better world.

Adieu best of wives and best of women. Embrace all my darling children for me.

Ever yours,

AH

On the night before the duel, Hamilton wrote several letters that attempted to explain why he had accepted Burr's challenge and how he planned to fight the duel. He was going to "throw away" his first shot, and possibly his second shot, if Burr's first bullet missed him. The "scruples of a Christian" supposedly prompted this decision, he claimed. The real reason was Hamilton's guilt for giving this advice to his son Philip. He felt compelled to take the same risk. But he also hoped his *delope* would enable him to triumph over Burr. If Hamilton survived, Burr would leave the dueling ground a humiliated, politically neutered man.

That same night Hamilton wrote another letter to Eliza, asking her to be generous to his mother's aging needy cousin Ann Mitchell, who had tried to help him and his brother in St. Croix. To this request he added an impulsive postscript:

The scruples of a Christian have determined me to expose my own life to any extent rather than subject myself to the guilt of taking the life of another. This must increase my hazards & redoubles my pangs for you. But you had rather I should die innocent than live guilty. Heaven can preserve me and I humbly hope will, but in the contrary event, I charge you to remember you are a

Christian. God's will be done! The will of a merciful God must be good.
   Once more adieu my darling
   darling wife
   AH

As the sun rose above the Weehawken dueling ground, Aaron Burr's first shot struck Hamilton just above his hip and tore through his body to lodge in his spine. Hamilton convulsively clutched his trigger when the bullet hit, and his shot struck a tree limb high over Burr's head. "This is a mortal wound, Doctor," Hamilton gasped when Dr. Hosack rushed to his side. His friends brought him back to New York, where Hosack and several other doctors examined him; all agreed that Hamilton was right, the wound was mortal.

Hamilton asked his friend, merchant William Bayard, who had offered his riverside house on Jane Street to the dying man, to send for Benjamin Moore, Episcopal bishop of New York. The bishop responded, but he was unhappy to learn that Hamilton had been wounded in a duel, an activity Moore considered immoral. He was even more troubled when he learned that Hamilton had never joined the Episcopal Church, yet he wanted to receive holy communion. Moore declined to administer this central act of the Christian faith in such a situation. He feared it would imply that he condoned dueling. The bishop departed, declaring he wanted to give Hamilton "time for reflection."

Hamilton sent for another clergyman, the Reverend John M. Mason, pastor of the Scotch Presbyterian Church, and asked him to administer communion. Mason regretfully informed Hamilton that his church did not believe in private communion. For them it was a public ceremony, available only on Sunday. He tried to reassure Hamilton that communion was not necessary to win God's forgiveness for his sins if his faith in Christ was sincere. Hamilton shook his head frantically; he said he was aware of the power of Christ's benevolence and that he wanted communion "only as a sign."

Hamilton could not say more without revealing his real reason: he wanted to reassure and comfort Eliza. He knew that Mason would not consider this a sufficient reason to alter the regulations of his church; Mason might even have concluded Hamilton's profession of faith was insincere.

A distraught Hamilton begged Bayard to persuade Bishop Moore to change his mind. At this point, Eliza arrived at the house. Knowing her emotional fragility, Hamilton had ordered the friend who brought her from The Grange to say nothing about his condition. She hurried to his bedside thinking he was suffering from stomach spasms. When Dr. Hosack told her the truth, she started to sob and gasp for breath. Hamilton writhed on the bed, wondering why either or both of those clergymen had refused him communion so he could offer Eliza proof that he had made his peace with God. Now all he could utter was a desperate reminder for her to seek the solace that had sustained her marriage: "Remember, my Eliza, you are a Christian!"

At first the words did little good. Elizabeth began to unravel. Dr. Hosack and others tried to comfort her. Hamilton repeated the words several times as Elizabeth sobbed and shuddered with grief. Suddenly there was another woman in the room, tears streaming down her tormented face: Angelica Schuyler Church. Eliza flung herself into the arms of her older sister. Only she understood the pain of losing this unique, incomparable man. Whether the love between Angelica and Hamilton had ever been physical was irrelevant now.

A few hours later, Bishop Moore returned to the Bayard house. Hamilton's friends pleaded with the prelate to change his mind. The bishop had another talk with the dying man, who told him he had been planning to join the Episcopal Church "for some time past." He declared that he detested dueling and bore "no ill will" toward Aaron Burr. He had gone to Weehawken "with a fixed resolution to do him no harm." The bishop gave him holy communion, and Hamilton's head fell back on the pillow. In spite of his agonizing pain, a faint smile played across his lips. He had achieved the only gift he could offer Eliza now: his reborn faith in the mysterious God who had allowed so much grief to engulf their lives.

IV

Alexander Hamilton died peacefully the following day with a weeping Eliza and Angelica at his bedside, along with at least a dozen other friends. Gouverneur Morris found the scene so unendurable that he fled to the garden of the Bayard House. New York City gave Hamilton a public funeral replete with orations and a huge crowd of spectators. When his friends examined the will Hamilton had left among his papers, they realized he was penniless. His only meaningful investments were in western lands that would take years, perhaps decades, to appreciate in value. Meanwhile, he had debts that amounted to almost a half million dollars in modern money. It was unarguable proof that Hamilton had never dipped into the millions of dollars he had handled for the federal government or used his insider's knowledge to speculate on Wall Street.

The friends, led by wealthy Gouverneur Morris, raised enough money to enable Eliza to keep The Grange. But contributions lagged, because many people thought she could turn to her wealthy father for support. Four months later, when Philip Schuyler died, they learned that he, too, was close to bankruptcy. He had lost a great deal of money in unwise investments during the 1790s.

Over several years, Hamilton's friends managed to raise about $80,000 for Eliza—more than a million and a half dollars in the inflated currency of our time. She was able to live in reasonable comfort and dignity. In the first few years, she sometimes expressed near despair. She spoke of her "wounded heart" being unequal to the burdens cast upon it. But she found strength in Hamilton's deathbed conversion. She spoke of him as "my beloved sainted husband and my guardian angel."[3]

The words were not chosen casually. She believed Hamilton's faith had been reborn and that he had joined the ranks of the "sainted" in eternity. She also believed, like Episcopalians and Catholics today, in the "communion of saints," which enabled the dead to protect and nurture those they loved who were still on earth—the role of the guardian angel. Knowing that her prayers and example had played a part in this transformation was enormously meaningful to her. Gradually, she began to see that Hamilton's early death had given her a mission in life. She would devote herself to protecting and even enlarging Alexander Hamilton's reputation as one of the founders of the American republic.

This was a task that needed doing. Hamilton's tragic death did not change the Jeffersonian Republicans' opinion of him. John Beckley, the man who had given the Reynolds story to James Thomson Callender, gloated openly at Hamilton's fall. He mocked the magnificent funeral and the eulogies that Hamilton had received. "Federalism has monumented and sainted their leader up to the highest heavens. . . . The clergy, too, are sedulously trying to canonize the double adulterer as a moralist, a Christian and a saint."[4]

As the Jeffersonians saw it, Hamilton was a menace, a perpetual threat to the republic's political purity, and Burr could and did gun him down without an iota of blame or remorse. Proof that this conclusion was not confined to a few politicians of Beckley's stripe was the way Burr was treated by Jeffersonian Republicans, especially in the South, after the duel. He fled to Saint Simons island off the coast of Georgia to escape the threat of a show trial concocted by his New York enemy, Governor George Clinton. Burr's host, wealthy Republican Senator Pierce Butler of South Carolina, lavished hospitality on the fugitive. On his way back to Washington, D.C., for the next session of Congress, the vice president was feted in Savannah, Georgia, and Petersburg, Virginia, by Republican admirers as if he were a general returning from a triumphant campaign.[5]

Other ex-revolutionaries were equally unsympathetic to Hamilton. Thomas Paine wrote a mocking satire asking Christian believers why their God had arranged to have Hamilton killed in a duel. John Adams, brooding in Quincy, wrote that "a caitiff had come to a bad end." He would only concede a hope that Hamilton was "pardoned in his last moments."[6]

V

Eliza Hamilton set herself the task of collecting her husband's papers. She devoted many hours to retrieving copies of the hundreds of letters he had written to friends and supporters. She even persuaded the Washington family to let her borrow and copy Hamilton's letters to the president. She interviewed politicians who had worked with Hamilton and added memoranda of their memories to her files. For the next twenty years, she tried to persuade one of Hamilton's friends or admirers to write his biography. She met with one frustration after another. Various prominent politicians and writers studied the papers for years at a time and then returned them, pleading ill health or advancing age.

A climactic disappointment was the failure of former secretary of state Timothy Pickering. He kept the papers for almost a decade. When he died in 1829, his heirs found only a few disjointed, unfinished chapters. By this time Eliza's second-oldest son, John Church Hamilton, was ready to undertake the task, and he began his seven-volume *History of the Republic as Traced in the Writings of . . . Hamilton,* which would consume the next thirty years of his life.

This devotion to her husband's memory was by no means the only thing that occupied Eliza Hamilton. Long regarded as a fragile, dependent woman, she began demonstrating an independence and originality that amazed everyone. All her life Eliza had been dominated by strong-willed men, first her father, then her husband. Now that she was free to act on her own, she joined a group of equally religious women who founded The New York Orphan Asylum Society.

Eliza was perpetuating for friends and family another memory of Alexander Hamilton, the orphaned West Indian boy. She became a hardworking member of the board, with the title of deputy director. Over the years she devoted an immense amount of time to the rescue of New York children who had met a similar fate. In 1821, with her own children grown and launched on respectable careers, Eliza became the asylum's director. She learned to know and care about each of the 158 children then in residence in the building the society had erected in Greenwich Village. She helped them get jobs and persuaded a New York politician to recommend one boy for West Point. Later she persuaded the New York state legislature to give the school annual grants. In 1836, she presided over a ceremony that began the construction of a larger and more permanent orphanage at Riverside Drive and 73rd Street.

On the side, in 1818 Eliza founded the Hamilton Free School on land she owned between 187th and 188th streets in Washington Heights. It was the first school in that developing part of Manhattan Island. She enjoyed all this hard work immensely. Once, she told a son how grateful she was to God for "point[ing] out this duty to me and giv[ing] me the ability and inclination to perform it."[7]

A portrait of Eliza Hamilton in her later years is the most telling evidence of how profoundly she had changed. The mouth is now a strong emphatic line and the dark eyes are bright with self-confidence—and kindness. The anxiety and uncertainty of her married years have vanished. A woman friend described her face as "full of nerve and spirit." As she grew older, the same friend marveled at how she "retains to an astonishing degree her faculties and converses with . . . ease and brilliancy."[8]

In 1848, at the age of ninety-one, Eliza Hamilton moved to Washington, D.C., to live with her widowed daughter, Eliza Holly, who had a comfortable house only a few doors from the White House. Here, Mrs. Hamilton launched a new career as a celebrity. Politicians of all ages and parties rushed to meet her. Southerners may have been dismayed by her brisk interest in the politics of the day: she was a critic of slavery. But everyone was charmed by the "sunny cheerfulness of her temper and quiet humor." She was a frequent guest at the White House, where presidents from Polk to Fillmore were fascinated by her still-vivid memories of chatting with George and Martha Washington and John and Abigail Adams.[9]

Perhaps the most startling of Eliza's Washington, D.C., activities emerged from her friendship with another woman who shared many of her memories of the republic's early years—Dolley Madison. Dolley came to her one day proposing to do something about the stalled monument to George Washington. For the better part of a decade, the proposed 855-foot obelisk had remained nothing more than an embarrassing idea. Eliza Hamilton called on the know-how she had acquired in her forty years of lobbying and fundraising for New York's orphans.

Together these two remarkable women loaned their names to a campaign to raise enough money to begin the huge task. Their appeal inspired startling numbers of people to open their wallets, and on July 4, 1848, the city fathers laid the cornerstone of the great marble pillar that would declare George Washington's singular greatness. Among those at the ceremony were President James K. Polk and his wife, Martha Washington's grandson George Washington Parke Custis, and an obscure congressmen named Abraham Lincoln.

Elizabeth Hamilton died in 1854 at the age of ninety-seven. In a small pouch she wore around her neck, her daughter found the letter Alexander Hamilton had written to her on July 4, 1804, testifying to his wounded, wounding, but ultimately transcendent love.

BOOK FIVE

—

 Thomas Jefferson

ROMANTIC VOYAGER

The first woman close to his own age who stirred Thomas Jefferson's affection was his older sister, Jane. She shared his enthusiasm for music and books, which was not widespread in the Jefferson household. In later years, Jefferson described Jane as "a singer of uncommon skill and sweetness." Often on summer nights, he would play songs on his violin and the two would sing together. Even in his old age, Jefferson spoke of her to his granddaughters "in terms of warm admiration and love."

Jane remained a spinster until she died in 1765 at the age of twenty-five. One suspects she had fastidious tastes, like her younger brother. No one mourned her more than Jefferson. Years later, when he began planning to build a house on a nearby small mountain, one of its features was going to be a family cemetery; at its center would be a small stone altar, dedicated to Jane's memory. In one of his account books, after describing this memorial, Jefferson wrote a touching epitaph:

Ah! Joanna puellarum optima!
Ah! Aevi virentis flore praerpta!
Sit sibi terra laevis!
Longe, longeque valeto![1]

(Ah! Jane, best of girls!
Ah! Plucked too soon from your blooming youth!

Why was your native soil so unfavorable!
Long, long shall I bid you farewell!)

II

Jefferson's father, Peter Jefferson, was another rural strong man, on the model of Augustine Washington. His main interest in his son seems to have been figuring out how to toughen the skinny, dreamy youth to qualify him for manhood, Virginia style. Peter died when Tom was fourteen, probably suspecting he had not succeeded. At an early age, Tom was sent into the woods with a gun to bring back a wild turkey. He blazed away but hit nothing until he found a turkey trapped in some sort of pen. He pinned the bird in place with a garter and shot him at point-blank range. We can be fairly sure that his father was not pleased by this performance. Tom continued to spend most of his time with his head in a book—or practicing his violin as many as three hours a day.

With his mother, Jane Randolph, Jefferson seems to have had an uneasy relationship. It was not as overtly turbulent as George Washington's conflicts with Mary Ball Washington. But there are hints in Jefferson's papers that he chafed under his mother's role—or rule—as overseer of the family's finances until her death in 1776. Even when he was twenty-five and a practicing attorney, she seems to have expected him to submit all his accounts to her for approval.

There are also hints that Jane Randolph tended to play up her family's heritage, which supposedly went back to Scottish and English nobility, at the expense of the Jeffersons' more mundane Welsh origins. Jefferson had a lifelong habit of making disparaging remarks about coats of arms and noble ancestors. Jane was the daughter of one of the seven sons of William Randolph and Mary Isham, who are sometimes called the Adam and Eve of Virginia. Their numerous descendants married into virtually every notable family in the colony.

Jane Randolph Jefferson was born in London, where her father was living as a merchant, and the Jefferson home, Shadwell, was named for the London parish in which she had lived. At least one psychobiographer has speculated that Jane Jefferson disapproved of her son's participation in the Revolution. One of Jane's brothers was so disgusted by the upheaval that he sold his Virginia lands and moved to England. Another brother was already living there.

As the older son (his brother Randolph was twelve years younger), Tom inherited the pick of Peter Jefferson's lands along the Rivanna River and elsewhere, totaling more than five thousand prime acres. The income from these fruitful fields left him free of financial worries. At seventeen, Jefferson enrolled in the College of William and Mary in Williamsburg, the little town that served as colonial Virginia's capital.

The year was 1760. A new king, George III, had just ascended the British throne. The British Empire was in the process of winning the Seven Years' War. The American side of it, usually called The French and Indian War, had already ended in victory with the capitulation of Canada. Ex-Colonel George Washington was discovering unexpected happiness in his marriage to Martha Dandridge Custis and working hard to wring a profit from Mount Vernon's mediocre soil. Benjamin Franklin was in London, enjoying the international celebrity he had achieved for his epochal discoveries in electricity.

As a country boy, Jefferson approached the scions of Virginia's first families—and their sisters—with not a little diffidence. He was thin-skinned and rather shy, and he feared rebuffs. He concealed his inclination to be sociable behind a facade of studiousness. "He used to be seen with his Greek grammar in his hand while his comrades were enjoying relaxation," one friend recalled.

Family legend has him studying fifteen hours a day, but other evidence indicates that he knew how to have a good time, like most Virginians. He joined the Flat Hat Club, a college society that recorded its zany doings in mock-Latin verse. He made lifelong friends such as John Page, a descendant of "King" Carter, who invited him to Rosewell, the magnificent three-story mansion in which Page had grown up. There and in other great houses to which his Randolph kinship opened doors, Tom discovered the fascinations of the female sex. Their names still twinkle in his youthful letters: Rebecca Burwell, Susanna Porter, Alice Corbin, Nancy Randolph.

Tom spent the years between nineteen and twenty-three adoring Rebecca, an heiress from Yorktown whose parents were long dead. "Enthusiasm" was the word her contemporaries used to sum up Rebecca in later years. In the eighteenth century this meant a strongly emotional personality. In the early decades of the century, when the severe, controlled classicism of Joseph Addison and Samuel Johnson dominated English taste and wild-eyed revivalists had Americans leaping and shouting in their churches, the word had a negative timbre. But by the 1760s, romanticism was in the air, and people with vivid emotions were not only tolerated, they were admired.

For two years, the romance percolated while Jefferson struggled to bring himself to ask the ultimate question. He rhapsodized in letters about "Belinda"—the exotic name he gave Rebecca. But something about her tied his tongue in knots. Perhaps it was the simple knowledge that marriage meant the end of youth, farewell to the bachelor's freedom. Perhaps it was another hint that his parents' marriage had not been very happy and he was wary of choosing a member of Virginia's aristocracy as his wife.

For Jefferson, marriage also meant the extinction of his ambition to travel to Europe to see the monuments and palaces and paintings of England and France and Italy, about which he had read so much. Once, in a garden with Rebecca, he hinted that he would begin his grand tour at once if she promised to wait two years for his return. But her frown cast a shadow on this idea, and Jefferson returned to his studies. By this time he had passed from college Greek and Latin to the hard work of mastering "Old Coke"—Sir Edward Coke, the English jurist whose commentaries on the laws of England were famed for their crabbed style and "uncouth but cunning learning." The young man had decided to become a lawyer.

Retreating to Shadwell, he lamented that he was certain to spend his time thinking of Rebecca "too often, I fear, for my peace of mind, and too often, I am sure, to get through Old Cooke [Coke] this winter." A month later, he was writing plaintively to his friend John Page: "How does RB do? What do you think of my affair, or what would you advise me to do? Had I better stay here and do nothing or go down [to Williamsburg] and do less?" He decided on the first choice, and spent his days and nights struggling with Coke and planning a voyage in an imaginary ship called *The Rebecca* in which he would visit "England, Holland, France, Spain, Italy (where I would buy me a good fiddle) and Egypt."[2]

Jefferson was not the first lover in history to be tongue-tied at the sight of his beloved. But the duration of his reluctance is significant. He seemed to revel in the idea of love, the beauty and transcendence of it, and to recoil or at least hesitate from its physical expression. He seemed to enjoy the brooding, the semisweet despair of his frustration. This experience was becoming familiar in the emerging romantic movement in Europe. Lovers devoted themselves to unattainable women and sometimes shunned consummation to prove the depths of their devotion. It would gradually become apparent that this division was an important part of Jefferson's psyche.

Another college friend, Jacquelin Ambler, began pursuing Tom's unattainable damsel. All through the following spring and summer, Jefferson stayed home, philosophizing: "If she consents I shall be happy, if she does not I must endeavor to be so as much as possible. . . . Perfect happiness, I believe, was never intended by the Deity to be the lot of any one of his creatures in this world."

His friend Page, acting as both adviser and ambassador, warned Tom that Ambler was making ominous progress. So the philosophic lover came to Williamsburg for the social season. He was soon giving Ambler strong competition—until a climactic night at the Raleigh Tavern, the favorite gathering place of the young bloods and their belles. Arriving for a ball, the ladies were dressed in that "gay and splendid" style that made Virginia famous, their hair "craped" high with rolls on each side, topped by caps of gauze and lace. The men looked almost as splendid in clockwork silk stockings, lace ruffles, gold- and silver-laced cocked hats, and breeches and waistcoats of blue, green, scarlet, or peach.

Jefferson had spent the hours before the ball composing a whole series of romantic compliments, witty remarks, and bright observations for Rebecca. "I was prepared to say a great deal," he told his friend John Page. "I had dressed up in my own mind such thoughts as occurred to me in as moving a language as I knew how, and expected to have performed in a tolerably creditable manner."

But when the lover came face to face with Rebecca in her finery, "Good God! . . . a few broken sentences uttered in great disorder and interrupted with pauses of uncommon length were the too visible marks of my strange confusion!"

"Last night," the despairing suitor groaned, "I never could have thought the succeeding sun would have seen me so wretched as I now am!"[3]

The following year, Rebecca became engaged to Jacquelin Ambler, and Jefferson was stricken with the first attack of what he called "the headache"—a disabling migraine-like disorder that would periodically torment him for the next forty years.[4]

III

Jefferson retreated once more to philosophizing—and making matches for his friends. He proposed to Sukey Potter for his fellow lawyer William Fleming. A few weeks later, he was cheerfully reporting to Page the fate of their friend Warner Lewis: "Poor fellow, never did I see one more sincerely captived in my life. He walked to Indian camp with her yesterday, by which means he had an opportunity of giving her two or three love squeezes by the hand, and like a true Arcadian swain, has been so enraptured ever since that he is company for no one."[5]

These examples of romantic bliss did not impress Jefferson. After his disaster with Rebecca Burwell, he settled down to practicing law and enjoying life. He dated his Williamsburg letters "Devilsburgh" and needled friends such as Fleming for falling in love. Soon he was making cynical comments on the matchmaking all around him. When William Bland won Betsey Yates, Jefferson wryly remarked, "Whether it was for money, beauty, or principle, it will be so nice a dispute that no one will venture to pronounce."[6]

The beleaguered Jefferson began copying from his favorite books dour quotations that traced a frequent evolution of the bachelor's psychology:

> *Wed her?*
> *No! Were she all desire could wish, as fair*
> *As would the vainest of her sex be thought*
> *With wealth beyond what woman's pride could waste*
> *She could not cheat me of my freedom*[7]

Here, too, we can see a recoil from the power of the erotic, which compels a man to surrender to his physical desire for a woman. In the eighteenth century, this meant a surrender of not only sexual freedom, but a host of other freedoms. Jefferson placed more and more value on the company of his men friends, conveniently ignoring the fact that most of them had married. One day he was writing John Page about how much he enjoyed "the philosophical evenings" at Roswell. The next day he was copying, *I'd leave the world for him that hates a woman, woman the fountain of all human frailty.*

IV

The remarks about frailty may have special significance. Around this time, the roving bachelor became involved in an affair that he would bitterly regret in later years. One of his close friends in Albermarle County was Jack Walker. He had married a buxom miss named Betsey Moore and was living only a few miles from Shadwell. Young Walker was offered the job of clerk to a Virginia commission negotiating a treaty with the Indians at Fort Stanwix in northern New York. He made a will before he departed, making Jefferson his executor if some warrior planted a hatchet in his skull or he drowned in the cold, swift waters of the Mohawk River. During Jack's four-month absence, Jefferson frequently visited Betsey and her baby daughter to make sure all was well.

What began as a favor to a friend suddenly erupted into desire. Jefferson had no interest in idealizing, much less marrying, Betsey Walker. Perhaps the sophisticated lawyer began assuring Betsey that there was nothing wrong with a little fling—everyone did it. To back him up, he may have quoted Ovid and other poets on the delights of illicit love. Perhaps Betsey encouraged him, either deliberately or inadvertently. Although they were supposed to feign disinterest, many women in the eighteenth century enjoyed sex, and Betsey had been married for well over a year. She may have resented her husband abandoning her for a jaunt to the northern woods.

We know only this much: after tantalizing him long enough to send the bachelor into a frenzy of frustration, Betsey said no at the crucial moment, and Jefferson retreated with nothing to show for his efforts but a wounded ego. Betsey waited twenty years to tell her husband about the episode, which did not prevent Jack Walker from making Jefferson's next ten years miserable with public and private accusations. By that time Walker was an ally of Jefferson's political enemies.[8]

Fortunately, at this point a healthier influence entered Jefferson's life. He had another friend from Albermarle. His name was Dabney Carr, and he was also pursuing the law as a career. In 1765, Carr married Jefferson's younger sister, Martha. Their happiness made a deep impression on the disillusioned bachelor. Writing to John Page, Jefferson confessed his amazement at the way Carr "speaks, thinks and dreams of nothing but his young son. This friend of ours . . . in a very small house, with a table, half a dozen chairs, and one or two servants, is the happiest man in the universe."[9]

Seeing Page himself happily married, with two growing children and a wife who shared his love for books and good talk, also impressed Jefferson. After a 1770 visit to Roswell, the Page estate, he wrote, "I was always fond of philosophy even in its dryer forms, but from a ruby lip it comes with charms irresistible. Such a feast of sentiment must exhilarate life . . . at least as much as the feast of the sensualist shortens it."

V

Seven months after he wrote these words, the fast-fading bachelor was singing love songs. In Martha Wayles Skelton, Jefferson found a woman who routed the cynic in his nature and renewed his yearning for ideal love. She lived in The Forest, a tall, rather ungainly wooden mansion on a knoll overlooking the broad James River, the main highway of Virginia. A fragile, delicately boned beauty of twenty-two, she was the widow of a college friend, Bathurst Skelton, by whom she had a young son, John. Martha had in abundance the virtue that Jefferson put first in a list of wifely attributes he compiled for one of his many notebooks: sweetness of temper. Again, there is a suggestion that this was a virtue Jane Randolph Jefferson lacked.

Two other qualities were attractive to this most romantic of the founders: "spriteliness and sensibility." The latter had become especially important to Virginians. It, too, was an offshoot of the growing fondness for emotion in poetry and novels. A woman, especially, felt inferior if she was deficient in sensibility. One belle reported proudly to a friend her reaction to the latest work of the popular author Lady Julia Mandeville: "I never cried more in my life reading a novel."

To Jefferson's delight, Martha shared his enthusiasm for the younger generation's favorite novel, *Tristram Shandy* by Laurence Sterne. This comic masterpiece was the black humor of its day, considered a little naughty by the straitlaced but fervently admired by the rising generation for the way it spoofed musty academic writing, windy doctors of medicine, boring old soldiers, and a dozen other favorite targets of the young, while bathing in sentiment such inevitabilities as true love and premature death.

Martha Wayles Skelton's fondness for *Tristram Shandy* was not her only recommendation. She also shared Jefferson's love of music, and played beautifully on the spinet and harpsichord. Add to these accomplishments a natural grace when walking and riding, large, expressive eyes of the richest shade of hazel, and luxuriant hair of the finest tinge of auburn, and you know why Thomas Jefferson was soon wearing out his horseflesh on the road to The Forest.

To his alarm, Jefferson discovered there were other suitors. Martha was not only beautiful, she was rich. Her father was, in Jefferson's words, "a lawyer of much practice" who also made hefty profits as a slave trader. Martha had inherited still more wealth from her two-year marriage to the late Bathhurst Skelton. Martha's industrious father had trained her "to business" and urged her to find a good manager for her estate. Jefferson certainly qualified in this category. At twenty-seven, he was already a member of the House of Burgesses and a well-established lawyer with a handsome income from his practice and his flourishing farms.

Jefferson soon had not the slightest doubt that he and Martha were in love. But he was confronted by an unexpected obstacle: John Wayles did not want him as a son-in-law. Wayles may have had higher ambitions for his daughter. Socially, the Jeffersons were comparative nobodies; their Randolph connection had been diluted by that family's prodigious fecundity.

John Wayles may have contributed to Jefferson's hostility to "the cyphers of the aristocracy" who had immense social influence in the Virginia of his youth. Not that Wayles himself was an aristocrat: he had come to Virginia as a servant boy and risen high thanks to his brainpower and industry. Unfortunately, that sort of man was even more likely to seek confirmation of his status by a connection with one of Virginia's great families. On February 20, 1771, Jefferson complained to a friend about how "the unfeeling temper of a parent" could obstruct a marriage.[10]

There is another possible, even probable, explanation for John Wayles's hostility. Like most first-generation English emigrants to America, Wayles may have retained a strong loyalty to the king. Jefferson was already an outspoken advocate of American rights in the imperial quarrel that had been simmering since 1765.

The love-smitten patriot did not let John Wayles's opposition discourage him. Seldom did two weeks go by without seeing Jefferson and his traveling companion, the slave Jupiter, ride up the hill to The Forest. His suit may have been helped by a temporary suspension of transatlantic hostilities that began in 1770 and made some people think the dispute with the mother country would simply go away.

VI

Jefferson spread his devotion to Martha Wayles Skelton all over Virginia. In August 1771, he asked Robert Skipwith, who was courting Martha's younger sister, Tabitha, to "offer prayers for me at that shrine to which though absent I pray continual devotions. In every scheme of happiness she is placed in the foreground of the picture, as the principal figure. Take that away, and there is no picture for me."[11]

From Williamsburg in the spring of 1771 came a letter from Mrs. Drummond, an older woman friend for whom Jefferson had described Martha in extravagant terms. "Let me recollect your discription," she wrote, "which bars all the romantic poetical ones I ever read . . . Thou wonderful young man, indeed I shall think spirits of an higher order inhabits yr aery mountains—or rather mountain, which I may contemplate but never aspire to . . . Persevere, thou good young man, persevere—she has good sence, and good nature and I hope will not refuse (the blessing shal I say) why not as I think it,—of yr hand, if her heart's not ingaged allready."[12]

The "aery mountain" Mrs. Drummond mentioned was a conical 857-foot peak little more than a mile outside Charlottesville. In their teens, Jefferson and Dabney Carr used to ramble the slopes of this oddly isolated little elevation, which Jefferson had inherited from his father. Jefferson found the crest of the mountain exhilarating. One day he told Carr that he planned to build a house on it. The idea might have remained an adolescent dream—but fire destroyed Shadwell in 1770, and his mountaintop house suddenly became a real possibility.

After a year of labor by slaves and white artisans, Jefferson had managed to construct only a one-room brick cottage. But he was so much in love with his mountain that he moved into the tiny house, vowing he would get "more elbow room this summer." Only Martha Wayles Skelton and a handful of other people had any idea of the magnificent mansion Jefferson had already sketched and planned down to the precise proportions of every room. The gifts that would make him the father of American architecture were beginning to flower. In his college years he had pored over the sketchbooks of the great Renaissance architect Andrea Palladio, and from them he had conceived an American style that would have the same chaste lines and carefully calculated symmetry.

In his bachelor days, Jefferson had called his mountain mansion "The Hermitage." Now, with the prospect of Martha Wayles Skelton joining him, it became Monticello—a name with a sweetly romantic ring. But the name would mean nothing unless Jefferson remembered Mrs. Drummond's advice to persevere. From England he ordered an expensive "forte-piano," the very latest in musical instruments. It was obviously intended for only one player. He also asked his British purchasing agent to search the herald's office for the coat of arms of the Jefferson family. "It is possible there may be none," he wrote. "If so, I would with your assistance become a purchaser, having [Laurence] Sterne's word for it that a coat of arms may be purchased as cheap as any other coat."[13]

The suitor was straining to persuade John Wayles that the Jefferson family lineage was not a wholly worthless Welsh concoction, only a few millimeters above the subterranean Irish. All these efforts, plus more visits to The Forest, had their inevitable effect. John Wayles realized that his daughter had no intention of loving anyone but Thomas Jefferson. On November 11, 1771, the no longer unfeeling parent gave his permission. The young couple set the wedding date for January 1, 1772—visible proof of their impatience. An exuberant Jefferson scattered two- and three-pound tips to The Forest's servants and galloped back to Albermarle to prepare his family for the wedding.

VII

As the happy day approached, Jefferson put his legal training to good use. He wrote out the license-bond for the wedding, in which he and his best man, Francis Eppes (his future wife's brother-in-law), pledged fifty pounds to support their joint declaration that there was no known cause to obstruct a marriage between "the aforementioned Thomas Jefferson and Martha Skelton." After *Skelton,* Jefferson wrote a word that again suggested a division between spiritual and sexual love in his soul: *spinster.* Someone else, probably his best man, crossed it out and wrote *widow.* Why was Jefferson unconsciously denying that Martha had already submitted to another man's desire? Apparently he could not tolerate the thought—

even though there had been living proof—her son, John Skelton, who had died six months earlier, in June 1771.[14]

The wedding celebration lasted two and a half weeks—not unusual, and proof of how much Virginians loved a party. Not until January eighteenth did the newlyweds set out for Monticello. On the way they made a sentimental stop at Tuckahoe, where Jefferson had spent some of his boyhood while his father managed the estate of his friend and in-law, William Randolph. From there they set out on the final miles to Monticello, undeterred by a veritable blizzard, which forced them to shift from phaeton to horses.

It was midnight when they reached the whitened mountaintop. Jefferson led Martha to the one-room brick cottage where he had pictured himself whiling away his days as a bachelor hermit. While Martha shivered beneath her cloak, the bridegroom built a fire. Soon a roaring blaze sent waves of warmth and light against the walls of their refuge, turning the blizzard's howl into a curiously comforting sound. They lay down before the blaze wrapped in each other's arms.

Suddenly Jefferson leaped up, remembering a hidden treasure. From behind a shelf of books he flourished a half bottle of wine. With bodies warmed and glasses full, they lolled before the fire. Martha's auburn head bent low, her hazel eyes shining over the latest sketches of the magnificent house in which Jefferson vowed they would grow old together. It was a night they would remember for the rest of their lives.[15]

THE TRAUMAS OF HAPPINESS

The great world of Virginia—the courthouses, the mansions—saw nothing of Thomas and Martha Jefferson for two months after they arrived at Monticello. The House of Burgesses met in Williamsburg without the twenty-seven-year-old delegate from Albemarle County. The bridegroom did not bother to jot a single note about money or other matters in his usually busy pocket diary. Not until April did the Jeffersons end their honeymoon and descend their mountain for a journey to Williamsburg.

They enjoyed the capital's lively spring season, going to the theater frequently and riding out to visit friends in the vicinity. They also visited a Dr. Brown. On the way back, they stopped for a month-long visit with John Wayles, Martha's father. By the time they reached Monticello it was almost summer; the flowers and fruit trees planted by Jefferson before his marriage were blooming, and so was Martha Jefferson. She was expecting their first child.

On September 27, 1771, an hour after midnight, Martha gave birth to a daughter. Jefferson promptly named the child after her mother. The next months were an anxious time. The baby was underweight, and she did not seem to thrive at Martha's breast. Not until someone, perhaps Jefferson, suggested letting one of the Monticello slaves nurse her did tiny Martha begin to grow plump and healthy. The ordeal left her mother weak and virtually bedridden.[1]

Her husband relieved his tension by concentrating on the mansion he was in the process of building. By now its basic structure was visible. It was organized around a spacious central room with an octagonal section, facing west, that Jefferson called the parlor. This was entered from a hall on the east side, through a classic white pillared portico. Flanking the parlor was a smaller, square dining room to the north, balanced by a second square room to the south, which could also be used as a dining room. Off the hall was to be a large staircase to the roomy library, above the parlor. On either side of the library were two bedrooms. The ceilings of the downstairs rooms were eighteen feet high—twice the height of an ordinary plantation house. The decorations on the portico, on the mantels and indoor friezes, were carefully selected for variety and beauty from the classic architectural "orders"—Doric, Ionic, and Corinthian. The goal was a house that elevated the soul and comforted the body.

For Martha, perhaps the best proof of her husband's originality was Jefferson's decision to put all the outbuildings that marred the appearance of many Virginia mansions below the ground under two L-shaped terraces that ran from the house to their honeymoon cottage in one direction and to a matching stone cottage in the other direction. A kitchen, a laundry, a pantry, a dairy—there was room underground for all the house's necessities in this ingenious plan. Soon Martha was conferring with their cook and other servants about who would work in these places, and ordering furniture and rugs to decorate the mansion that was emerging before her delighted eyes.[2]

II

In the summer of 1773, Martha Jefferson became pregnant again. Theoretically, she and her husband considered it good news. She may even have rejoiced with him. Like most males, Jefferson wanted a son to continue the family name and lineage. In days when three out of every four babies failed to survive childhood, a woman accepted the need to have many children as a fact of life. But there was probably an undercurrent of worry in the Jeffersons' joy. A man as sensitive as Jefferson could not help noticing that for Martha, childbirth was a source of more than normal anxiety. Her own mother had died giving birth to her, and her father's next wife seems to have been notably unkind to Martha. This doubled her fears—her death could mean future unhappiness for her infant daughter.

By now, Jefferson had returned to the practice of law. This meant he was away from Monticello for weeks at a time, defending clients in various county courtrooms around Virginia. Only one of his cases compels our attention in this study of his private life—a 1770 attempt to free a mulatto named Samuel Howell. Jefferson represented him without charging a fee. Howell's grandmother had been the product of a marriage between a black man and a white woman. According to a Virginia law, she remained a slave until she was thirty-one. (The law had been passed to discourage such liaisons.) During her years in bondage, she had given birth to Howell's mother, who also was required to remain a slave until thirty-one. Howell was similarly enslaved until he reached the same age. He claimed he was a free man.

Jefferson agreed with him. Under the law of nature, he told the startled judge, "we are all born free." He contended it was bad enough that the law required a person born of a slave to remain in servitude for thirty-one years, and then extended this statute to her children. But only a new law passed by some future legislature, "if any could be found wicked enough," would inflict this fate on a grandchild. The judge found no merit in Jefferson's argument, and Howell remained enslaved. The defeated lawyer proceeded to do something highly irregular: he loaned his client $10. There may have been some verbal advice included with the money. Howell soon ran away and was never seen in Virginia again. The case not only revealed Jefferson's dislike of slavery but his sympathy for mulattoes whom he considered especially victimized by the system.[3]

II

Another side of Jefferson's career also kept him away from home and more than a little distracted while he was in residence at Monticello: politics. The British were determined to enforce the empire's laws against smuggling with new rigor. They decreed that anyone who interfered with the royal navy in its execution of this policy would be punished by death—and the trial would take place in England, rather than America. The House of Burgesses was enraged and bombarded the British with fiery protests.

Jefferson was an admirer of the leading protestor against this British policy—the backwoods lawyer Patrick Henry. He had been a law student in 1765, paying a visit to the Houses of Burgesses, when Henry roared his defiance of the Stamp Act: "Caesar had his Brutus, Charles The First his Cromwell—and George The Third may profit from their example! If this be treason, make the most of it!"

Jefferson and his best friend and brother-in-law, Dabney Carr, continued to support Henry. In 1773, Jefferson wrote a series of resolutions warning that Virginians would not tolerate the loss of their "ancient legal and constitutional rights." They further resolved to create a committee of correspondence to keep in touch with fellow protestors in other colonies. They knew all too well that this move would give the British a bad case of the jitters. Jefferson, aware that oratory was not one of his gifts, encouraged Dabney Carr to present the proposal to the House of Burgesses. Carr performed magnificently, making more than one listener wonder whether Patrick Henry had a serious rival to his claim to be Virginia's Demosthenes.

Thirty-five days after his brilliant debut, Dabney Carr died of a "bilious fever" in Williamsburg. No one had any clear idea what the medical term meant—beyond its deadly effect. The fever was probably an acute form of malaria or typhus. By the time Jefferson, away on legal business, returned to Monticello, Carr was already buried at Shadwell. His grief-stricken friend brought the body to Monticello and wrote his epitaph, making clear it was a tribute from "Thomas Jefferson, who of all men living loved him the most." To his grieving widow (Jefferson's sister, Martha) and her six children, Jefferson opened the doors of Monticello. Henceforth, he declared, he would regard the children as his own.[4]

III

Two weeks later in 1773 came more grim news: John Wayles was dead at fifty-eight. His death did not cause the pain of Dabney Carr's loss, but Martha and her husband were saddened nonetheless. Jefferson had come to like this genial man, even if he did not respect him very much as a lawyer. Wayles had returned the friendly feelings, making his new son-in-law the executor of his will. Jefferson soon realized that he and Martha were rich. They had inherited 11,000 additional acres of land and 142 slaves. There were heavy debts to British merchants, but Jefferson sold almost half the land to settle them.

Most of John Wayles's slaves stayed on the three plantations the Jeffersons had inherited from him. Some came to Monticello. These included a family of mulattoes who, like Samuel Howell, had a last name: Hemings. They reportedly were John Wayles's children by Elizabeth Hemings, who was the daughter of a white British sea captain and a black slave mother. After his third wife died, John Wayles had apparently taken Elizabeth (usually called Betty) as his mistress, and she had six children by him, three sons and three daughters. Jefferson was probably unhappy with this inheritance. As we have seen in Howell's case, he did not believe that mulatto grandchildren should be raised as slaves. But the laws of Virginia remained inflexible on this point, and the Hemingses came to Monticello as Martha's property.[5]

Who was responsible for bringing them there? As we have seen in the case of George Washington's supposed son, West Ford, white parents or relatives of illegitimate mulatto children were often inclined to feel a special concern for them. In the eighteenth century, blood was a powerful bond between people of the same family, including nieces, nephews, and often distant cousins. There is a strong likelihood that Martha Jefferson persuaded her husband to bring the Hemingses to Monticello. The children were her half sisters and brothers, and she may have hoped that eventually she could take steps to move them beyond slavery—or at the very least to offer them her protection from sale, which often broke family ties. Another motive may have been a desire to conceal her father's liaison with Elizabeth Hemings. Although it was not at all uncommon for white masters (and their sons) to take enslaved women as mistresses, it was not a practice that was publicly condoned, much less approved.

IV

Next came a decision that flowed from the unexpected wealth bestowed by John Wayles's will. It also revealed something about the Jefferson marriage. The man who had spent so many hours wrestling with "Old Coke" and worn out horses up and down Virginia's wretched roads to build a thriving law practice abandoned his profession. He sold his client list to a cousin, twenty-one-year-old Edmund Randolph, and henceforth became a man with only two pursuits, husband and farmer.

Giving up his law practice may have been a sacrifice Jefferson made to prove to Martha Wayles that there was nothing in his life more important than her happiness. Martha wanted Jefferson within reach of her hand. She needed the touch of his reassuring lips to banish the anxiety swirling in her soul as her pregnancy advanced. For the next months, he devoted himself to her and Monticello.

By now the house was reaching the first stage of its perfection. No one, including Jefferson, imagined that in a few years he would tear it down and build another, even more remarkable version. He cut winding trails that he called roundabouts through the little mountain's forested slopes. Ultimately he created four of these paths for woodland wandering, connecting them by oblique roads. He planted more fruit trees and a vegetable garden on the southeastern slope. He was like a man attempting to create his own Garden of Eden for the woman he loved.

Martha gave birth on April 3, 1774. Everyone, above all the worried father, was hoping for a boy. But it was another girl, whom Jefferson named Jane Randolph in honor of his mother. One suspects that the name was really a tribute to his dead sister, with the "Randolph" a gesture of respect to his mother, still very much alive only a few miles away in his rebuilt boyhood home, Shadwell.

This was a minor matter, compared with Jefferson's concern for Martha. Once more she recovered slowly from the ordeal. Jefferson fussed over her, repeatedly assuring her he was the happiest man on the planet, with a loving wife and two thriving daughters. He played with little Martha, toddling at eighteen months, and began schooling Peter Carr, Dabney's oldest son.

In the evenings, he and Martha read to each other from their favorite books, *Tristram Shandy* or Jefferson's other favorite, *The Poems of Ossian* by James McPherson. These were supposed to be the works of a Celtic Homer who flourished in the dawn of Scotch-Irish history, discovered by the learned McPherson in the Scottish highlands. In fact, McPherson had created Ossian and the poems in his imagination. Today the fakery is a minor matter. The poems are a vivid example of the extravagant style of the romantic movement in literature. Here is one of Jefferson's favorite passages from "The Songs of Selma":[6]

> *Star of descending night!*
> *Fair is thy light in the West!*
> *Thou liftest my unshorn head from thy cloud*

Thy steps are stately on thy hill
What does thou behold in the plains?
The stormy winds are laid
The murmur of the torrent comes from afar
Roaring waves climb the distant rock.
The flies of evening are on their feeble wings
The hum of their course on the field
What dost thou behold, fair light?
But thou dost smile and depart.
The waves come with joy around thee:
They bathe thy lovely hair
Farewell thou silent beam!
Let the light of Ossian's soul arise!

The goal of such poetry was the emotion Jefferson called "the sublime." The word reveals the mystic side of Thomas Jefferson's soul. On Monticello's summit his spirit soared in every direction. The days and nights became an almost continuous enjoyment of beauty in books, in the woman he loved and the children she had given him, in the bountiful blooming world of nature that fascinated both his eyes and his mind. It is easy to see how this sensitive man treasured almost every hour he lived on his mountain, and would willingly have spent the rest of his life there, yielding with reluctance the few days a year he was required to sit in the House of Burgesses. But events beyond the horizon of the world Jefferson commanded from his hilltop were about to force him to surrender that dream.

V

The Revolution that exploded in Boston soon swept Virginia and the rest of the American colonies into the growing upheaval. Jefferson's talented pen was pressed into service by his colleagues in the House of Burgesses. His fierce essay "A Summary View of the Rights of British America" was soon being read by Americans in other colonies and even in London, where Britons sympathetic to the Americans' grievances distributed it with enthusiasm. Jefferson was not among the delegates chosen for the First Continental Congress in 1774; they were all a decade or more older. But in 1775, as the clash veered toward full-scale war, some of the older men such as Patrick Henry stayed home to guide Virginia, and Jefferson was selected for the Second Continental Congress.

There he was engulfed by the excitement of being at the center of the Revolutionary ferment. This is what Congress became, after the news of the blood spilled at Lexington and Concord reached Philadelphia. He met Benjamin Franklin and George Washington, who was already a hero to Jefferson. He had been in his early teens when Colonel Washington became Virginia's most admired soldier.

New Englanders came within Jefferson's orbit for the first time. Several Yankees who had read the "Summary View" referred to its writer as "the famous Mr. Jefferson." John and Samuel Adams did not hesitate to say to people they trusted that they were for independence—the sooner the better. They met no opposition from Thomas Jefferson. John Adams later recalled that Jefferson was so frank, explicit, and decisive on committees and in private conversation, he soon "seized upon my heart."[7]

When Congress adjourned for a few weeks in August 1775, Jefferson rushed back to Virginia. His letters from Martha made it all too clear that she missed him desperately. Worse, their little daughter, Jane Randolph Jefferson, was ailing. En route, Jefferson bought a rare violin from his cousin, John Randolph, father of the man to whom he had sold his law practice. Randolph had decided his loyalty to the king required him to resign as attorney general of Virginia and sail for England.

In an exchange of notes, Jefferson revealed how much politics was dividing his soul. He told Randolph he yearned for the day when "this unnatural contest" with Great Britain would end with "a restoration of our just rights." That would enable him to achieve his deepest wish—to "withdraw myself totally from the public stage and pass the rest of my days in domestic ease and tranquility."[8] The emotional intensity in these words suggests the strong possibility that there were people in Jefferson's own family who were making him feel the revolutionary contest was unnatural. The two prime candidates are his mother, Jane Randolph Jefferson, and his wife, Martha Wayles Jefferson. The revolution had been percolating for six years before Jefferson married Martha—more than enough time for her to hear her father express his disapproval of troublemakers like Samuel Adams and Patrick Henry.

Alas, there was little tranquility for Jefferson at Monticello. A few weeks after he returned, he watched little Jane Randolph Jefferson die in her weeping mother's arms. Unable to bear the thought of leaving Martha alone, Jefferson persuaded her to pay a visit to her half sister, Elizabeth Eppes, and her husband, Francis, while he headed back to Philadelphia in his phaeton. Martha found the separation unendurable. She sank into a depression and was too ill to write a single letter.

Jefferson devoted one day a week to writing letters home to Martha and his numerous friends. When a month passed without a response from Martha, he grew almost frantic. He wrote to Francis Eppes, begging him for news. Eppes assured him Martha was not seriously ill, but the distraught—and perhaps guilt-ridden—Jefferson replied that he wanted to hear that news from her own hand.

VI

In the last week of December 1775, Jefferson abruptly abandoned Congress and returned to Monticello for the next four months. He supervised the continued work on the house and did his utmost to restore Martha's health and happiness. On March 31, 1776, another woman in his life gave him a shock: Jane Randolph Jefferson died suddenly at the age of fifty-seven. Within a few hours, Jefferson experienced a violent headache that incapacitated him for the next five weeks. It was far worse than the pain he had experienced when he lost Rebecca Burwell.

What was the source of this migraine-like agony? Jefferson never revealed a trace of sorrow about his mother's death in a letter or notebook. It adds weight to the suspicion that she did not approve of his revolutionary activities. We have noted a hint of this disapproval in his emotional letter to John Randolph. Martha's health no doubt complicated Jefferson's feelings. If he was doing the "wrong" thing (in his mother's view) by joining the Revolution, was his wife's illness a kind of punishment? Rationally, such thinking makes no sense. But in the subconscious, reason has to compete with the heart's unreasoning demands. Jefferson's guilt only intensified his wish to do everything in his power to make Martha well. He had planned to return to Congress at the end of March. Now he postponed his departure for another month.

VII

By May 13, 1776, Jefferson was back in Philadelphia. The mood in Congress had changed dramatically. Much of the reason for it was a tough talking pamphlet called "Common Sense" by a recently arrived Englishman, Thomas Paine. A friend had sent Jefferson a copy in February. Paine heaped scorn on George III and urged the Americans to declare their independence. Two days after Jefferson arrived in Philadelphia, the Virginia Convention under the leadership of Patrick Henry voted in favor of the explosive word. The news soon reached the colony's delegates in Philadelphia, and their chief spokesman, Richard Henry Lee, reported the decision to Congress in sonorously historic words: "These united colonies are and of right ought to be free and independent states."

In the midst of this turmoil, Jefferson's ongoing anxiety for Martha almost derailed his rendezvous with history. He wrote two letters to friends in the Virginia Convention saying he did not want to be reappointed as a delegate to Congress—and requested permission to return home as soon as possible. Edmund Randolph, the purchaser of Jefferson's law practice, presented his request to the Convention and found himself "with a swarm of wasps around my ears." Dismissing Randolph's plea, the Convention elected Jefferson a delegate for another year.

A few weeks later, John Adams chose Jefferson to draft the Declaration of Independence, a task that became the foundation of his fame. As the unforgettable words flowed from his pen, his heart and mind remained clotted with dread. Post rider after post rider had arrived from Virginia with no letter from Martha. The declaration was proclaimed and read throughout America, changing the war from a colonial quarrel to a world-transforming upheaval. But for the moment, Jefferson's role was known to only a handful of his friends and fellow congressmen. Not until 1784 would he be mentioned in a newspaper as its author.[9]

A letter finally arrived from Martha, begging him to come home as soon as possible. Whether she was seriously ill or simply thought she was remains a mystery. Jefferson impulsively assured her that he would be at Monticello by the middle of August. But he soon discovered that the older members of the Virginia delegation were deciding to go home without asking anyone's permission. Soon, Jefferson was the only Virginian in Congress. If he departed, the state would lose its vote. He wrote to Richard Henry Lee, begging him to return to Philadelphia as soon as possible. "The state of Mrs. Jefferson's health" made it "impossible for me to disappoint her expectation of seeing me." At the bottom of the letter, after a few sentences about congressional business, he scribbled, "I pray you to come. I am under a sacred obligation to go home."[10]

Lee, whose own wife was in poor health, refused to be hurried, and left Jefferson wracked with worry for the entire sweltering month of August. On September 1, 1776, he could stand the agony no longer and set out for Monticello, leaving his state unrepresented in Congress. As he departed, he wrote an almost hysterical letter to Edmund Pendleton, the chairman of the Virginia Convention, telling him he wanted to retire from politics.

VIII

Back at Monticello, Jefferson's neighbors urged him to participate in the Virginia Convention, which was rewriting the new state's laws. He took Martha to Williamsburg with him. They lived in the comfortable house of his friend and mentor George Wythe, who was serving in Congress. Soon they were lovers again and Martha became pregnant. Jefferson consulted the best doctor in Williamsburg on Martha's behalf, but he could only tell the anxious husband that she was a fragile woman whose health would always be uncertain.

A few weeks later, the president of Congress, John Hancock, asked Jefferson to join Benjamin Franklin in an embassy to Paris to seek French aid. It was flattering evidence of how much Jefferson had impressed his fellow congressmen. It also confronted him with an agonizing decision. Here was a chance to see Europe, something he had yearned to do since he was a student at William and Mary. Hancock told him his "great abilities and unshaken virtue" made him Congress's choice for this task, which was crucial to the future survival of America.

For three days, Jefferson was a man in torment. Could he take Martha with him? The way she leaned on his arm like an invalid when they walked in George Wythe's garden scotched that idea. Only someone in the best physical health could endure six or eight weeks on the Atlantic in late autumn. Could he go alone? Martha would give him her permission of course. But she had been unable to tolerate a separation of a few months while he was in Philadelphia. This ambassadorship meant an absence of at least a year and probably two. Jefferson could not even bring himself to discuss it with her.

Instead, he wrote the most painful letter of his life, turning down the appointment. He tried desperately to defend himself against the imputation of cowardice or self-interest. "No cares of my own person or yet for my private affairs would have induced one moment's hesitation" to accept the appointment, he avowed. "But circumstances very peculiar to my family, such as neither permit me to leave nor to carry it," compelled him to refuse the offer. Those tormented words won him nothing but a scathing rebuke from Richard Henry Lee, who almost certainly spoke for other Virginia congressmen. "No man feels more deeply than I do, the love of and the loss of, private enjoyment," Lee wrote. But if everyone followed Jefferson's example, America was "beyond redemption, lost in the deep perdition of slavery."[11]

IX

Meanwhile, Martha's pregnancy advanced, and her anxiety mounted with it. Once more there was the swirling dread as the day of delivery approached. On May 28, 1777, an all but distracted Jefferson waited on the first floor of his still unfinished mansion while Martha gasped and sobbed in labor. At last, down the stairs came his friend, Dr. George Gilmer, with a broad smile on his face. In his arms was a tiny red body swathed in blankets. It was a boy! Thomas Jefferson had a son!

Seventeen days later, Jefferson scrawled in his pocket diary: "Our son died 10 H. 20 M P.M." Letters came to Monticello from John Adams and Richard Henry Lee, telling him how badly he was needed in Congress. Lee's missive was unmistakably insulting. Jefferson did not bother to answer either letter. He retreated from national politics, paying only sporadic attention to the ongoing war, with its unnerving mixture of victories and defeats. He continued to accept election as a delegate to the Virginia Convention, but his attendance was erratic, depending largely on Martha's health.

On August 1, 1778, Martha gave birth to another child, a girl whom Jefferson named Mary. She survived the first months, when so many babies died, and Martha also made a rapid recovery. An exultant Jefferson plunged into a near frenzy of planting and building. Determined to finish

Monticello, he ordered no less than 190,000 red bricks. Some of this energy was inspired by good news about the war. France had signed a treaty of alliance, and the British army had abandoned its grip on the American capital, Philadelphia. General Washington pursued the retreating redcoats and claimed a victory in a brutal clash at Monmouth, New Jersey. Jefferson and many other Americans assumed peace and independence were imminent. In May 1779, Jefferson informed his friend Edmund Pendleton that he was planning to retire from both state and national politics.

Instead of seeking peace, the British were goaded to fresh fury by the intrusion of their ancient French enemy and embarked on a new strategy. They shifted the war to the South. An army rampaged up from Florida and subdued Georgia. Alarmed patriots in South Carolina begged for reinforcements. In this threatening atmosphere, Jefferson's friends elected him governor of Virginia. The news eased the pain of the letters he had received implying he was letting "domestic pleasures" impede his patriotism, and he accepted the post. It soon became the major misfortune of his life.

X

Jefferson took Martha and the two children with him, first to the old colonial capital, Williamsburg, and then to Richmond, to which the legislature decided to move because it was less vulnerable to seaborne assaults from the enemy. But the British landed troops from their ubiquitous fleet with almost ridiculous ease, and Governor Jefferson could do little to stop them. Virginia's militia law was so lax, a man could refuse to turn out by giving almost any excuse. In the fall of 1779, Martha and the children had to flee Richmond to a nearby plantation until British raiders withdrew.

In 1780, Martha became pregnant again, adding another worry to Governor Jefferson's lengthening list of woes. In November of that year, in a small Richmond house the governor had rented from Martha's uncle, she gave birth to another daughter, whom Jefferson named Lucy Elizabeth. Both the baby and her mother recovered quickly—the first good fortune to come Jefferson's way in months.

Little more than a month later, another British raid, led by the traitor Benedict Arnold, now a British brigadier general, sailed boldly up the James River, and Martha and her children again were forced to flee, this time in foul winter weather, while Governor Jefferson frantically tried to rally the state's militia. Barely two hundred men turned out, and he had to sit helplessly on his horse on the south side of the James River and watch Arnold burn millions of dollars worth of cotton, tobacco, and other property in Richmond.

Governor Jefferson brought Martha and the children back to the ravaged capital and struggled to keep a semblance of government alive. As spring approached, his spirits seemed to be reviving. But on April 15, a raw, rainy day, came a devastating personal blow. "Our daughter Lucy Elizabeth died about 10 o'clock a.m. this day," Jefferson wrote in his account book. Martha was inconsolable. Jefferson did not even dare leave her to walk a few dozen steps to the nearby house where his council met. He sent these gentlemen a note, saying that Mrs. Jefferson's "situation" made it impossible for him to attend their daily session. Three days later, Jefferson wrote to a friend, "I mean shortly to retire."[12]

For the next months, Martha was apparently too ill to supervise Jefferson's household. The house slaves took charge of buying food and caring for the two surviving children. The harassed governor had little or no time to spare for his family. A few weeks later, the main British army in the south invaded Virginia with seven thousand men. In spite of this crisis, Jefferson informed his council and the state legislature that he was not going to accept a third one-year term as governor. Not a few people, unaware of his reason for retiring, saw this decision as a shameful abandonment of his post when he was most needed. Among the most scornful critics was Jefferson's old friend, former governor Patrick Henry.

The legislature and the governor decided to transfer their operations to Charlottesville, which they thought was deep enough in Virginia's interior to be immune from British attack. Jefferson took Martha and the children to nearby Monticello and joined the lawmakers in the little town, where he reiterated his intention to resign.

At dawn on June 4, a huge horseman came pounding up the winding road to Monticello's portico. Jack Jouett was the bearer of bad news. A strike force of 180 British dragoons commanded by Lieutenant Colonel Banastre Tarleton, the most feared cavalry leader of the war, was heading for Charlottesville to capture the retiring governor and the legislature.

Jefferson gave Jouett a glass of Madeira and told him to head for Charlottesville to warn the legislators. No one, including Jouett, knew when Tarleton and his horsemen might arrive. But it seemed probable that Virginia's Paul Revere had made far better time than the dragoons, riding in formation and limited to the pace of the slowest horses. In Charlottesville, Jouett had barely sounded the alarm when another Virginian reported Tarleton was only minutes away. The lawmakers scattered in all directions, many barely dressed, all shorn of any semblance of dignity.

On Monticello's crest, Jefferson remained calm, studying the empty road through a telescope. He decided he had time to collect important personal papers and hide them in the woods. As he began to give orders to his house slaves, a man named Hudson came pounding to the portico to shout that part of the British raiding force was at the foot of the mountain.

Underscoring his desire to bag Jefferson—and his contempt for Virginia's ability to stop him—Lieutenant Colonel Tarleton had divided his small force and ordered Captain Kenneth McLeod to take a detachment of dragoons to Monticello. A wild scramble ensued. Jefferson had to hurry Martha and their daughters into a carriage and send them off to nearby Blenheim Plantation.

It is not hard to imagine Martha's terror. Not even at Monticello were she and her children safe! If they captured her husband, the British might hang him on the spot or transport him to England for a degrading show trial as a traitor followed by an even more grisly execution. At the very least, these rampaging dragoons were likely to loot Monticello and burn it to the ground.

As Martha and the children vanished down the road, Jefferson gave orders to the house servants to hide as much of the silver and other valuables as they could grab. The mortified ex-governor (his term had legally expired two days earlier) told other servants to walk his horse from the blacksmith's to a point in the road between Monticello and adjoining Carter's Mountain. He legged it ignominiously into the woods, cutting across his own property to rendezvous with the horse.

On Carter's Mountain, Jefferson resorted to his telescope again and saw no trace of Tarleton's green-coated horsemen. He was about to return to Monticello when Charlottesville's main street swarmed with sabre-waving horsemen in exultant pursuit of scurrying legislators. Jefferson rode hastily down the other side of Carter's Mountain and soon joined Martha and his daughters for midday dinner at Blenheim Plantation, their temporary refuge.

Martha's anxiety remained acute, and so did Jefferson's mortification.

Tarleton's incursion was a savage final commentary on his failed governorship. With more than fifty thousand militiamen on its rolls, Virginia under his leadership was unable to stop 180 British dragoons from riding into the heart of the state. No one had fired a shot at them. Jefferson decided to take Martha and the children to Poplar Forest, his small plantation in Bedford County, seventy miles away, where the British were unlikely to come—and he did not have to face angry legislators when they returned from their hiding places in the woods.

Almost as if evil spirits were pursuing him, at Poplar Forest Jefferson cantered out for a morning ride on his favorite horse, Caractacus. The steed reared and pitched his master from the saddle, breaking his left wrist. Badly shaken up by the fall, he took six weeks to recover. Then came a letter from a friend, telling him that the Virginia legislature had approved a resolution calling for an investigation of his governorship. Behind this nasty move was Jefferson's former friend Patrick Henry. He may well have been sincerely disgusted with Jefferson's performance, but it was also a chance to ruin a potential political rival. The assembly had adjourned until the fall, leaving Jefferson dangling between guilt and innocence.

XI

During these months of Jefferson's woes, something close to a political and military miracle occurred in Virginia. A huge French fleet appeared off the coast and blockaded Chesapeake Bay and the port of Yorktown, which the British army was using as its headquarters. Simultaneously, George Washington led the American army and a French expeditionary force on a forced march from New York, trapping the British in a deadly grip. On October 19, 1781, the British army surrendered. Suddenly peace and independence seemed real possibilities.

Shortly after the good news reached Monticello, Jefferson wrote George Washington a letter, congratulating him and explaining why he had not come to Yorktown to say this in person, "notwithstanding the decrepitude to which I am unfortunately reduced." He was referring to his broken wrist. It may also have been an oblique comment on his political status. He was a mere "private individual" now, and there was no reason why Washington should waste time talking to him. This almost incoherent letter was written by a man in the grip of acute anxiety.[13]

News of the resolution to investigate Jefferson's governorship had plunged him into depression. This state of mind made him doubly vulnerable to a familiar but ever new anxiety: Martha was pregnant again. It was her sixth pregnancy, and the shocks of the war years—three frantic flights from marauding British and the deaths of her son and little Lucy Elizabeth—had dangerously weakened her already delicate health. Soon, Jefferson was telling friends that the investigation of his governorship and its implication that he had been an inept and incompetent chief executive justified his total withdrawal from public life.

"I have retired to my farm, my family and my books, from which I think nothing will evermore separate me," Jefferson wrote to his former law colleague Edmund Randolph. The astonished Randolph replied, "If you can justify this resolution to yourself, I am confident you cannot to the world."[14] When the legislature met late in the fall of 1781, Jefferson attended and declared he was ready to defend himself against any and all accusations. The ex-governor read a list of the charges that his chief accuser (a Patrick Henry pawn) had sent to him, and his answers to them. Without a word of debate, the assembly passed a resolution declaring their high opinion of Jefferson's "ability, rectitude and integrity as chief magistrate of this commonwealth." One might think this would have satisfied Jefferson, but he went back to Monticello still vowing the experience had forever soured him on future public service. He resigned from the legislature and refused an appointment to the Continental Congress.

On May 8, 1782, Martha Jefferson gave birth to another baby girl. Perhaps hoping to raise her spirits, Jefferson named the child Lucy Elizabeth, a kind of defiance of the fate that had deprived them of the daughter with that name while they were "refugeeing" around Richmond. According to a family legend, the baby was huge—perhaps sixteen pounds. If so, it suggests that Martha Jefferson may have suffered from diabetes. Women with that disease tend to have ever larger children until childbirth finally overwhelms their bodies.[15]

Now began the four worst months of Jefferson's life. Martha did not rally after this exhausting birth. Day by day she became weaker and more wasted, in anguishing contrast to the blooming spring and summer outside her bedroom windows. Jefferson's mounting anxiety was worsened by a summons from the Virginia Assembly. He had been elected against his wishes as a delegate from Albermarle and refused to serve. James Monroe, an ex-soldier who had become a Jefferson disciple, warned him that many people were criticizing him. The speaker of the House of Delegates informed Jefferson that he might be dragged to Richmond under arrest if he persisted in his refusal to serve.

Jefferson wrote Monroe a frantic letter, claiming that the investigation of his governorship had been an injury that would be cured only "by the all-healing grave." At the close of this wild diatribe, he blurted out the real reason for his refusal: "Mrs. Jefferson has added another daughter to our family. She has been ever since and still continues dangerously ill." Monroe wrote a compassionate and understanding reply, which Jefferson never answered. He spent the rest of the summer watching Martha slip away from him.[16] He had summoned his sister, Martha Carr, and Martha's sister, Elizabeth Eppes, to help him. But he did most of the nursing himself. He sat beside Martha's bed for hours, reading to her from her favorite books when she was awake. When she slept, he retreated to a small room adjoining the bedroom, where he tried to work on a book he was writing, *Notes on Virginia,* a response to a set of queries sent to him by an inquisitive French writer. Years later, his daughter Martha, who was ten at the time, remembered that her father "was never out of [her mother's] calling" during these four scarifying months.

Husband and wife could not conceal from each other, no matter how hard each tried, that both knew what was happening. One day, Martha took a pen from her bedside and wrote on a piece of paper words from their favorite book, *Tristram Shandy:*

> *Time wastes too fast: every letter*
> *I trace tells me with what rapidity*
> *Life follows my pen. The days and hours*
> *Of it are flying over our heads*
> *Like clouds of windy day, never to return*
> *More everything presses on—*

Her handwriting faltered and stopped with those words. But she knew the rest of the passage as well as Jefferson. A few hours later, he took the paper and completed it in his strong hand:

> *—and every*
> *Time I kiss thy hand to bid adieu,*

Every sentence which follows it, are preludes to
that eternal separation
Which we are shortly to make!

Jefferson saved the paper for the rest of his life, adding to it a lock of Martha's hair.[17]

On Friday, September 6, Jefferson watched Martha's breath grow more and more labored. Beside him was his grieving sister, Martha Carr, and sister-in-law, Elizabeth Eppes. Nearby stood Martha's favorite household slaves— Monticello's cook, Ursula, Elizabeth Hemings, and some of her daughters.

Martha Jefferson knew the end was near. She began giving her husband instructions about things she wanted done after her death. When she came to the children, tears streamed down her sunken cheeks. Finally she held up her hand and told her husband she "could not die happy" if she thought her daughters "were ever to have a stepmother brought in over them." It was testimony to the still painful memory of her own mother's death and her unhappy childhood with her father's second wife. "Holding her other hand in his," the witnesses later recalled, "Mr. Jefferson promised her solemnly that he would never marry again."[18]

This is a tragic moment—so tragic, it is difficult to think rationally about Martha's request. It was an extraordinary demand to make on a thirty-nine-year-old man, especially when we consider how frequently Virginia men— and women—remarried after losing a spouse. It revives the suspicion that Martha Jefferson never approved of her husband's revolutionary career. If we factor in Martha's virtual summons for Jefferson to abandon his post in Philadelphia and return to her in 1776, and contrast it with Abigail Adams's relinquishment of her husband in the name of patriotism, the motive grows even darker. Beneath Martha's concern for her children, was there a bitter farewell message? *I don't want you to neglect them the way you neglected me.* Few men would be more vulnerable than Thomas Jefferson to such an accusation.

As Martha sank into a coma and her breath became the shallow gasps of the dying, Jefferson blacked out. Martha Carr seized him before he toppled to the floor and with some help from Mrs. Eppes and the household slaves, they half dragged, half carried him into the library, where he lost consciousness. For a while the agitated women thought he, too, was dying. It took the better part of an hour to revive him. His grief was so terrifying, their fear of his death was replaced by fear of madness. For three weeks he did not come out of the library. Hour after hour he paced up and down, collapsing onto a couch only when, in the words of his daughter Martha, "nature was completely exhausted."[19]

Two weeks after Martha died, Edmund Randolph visited Monticello. A few minutes with Jefferson convinced him that the rumors he had heard of his friend and mentor's "inconsolable grief" were true—to the point of "his swooning away whenever he sees his children." Other friends, such as James Madison, refused to believe Randolph's eyewitness account. They could not connect the intensity of his grief to Jefferson's "philosophical temper."

None of these people were psychologists; rather the opposite. Today we would say that Jefferson was experiencing a trauma, a psychic wound so intense it affected his relationship with women for the rest of his life. Martha's death was the climax of a series of personal losses, compounded by his failures as a political leader during his governorship. The two streams of anguish blended into a terrible regret for the time he had given to pursuing his public career. He could not avoid thinking that the separations, the frantic flights from British raiders, were the reasons for Martha's death. He had sacrificed her to the Revolution, and for what? All he had gotten in return was a blotted name and sneers from foes, and even from friends, that he was a man who preferred "domestic pleasures" to serving his country.

No one watched or remembered these terrible days and weeks with more anguish than Jefferson's oldest daughter, Martha. Writing fifty years later, she confessed, "The violence of his emotion, of his grief . . . to this day I do not trust myself to describe." When Jefferson emerged from his library like a ghost from a tomb, he was still a haunted man. All he could do was ride around the countryside hour after hour, swaying in the saddle like a corpse. Ten-year-old Martha rode beside him, reaching out to this reeling, incoherent man to hold him erect, offering herself as a mostly wordless companion on their aimless rambles along unfrequented roads.

By now Martha knew from the household slaves or from her aunts that Jefferson had promised never to marry again, for her and her sisters' sakes. It was the beginning of a bond between father and daughter that became stronger and more meaningful to both of them with the passage of the years.

On October third, Jefferson wrote to Elizabeth Eppes, who had returned to her home, Elk Hill. He told of Patsy (Martha) riding with him and her determination to accompany him to Elk Hill when he made a visit that he had apparently promised Mrs. Eppes. "When that may be . . . I cannot tell," he wrote. "Finding myself absolutely unable to attend to business." His grief burst uncontrollably onto the page: "This miserable kind of existence is really too burthensome to be borne and were it not for the infidelity of deserting the sacred charge left to me, I could not wish its continuance for a moment. For what could it be wished?" He did not write another letter for eight weeks.[20]

Martha Jefferson was buried beneath the great oak on the side of the mountain, near Jefferson's friend Dabney Carr and her lost children. Over her grave, on a plain horizontal slab of white marble, Jefferson placed an inscription:

TO THE MEMORY OF
MARTHA JEFFERSON
DAUGHTER OF JOHN WAYLES;
BORN OCTOBER 19TH 1748
INTERMARRIED WITH
THOMAS JEFFERSON
JANUARY 1ST, 1772;
TORN FROM HIM BY DEATH
SEPTEMBER 6, 1782
THIS MONUMENT OF HIS LOVE IS INSCRIBED

IF IN THE HOUSE OF HADES MEN FORGET THEIR DEAD
YET WILL I EVEN THERE REMEMBER YOU, DEAR COMPANION

Those last words were a quotation from the *Iliad,* in Greek.

The most deeply felt dream of Thomas Jefferson's life was over. The figure who stood "always in the forefront" of his vision of happiness was gone. Could he find another vision to replace it? It would take almost a decade for him to realize the words he had written in Philadelphia about everyone's right to freedom and equality required defense and interpretation if they were to become the guiding credo of the new nation. But in all the twists and turns of a renewed public career that would transform America, Martha Wayles Jefferson remained a presence in her husband's mind and heart.

HEAD VERSUS HEART

*I*n the months after Martha Wayles's death, Thomas Jefferson's friends launched a campaign to lure him away from Monticello, a place that could do nothing for the moment but deepen his despair. In the Continental Congress, James Madison persuaded Congress to reappoint Jefferson as a commissioner to negotiate a peace treaty. Madison immediately wrote to Edmund Randolph in Virginia: "The resolution passed a few minutes ago . . . Let it be known to Mr. Jefferson as quickly as secrecy will permit. An official notification will follow . . . This will prepare him for it." Knowing Jefferson's sensitivity about his inglorious governorship, Madison added, "It passed unanimously, and without a single adverse remark."[1]

The news reached Jefferson at a plantation near Monticello, where he was having his three daughters inoculated against smallpox. In a letter to a French friend, he confessed he was "a little emerging from the stupor of mind which had rendered me as dead to the world as she was whose loss occasioned it." That same day he wrote to Madison and to the president of Congress, accepting the appointment. Leaving his daughters with Francis and Elizabeth Eppes, he journeyed to Philadelphia and then to Baltimore, where a French ship was supposed to take him to France. But before he could sail, word arrived in America that Benjamin Franklin and his fellow diplomats had signed a satisfactory peace treaty, and he returned to Monticello.

Madison, back in Virginia, persuaded the state legislature to appoint

Jefferson to Congress. He accepted, but soon found that politics did little to ease the gloom that shrouded his spirit. He was tormented by migraine headaches and a host of minor illnesses common to people suffering from depression. Nevertheless, he performed admirably and industriously as a drafter of committee reports and bills. In the spring of 1784, the delegates chose Jefferson to join Benjamin Franklin and John Adams in negotiating commercial treaties with other European states. The politicians were trying to loosen Britain's grip on America's import and export trade. Again, Jefferson accepted, and after two false starts, he arrived in France to begin a five-year exploration of the Old World that he had dreamed of making since the age of twenty.[2]

Now he was an older, wiser, and much sadder man. To lessen the pain of separation, he took his twelve-year-old daughter Martha with him. He left the two younger girls, Mary, whom Jefferson called "Polly," and Lucy Elizabeth, whom he called "Lu," with their aunt, Elizabeth Eppes. Jefferson also took one of Elizabeth Hemings's sons, James, to Paris with him. A bright, lively young man, James had welcomed his master's offer to apprentice him to a French *traiteur* (caterer), where he could learn the art of French cooking.

Benjamin Franklin and John Adams gave Jefferson the warmest of greetings and opened doors for him throughout Paris. Soon Jefferson was enjoying the French talent for charming visitors. "They were so polite," he remarked in one letter, "that it seems as if one might glide through a whole life among them without a jostle." He also liked their temperance. He seldom if ever saw anyone drunk. One of his favorite people was the Comtesse de la Rochefoucauld d'Anville, immensely dignified and sarcastic but with an amazing enthusiasm for America. Her son, Louis-Alexandre, the Duc de la Rochefoucauld d'Anville, lived in a magnificent mansion, where Jefferson met intellectuals such as the Marquis de Condorcet, one of the *philosophes* who were hoping to transform French society. Pretty, charming Madame de Corny, wife of one of the Marquis de Lafayette's closest friends, liked Jefferson so much that they began walking together in the leafy Bois de Boulougne. Jefferson was delighted by her exquisite femininity, which she combined with a penetrating intelligence.[3]

Young Martha Jefferson, still very much a country girl, was more amused than dazzled by French femininity. She and her father were barely settled in their lodgings, she told a friend, when "we were obliged to send immediately for the stay maker, the mantua maker, the millner and even a shoemaker before I could go out." She also submitted to a *friseur* (hairdresser) once, but "soon got rid of him and turned my hair down in spite of all they could say." Thereafter she put off Monsieur Friseur as long as possible, "for I think it always too soon to suffer."[4]

With the help of Lafayette's wife, Jefferson soon found a school for Martha, the Abbaye de Panthemont. The abbess, the father was assured, "was a woman of the world who understands young Protestant girls." Martha did not speak French, and none of her fellow pupils spoke English. But in the Abbaye lived fifty or more older women "pensioners" from good families, who quickly taught her the language. Soon everyone called her "Jeffy," and she was "charmed with my situation." Her father visited her often and found no fault with the education she was receiving.[5]

II

Outwardly, most people saw a serene, confident diplomat, vastly enjoying the architecture, the paintings, the plays, and operas of Paris, a city that was the artistic center of the civilized world. But for a year, Jefferson found it difficult and frequently impossible to shake off his depression. In November 1784, he wrote to a friend that he had "relapsed into that state of ill health, in which you saw me in Annapolis (where Congress met in 1783) but more severe. I have had few hours wherein I could do anything."

In January 1785 came a devastating letter: little "Lu" Jefferson was dead, "a martyr to the complicated evils of teething, worms and hooping cough." Jefferson relapsed into almost total gloom. He was sure his "sun of happiness" had clouded over, "never again to brighten." Throughout the winter, he was dogged by migraine, poor digestion, and lassitude. John Adams's wife, Abigail, who had become fond of him, reported he was "very weak and feeble" in March. Jefferson told his friend James Monroe he was "confined the greater part" of the winter.[6]

Spring sunshine—and new responsibilities—lifted his spirits. Benjamin Franklin returned to America, and Jefferson was appointed ambassador to France in his place. John Adams went to London as the ambassador to Britain. Jefferson soon realized his new role was "a lesson in humility." When a French man or woman asked if he was replacing

Dr. Franklin, Jefferson invariably replied, "No one can replace him. I am his successor."

This reply underscores an aspect of Jefferson's career that has escaped almost everyone's attention. In the Revolution, he had not achieved a degree of fame even close to the dimensions of George Washington and Benjamin Franklin. The leadership and political skills of these two men had sustained the Revolutionary struggle. Thanks to Lafayette, Jefferson's role as the author of the Declaration of Independence was better known in France than in America. His part in the following seven years of struggle with Britain had been negligible, and even tinged with failure, thanks to his poor performance as Virginia's governor.

III

After Jefferson had been in Paris almost a year, he began to pass judgment on the country he had so long yearned to visit. The power and privileges of the king and the ruling class of aristocrats troubled him. He told one American correspondent that "the great mass of the people were suffering under physical and moral oppression." Even the nobility did not possess the happiness "enjoyed in America by every class of people." The older nobles never stopped intriguing for political power. Younger aristocrats spent most of their time pursuing beautiful women. "Conjugal love" between a husband and a wife was virtually nonexistent. He contrasted this national tendency to the American ideal of a happy marriage.[7]

Perhaps influenced by William Temple Franklin's inglorious example, Jefferson declined to encourage young Americans to come to Europe. He was particularly emphatic on this point with his favorite nephew, Peter Carr, who corresponded with him throughout Jefferson's stay in France. Jefferson was full of advice to Carr on what he should be studying to prepare himself to become a man of distinction. He had asked his friend James Madison to become Peter's tutor, and he shipped him boxes of books from Paris. He had intimated when he left Monticello that he might invite him for a visit. But when Carr asked if the time had come, Jefferson informed him that he was now "thoroughly cured of that idea."

Jefferson explained why in scathing terms. If he came to Paris, Carr would probably pick up habits that would "poison" his spiritual and psychological health. Young Americans tend to succumb to "the strongest of all human passions" and become involved in "female intrigue[s] destructive of his and others' happiness." Or he would develop "a passion for whores destructive of his health." Either route taught the young American to consider "fidelity to the marriage bed as an ungentlemanly practice." These temptations were almost impossible to resist with "beauty begging on every street." Peter did not need foreign travel to make himself "precious to your country, dear to your friends, and happy within yourself."[8]

Young Martha Jefferson did not view French morals as gloomily as her father. This may have been a tribute to Jefferson's success in insulating her from the worst aspects of Parisian amorality—the prostitutes swarming on the boulevards, the brothels on the side streets. "There was a gentleman a few days ago," she told her father, "that killed himself because he thought his wife did not love him. They had been married ten years. I believe that if every husband in Paris was to do as much, there would be nothing but widows left."[9]

Jefferson was also uneasy with the Gallic fondness for racy jokes and overt sexual references. His secretary, William Short, reported he "blushed like a boy" when a French friend made an off-color remark. But he was not a puritan like John Adams, frowning disapproval on everyone who yielded or even admitted to sexual desire. Although he may have feared the worst, the new ambassador made no objection when William Short began a liaison with the beautiful young wife of the Duc de la Rochefoucauld. Eventually the affair plunged Short into a decade of misery— fulfilling Jefferson's remark to Peter Carr that such intrigues destroyed happiness on both sides of the erotic equation.

Although Jefferson commented acidly on infidelity as a way of life in most upper-class French marriages, he was far more disturbed by French women's passion for politics. He repeatedly deplored their intrigues and interference, calling them "Amazons" and contrasting them to American wives, who were "angels," faithful to their spouses and soothers of their husbands' nerves when they "returned [home] ruffled from political debate." He persisted in this opinion, even when he met and enjoyed the company of an extremely political American woman in Paris, Abigail Adams. The intensity of his feelings on this subject renews the suspicion that a woman in his own family did not approve of his revolutionary activities.[10]

In spite of his negative opinions, the ambassador had a lively social life.

He was constantly invited to dine with the Marquis de Lafayette and his charming wife, Adrienne. Jefferson was already admired as the author of the Declaration of Independence when news from Virginia added to his reputation. James Madison reported that Jefferson's proposal to establish freedom of religion had passed the state legislature. French intellectuals, eager to escape the grip of the Catholic Church, were enthralled.

Yet Jefferson's melancholy frequently returned to haunt him. "I am burning the candle of life without present pleasure or future object," he told one American friend. "I take all the fault on myself, as it is impossible to be among people who wish more to make one happy."[11]

IV

Lurking just around a bend in time was a cure for this recurring gloom. Her name was Maria Cosway. Jefferson met her one late summer day in 1786 strolling in the Halle aux Bles, Paris's great domed marketplace, on the arm of the American painter John Trumbull. A mass of golden curls crowned an oval face and exquisite rosebud lips. Liquid blue eyes shaded by an intriguing melancholy and a soft, fluttery voice that spoke English with a piquant Italian accent completed a style so meltingly feminine, the ambassador was mesmerized.

Maria had been the rage of fashionable London for several seasons. She had been born in Italy of Anglo-Irish parents and was in Paris with her husband, the miniature painter Richard Cosway, who stood beside her in an almost blinding array of colors. One of the great fops of the era, Cosway was a gnomish little man twice her age. He strutted around in "macaroni"-style outfits—mulberry silk coats embroidered in scarlet strawberries—and purple shoes.

Jefferson barely glanced at Mr. Cosway. Enthralled by Maria, he persuaded the Cosways and Trumbull to revise their social calendar for the day while he did likewise and declared himself ready and eager to be Maria's eyes and ears for a tour of Paris. (It was her first visit.) He hustled them into his ambassadorial carriage, and they rattled off to the royal park of St. Cloud, with its sun-dappled green lanes and magnificent fountains. They dined and strolled through the gallery of the Royal Palace, with its dazzling mythological murals. Back in Paris in the dusk, Jefferson led them to a pleasure garden designed by two ingenious Italians, featuring spectacular fireworks that created "pantomimes" in the night sky—Vulcan toiling at his forge, Mars in combat. Maria cried out with pleasure at these heavenly visions. The day ended with a visit to the most gifted harpist in Europe, who told Maria about his improvements in her favorite instrument while his wife played some of his exquisite compositions.[12]

For the next two weeks, the ambassador's carriage stopped at the Cosways' house almost every day to whirl Maria off for another six- or seven-hour tour of Paris or its environs. Richard Cosway was busy painting miniatures for a royal patron, the Duke D'Orleans and his family, and John Trumbull returned to London to paint (at Jefferson's suggestion) "The Declaration of Independence," which would make him famous. But the loss of these chaperones did not deter Maria and her enthralled admirer from spending whole days together, visiting new Parisian wonders such as the Bagatelle, a park containing exotic gardens and an elegant casino. They enjoyed cold suppers at a small inn near Marly-le-Roi, the favorite palace of long dead King Louis XIV, where pavilions were crowded with beautiful statues of gods and demigods.

Maria told Jefferson the story of her unhappy life. Educated in an Italian convent, she had seriously considered becoming a nun. Her Protestant mother had forbidden her even to think of it and when her father died, took her to London, where for several years she had scores of wealthy sons of noblemen and East India Company merchants panting after her. She was repelled by all of them, and her mother, perhaps to punish her, perhaps feeling it did not make much difference, ordered her to marry the physically repellent Cosway. Once more she obeyed, but her unhappiness only deepened. Cosway encouraged her to display her considerable talents as a painter and musician, but he was flagrantly unfaithful with lovers of both sexes and frequently rude. She was an ornament for his drawing room, nothing more.

Pity blended with the desire that Maria's beauty was stirring in Jefferson's psyche. There was an innocence about her that made her Italian-flavored coquetry seem harmless, unintentional, even when it was wreaking havoc on his emotions. He gazed into her mournful eyes as she told him that she yearned to do something important with her life. How she envied Jefferson, who had already helped to create a new nation and written some of its laws! As he listened and sympathized, Jefferson saw "music, modesty, beauty and that softness of disposition" that stirred memories of Martha

Wayles Jefferson. He began reading poetry and copying passages in which "every word teems with latent meaning." Some were clearly references to Martha:

> *Ye who e'er lost an angel, pity me!*
> *O how self fettered was my groveling soul!*
> *To every sod which wraps the dead . . .*

Other selections seemed to reflect his desire for Maria:

> *And I loved her the more when I heard*
> *Such tenderness fall from her tongue.*

But still other selections suggested a contrary emotion:

> *But be still my fond heart! This emotion give o'er'*
> *Fain would'st thou forget thou must love her no more.*

Jefferson continued to see Maria almost daily. He talked vividly about the natural beauties of Virginia, its rivers and mountains that would make perfect subjects for her brush—she was especially skilled in painting landscapes. A glimpse of Jefferson's high spirits is visible in a letter he wrote to Abigail Adams in London in which he declared that the French "have as much happiness in one year as an Englishman in ten." Everywhere he looked in Paris, he saw "singing, dancing, laugh[ing], and merriment."

This ebullience may explain what happened on September 18, 1786, as he and Maria strolled along the Seine. Full of exuberant energy, the no longer young ambassador tried to vault a fence or hedge, forgetting that Virginians, virtually born on horseback, were by no means agile on their own feet. He may have been telling Maria about the special thrill of the hunt, the soaring leaps over ditches and fences, and the crunching contact with earth on the other side. In Paris it was the ambassador, not his horse, who crashed to earth. When Jefferson staggered to his feet, his right hand dangled helplessly. He made light of it and strolled cheerfully to his own house, where he told Maria he had dislocated his wrist and had better call a surgeon. He would send her home in his carriage.[13]

For two weeks Jefferson was in constant pain. He slept little and no doubt wondered why he had made such a fool of himself. He was not a boy of twenty; he was forty-four years old. Maria sent him sympathetic notes and visited him several times, but a sickroom was hardly the place for further romance. Not until October fourth did Jefferson venture out with Maria again. It was her last day in Paris, and she had begged him to share it with her. The jouncing carriage jarred and possibly dislocated the damaged wrist again; Jefferson spent the night in agony. But another note from Maria, begging to see him one more time, nerved him to call his carriage and join the Cosways as they began their journey to Antwerp to board a ship for London. Maria had told him that her husband had promised to bring her back to Paris in April. "I . . . shall long for next spring," her note all but sighed. They had a last meal together in the village of St. Denis and exchanged wrenching farewells.

V

Jefferson spent the next two weeks writing one of the longest, most revealing letters of his life. He began by telling Maria that he had stumbled back to his carriage after saying goodbye to her, "more dead than alive." At home, "solitary and sad," he sat before the fireside and heard a dialogue begin between his head and his heart:

> *Head: Well, friend, you seem to be in a pretty trim.*
> *Heart: I am indeed the most wretched of all earthly beings. Overwhelmed with grief, every fiber of my frame distended beyond its natural powers to bear, I would willingly meet whatever catastrophe should leave me no more to feel or to fear.*
> *Head: These are the eternal consequences of your warmth and precipitation. This is one of the scrapes into which you are ever leading us. You confess your follies indeed: but still you hug and cherish them, and no reformation can be hoped where there is no repentance.*

So it went for twelve electrifying pages, written with Jefferson's left hand. The heart blamed all its troubles on Jefferson's head, which had taken them to visit the Halle aux Bles, the marketplace where he had met the Cosways, because the head wanted to sketch its magnificent dome. The head acerbically retorted that a chance encounter with a beautiful woman was no excuse for succumbing to a frenzy of love.

Opinions of this remarkable document have differed almost as violently as Jefferson's head differed with his heart. Some people have called it a great love letter. Others are put off by the way Jefferson frequently speaks of his fondness for both Maria and her husband. This complaint is easily dismissed. Jefferson was protecting both himself and Maria from scandal if a stranger or, worse, a newspaper got hold of the letter. But this dialogue between head and heart remains a very strange love letter.

Most love letters passionately avow devotion and adoration without any qualification. Those are the ones recipients save for the rest of their lives. But this dialogue between the head and the heart does not affirm that sort of passion. On the contrary, the heart's protestations of its rights and pleasures are repeatedly rebuked and checked by the head.

Here is the head telling the heart how to find tranquility:

> *The art of life is the art of avoiding pain: and he is the best pilot who steers clearest of the rocks and shoals with which it [life] is beset. . . . Those which depend on ourselves, are the only pleasures a wise man will count on: for nothing is ours which another may deprive us of. Hence the inestimable value of intellectual pleasures. Ever in our power, always leading us to something else and never cloying, we ride, serene and sublime above the concerns of this mortal world.*

Carried away by its own eloquence, the head goes much too far. It tells the heart to avoid friendships. Friends get sick, die, lose their money or their wives, and require exhausting amounts of sympathy. The heart replies eloquently that there is deep pleasure in consoling a friend or caring for him during an illness. Working itself into a fury, the heart decries "sublimated philosophers" and their "frigid speculations." If just once they experienced the "solid pleasure of one generous spasm of the heart," they would instantly change their arid minds. Defiantly, the heart tells the head it intends to go on loving people, especially Mrs. Cosway, who has promised to return in the spring. He is sure she (and her husband) will reappear in Paris's May sunshine. Even if she fails to come and fate places them on opposite sides of the globe, his "affections shall pervade the whole mass to reach them."[14] With those brave words, the heart wins a victory of sorts. But it is hardly a resounding one. Perhaps the best proof that this is a very special kind of love letter is the lady's reaction to it. "How I wish I could answer that dialogue," Maria wrote wistfully. "But I honestly think my heart is invisible and mute. . . ."[15]

Maria Cosway was not a stupid woman. She had no difficulty reading the fundamental message Jefferson was sending her: *My heart adores you as an ideal, as a woman who stirs my soul. But my wary, controlling head will never allow me to propose a flight to some hidden valley in Italy or the south of France, or a headlong escape to America, where we will defy your despicable husband and the rest of the supposedly respectable world in days and nights of rapturous love.* This was the sort of thing an impassioned lover would propose—but the widowed Thomas Jefferson was not this kind of man.

Using the same guarded style that Jefferson relied on to conceal his personal message, Maria wrote that her muted heart was bursting "with a variety of sentiments"—her sense of loss "at separating from the friends I left in Paris"—and her joy "of meeting my friends in London." It was enough to "tear my mind to pieces," but she would not go into it because he was "such a master on this subject"—presumably a mind torn to pieces—"Whatever I may say will appear trifling." This is not the language of a woman aflame with passion. If anything they are the words of a somewhat disappointed woman, who only dimly understands the reason for her dismay.

A year later Maria Cosway returned to Paris, without her husband. She stayed almost six months—and saw Thomas Jefferson only twice, both times at large dinner parties, one of which he gave for her. They corresponded off and on for the next forty years. Their letters were affectionate, but there was no attempt to rekindle the aborted passion of Paris.[16]

VI

Meanwhile, Jefferson became deeply involved in negotiations with a much younger member of the opposite sex. He decided that nine-year-old Mary Jefferson (now called Maria, probably at her insistence) must come to Europe. Her presence would complete the family circle and relieve him of the dread of receiving a message that she had followed little Lu into the shadows. Jefferson's letters to his daughter Martha suggested a streak of sternness in his parenting—he was constantly exhorting her to study hard, to become an accomplished woman. With Maria, Jefferson was the total opposite. Miss Polly, as she was often called, was to be persuaded, not ordered, to embark for Paris. It is more than a little interesting that Polly/Maria was often described as an almost exact replica of her mother both in looks and temperament.

The young lady turned out to be a challenge that taxed Jefferson's formidable rhetorical powers. In his first letter, he promised her innumerable French dolls and other toys in Paris, plus the chance to learn to play the harpsichord, to draw, to dance, to read and talk French, to see her sister Martha and, it need hardly be added, her lonely father. Maria replied: *I am very sorry that you have sent for me. I don't want to go to France. I had rather stay with Aunt Eppes.* Further attempts at parental persuasion got similar replies. *I cannot go to France and hope that you and sister Patsy are well.* The final riposte was: *I want to see you and sister Patsy but you must come to Uncle Eppes's house.*[17]

The baffled Jefferson finally resorted to deception. A ship was chosen, passage was booked, and her Eppes cousins joined Polly aboard the vessel for several days before it sailed. They cheerfully romped above and below decks. On the day of departure, Polly was allowed to play until night shrouded the ship and she tottered into a cabin and fell asleep. When she awoke, the ship was at sea. With her was Sally Hemings, a pretty mulatto girl of fourteen, the youngest daughter of Elizabeth Hemings. Sally had been pressed into service at the last moment, when an older, more reliable nurse became ill and could not make the voyage.

Polly (Maria) became the pet of the ship. She grew so attached to the captain and his crew that by the time the voyage ended, there was more trouble prying her off the vessel. The captain took her and Sally to London and handed them over to Abigail Adams, who was soon telling Jefferson that Polly "was a child of the quickest sensibility and the maturest understanding that I have ever met with for years. . . . I never felt so attached to a child in my life on so short an acquaintance." She described her escort, Sally Hemings, as "a girl of about 15 or 16" but "quite a child." Abigail reported that the captain of the ship had said Sally had been useless as Polly's nurse and he might as well bring her back to America on his return voyage. But Abigail thought she seemed fond of Polly and "appears good naturd."

Abigail—and Polly—assumed that Jefferson would rush from Paris to collect her. Instead he sent his French butler, Petit, whose English was primitive. When Petit arrived, Polly threw another tantrum and refused to leave Abigail Adams. In fact, she would not let that lady out of her sight, complaining bitterly of the way she had been deceived into leaving her Aunt Eppes.

Polly/Maria told Abigail she did not remember her father but had been taught to think of him with affection. Now she wondered whether that was another deception. He had forced her to leave all her friends in Virginia. She had expected him to come to England for her. Instead he had sent a man who could barely speak English! Her indignation inspired John Adams to write Jefferson a reproachful letter for failing to "come for your daughter in person." It took the better part of a week to persuade Polly to depart for Paris with Petit and Sally Hemings. At one point, she threw her arms around Abigail and cried, "Why are you sending me away—when I've just begun to love you!" Before the battle ended, Abigail was more distraught than Polly.[18]

Jefferson was totally delighted by this imperious young lady. He wrote to Abigail Adams about how Polly "flushed, she whitened, she flushed again" when she received a letter from Abigail. The pleasure Jefferson took in this performance leaves little doubt that in looks and manner, Maria was a vivid copy of her dead mother. After a week of showing her the sights of Paris, he enrolled her in the convent school with Patsy (Martha), where, he told Mrs. Eppes, she soon became "a universal favorite with the young ladies and their mistresses."[19]

VII

For the next year, Jefferson's personal life was overshadowed by another historical upheaval. Looming bankruptcy began to ravage the French government. Primarily the crisis was due to their antiquated tax system, which exempted most of the aristocrats from paying anything. The king was forced to summon a parliament called the Estates General to overhaul the system. The Estates numbered twelve hundred members, much too large an assembly to function efficiently as a governing body. Soon there were four distinct groups within the conclave, and Jefferson feared the possibility of civil war.

Meanwhile, the independent United States of America was entering a new phase of its existence. A convention met in Philadelphia in 1787, and Jefferson's friend James Madison had played a leading role in creating a new federal constitution for the republic, far stronger than the Articles of Confederation under which the Continental Congress had labored so ineffectually. Elections had been held, and a new government, with George Washington as the first president, had been chosen by the voters.

This transformation of the federal government made Jefferson decide it was time to return to America. He was also growing concerned about his neglected farms. The income from them had dwindled toward zero. Another reason was his daughters. They were becoming more French than American. A sort of climax in this department was Martha's announcement that she wanted to join the Roman Catholic Church and become a nun. Jefferson went to her convent school and had a talk with the abbess in charge. She agreed that it might be best if Martha and Polly withdrew from the school. Jefferson would supervise their education until they returned to America.

Another problem confronted Jefferson within the walls of his ambassadorial residence. James Hemings informed him that he did not want to go home. In Paris, slavery had been banned by the local Parlement (a semi-judicial ruling council), although it was tolerated elsewhere in France and was the economic backbone of the nation's overseas empire. James and his sister Sally were theoretically free. James did not want to relinquish this status for slavery in Virginia. Whether Sally Hemings also voiced a similar desire is less certain. James had a skill that would enable him to support himself. Both Hemingses had probably made contact with some of the estimated one thousand free blacks in Paris, where friends may have urged James and perhaps Sally to assert their freedom.[20]

Jefferson talked James out of this reach for independence with a combination of promises and appeals to his gratitude. James had become ill not long after he arrived in Paris, and Jefferson had spent a considerable amount of money on a doctor and nurse to help him regain his health. The ambassador had also paid for James's cooking lessons with one of the best chefs in Paris. If James returned to America, Jefferson promised him his freedom as soon as he trained one of his younger brothers to do the cooking for Monticello. Jefferson also promised to pay James a salary, which would enable him to save enough money to sustain himself when he went looking for work as a free man. Without James to support and protect her, Sally Hemings had little choice but to return to Monticello with her brother.

As Jefferson's agreement to free James Hemings made clear, he thought third-generation mulattoes should not be enslaved in America or in Paris, and he may have also told Sally this. In any event, he would never have allowed an attractive teenage girl to remain alone in Paris. She would very likely have become the mistress of a predatory young Frenchman, who would discard her after a few years and consign her to the ranks of Paris's sixty thousand prostitutes—the fate of many of the "opera girls" who had stirred Abigail Adams's sympathy.

The ambassador sailed for home on October 22, 1789, reserving cabins aboard the ship for his daughters and Sally Hemings, who probably functioned as their maid. He asked that Sally be given a cabin near them. Arriving in Virginia after a smooth twenty-nine-day voyage, he was astonished to read in the newspapers rumors that President Washington was going to appoint him secretary of state. He had planned to stay in America only long enough to get his daughters settled—probably with their Aunt Eppes, where Polly had been so happy—and restore his farms to prosperity with the help of expert overseers. He assumed he would return to France as ambassador. He felt that he was uniquely qualified to cement good relations between America and Revolutionary France.

At Eppington, the Eppes plantation, he found a letter from President Washington confirming the newspaper reports. Jefferson's admiration for Washington was so strong that he soon agreed to become secretary of state. By this time he was back at Monticello. As the Jeffersons' carriage appeared at the foot of the mountain, the slaves raced to welcome him in their brightest Sunday clothes. Cheering and shouting, they unhitched the horses, and the men hauled the carriage up the steep winding road to the summit. "When the door of the carriage was opened," Martha Jefferson later recalled, "They crowd[ed] around him, some . . . crying, others laughing." They lifted the protesting Jefferson in their arms and carried him to the portico. Martha and Maria Jefferson and James and Sally Hemings received equally warm greetings.[21]

There was much more than affection for Jefferson involved in this greeting. If Jefferson had died in Paris, or had been lost at sea, Monticello's slaves would have faced catastrophe. They would have been sold or handed over to Jefferson's heirs, with only minimal attention to preserving families or rewarding those who had established themselves as artisans or acquired other skills such as weaving cloth. Jefferson's safe return after five long years of uncertainty about their fates was more than enough reason to celebrate.

As Jefferson struggled to deal with the painful memories Monticello evoked, and make some progress on restoring the productivity of his farms, he had a surprise that gladdened his heart. Toward the end of December, Monticello had a visitor—a tall, dark-haired, swarthy young gentleman named Thomas Mann Randolph. He was warmly welcomed as the son of a man who had been Jefferson's playmate in his boyhood, when he spent seven years at the Randolph plantation, Tuckahoe. Thomas Mann Randolph's grandfather, William, had been Peter Jefferson's closest friend. When the elder Randolph died suddenly at age thirty-three, he had asked Peter in his will to take over the plantation and raise his orphaned children. Their mother had died a year or two before her husband.

The younger Randolph had conducted a lengthy correspondence with Jefferson while he was studying in Edinburgh. For a while, Jefferson had more or less taken charge of his education. After some pleasant small talk, Randolph informed Jefferson that he hoped to become his son-in-law. He had met Martha Jefferson not long after she arrived home, and the two young people had felt an instant attraction. One of Randolph's appeals for Martha was his European education. She had not been enthusiastic about returning to rural Albermarle County after five years of sophisticated Paris. Like Martha's father, Randolph was fascinated by politics, and he hoped to make it his career. He also had a strong interest in science and its Virginia subdivision, scientific farming. Almost as important was his six-foot-two-inch height. Martha had inherited her father's long-limbed body and dreaded the thought of marrying someone who was noticeably shorter than she. At least as influential was their families' long and intimate friendship.[22]

Jefferson gave his warmest assent to the match—and immediately began trying to arrange things so that Martha would remain within his paternal orbit. He urged Randolph to buy land near Monticello. The young man's father owned an excellent farm, Edgehill, only a few miles away. For the moment, Martha—and her husband—resisted his persuasion. Randolph's father had given him land in a distant section of Virginia and the young people, in a burst of independence, announced they were going to start their lives together far from both their homes.

Jefferson cheerfully acquiesced, but he by no means abandoned his determination to retain his daughter. For the moment, politics was absorbing his attention. He was about to depart for New York to join President Washington's cabinet; he would be gone for months, possibly years, which would make an objection to Martha's departure seem especially disagreeable. He contented himself with a son-in-law he liked and a daughter aglow with love.

In New York he wrote a revealing letter to Martha: "I feel heavily these separations from you. It is a . . . consolation to know you are happier and to see a prospect of its continuance in the prudence and even temper of both Mr. Randolph and yourself. . . . Continue to love me as you have done, and to render my life a blessing by the prospect it may hold up to me of seeing you happy."

Martha promptly replied: "I hope you have not given over coming to Virginia this fall as I assure you my dear papa my happiness can never be complete without your company." She assured him that "Mr. Randolph" was a wonderful husband and she was determined to please him in "every thing." All other aspects of her life would be secondary to that goal "*except my love for you.*"[23]

Meanwhile, Jefferson had decided to let Maria Jefferson, now thirteen, stay where she had been happiest—with her Aunt Eppes and her cousins at Eppington. He had taken Martha to Philadelphia with him when she was the same age, to advance her education. But Jefferson had long since realized that Maria was a totally different child who needed the companionship of loving friends and family to keep her contented. This solution worked so well that before the end of the decade Maria would marry her cousin, Jack Eppes.

With the two most important people in his life in happy situations, Jefferson headed for New York and its politics. He was a man with a mission. Conversations with James Madison, already his closest friend and advisor, had convinced him that there were tendencies in the United States that had to be exposed and defeated, lest the American Revolution end in betrayal of the ideals he had enunciated in the Declaration of Independence. In a speech in Alexandria, on the way to New York, he had told his audience the Republican form of government was the only one that was not "at open or secret war" with the rights of mankind.

On March 21, 1790, Jefferson reported to President Washington as a citizen soldier of the republic, ready for duty. He would soon discover that this duty was far more complex and emotionally abrasive than any task he had yet confronted. He would find himself virtually at war with men who had shared the task of achieving independence. His friendships with John Adams and George Washington would be ruined by vicious partisan politics. Worst of all, Jefferson would face devastating accusations about his personal life that threatened his growing fame as a founding father.

THE WAGES OF FAME

In ten years, Thomas Jefferson went from an untried, relatively unknown secretary of state in George Washington's cabinet to president of the United States. This amazing ascent owed a great deal to the way the Declaration of Independence became a major force in American politics. Almost from the moment Jefferson joined President Washington's administration, he clashed with Secretary of the Treasury Alexander Hamilton about their radically differing views of the French Revolution and America's political and economic future. As we have seen, from this conflict emerged two political parties, something the founders had never anticipated and at first deplored. When the Federalists flaunted the Constitution as their handiwork, the Jeffersonian Republicans retaliated with the Declaration of Independence as their sacred document and lavished praise on Jefferson as the author, adding cubits to his stature.[1]

As the political discussions grew more intense, anything Jefferson said or wrote became grist for the journalistic slander mills. One editor printed a letter he had dashed off to an Italian friend, Phillip Mazzei, which included a derogatory comment on Washington. Mazzei leaked it to a newspaper in Europe and it soon crossed the Atlantic. In the letter, Jefferson criticized America's declaration of neutrality in the war between Britain and Revolutionary France and described Washington as an "apostate" from the cause of liberty. He had been a "Samson" in the war for independence, but as president had allowed his head to be shaved "by the harlot England." The overheated comparison ended Jefferson's friendship with Washington. In Philadelphia, where Jefferson was serving as vice president during John Adams's presidency, people crossed the street to avoid speaking to him.[2]

Jefferson became involved with Scottish-born James Thomson Callender, a newspaperman who attacked George Washington, John Adams, and Alexander Hamilton with reckless accusations. Jefferson praised one of his effusions: "Such papers cannot fail to produce the best effect. They inform the thinking part of the nation," he told Callender. The journalist was a heavy drinker with a paranoid streak that widened appreciably when he was jailed under the Sedition Act, the law that made it a crime to criticize a president. Jefferson, who believed the law was unconstitutional, gave Callender money and sympathy. When Jefferson became president in 1800, he pardoned the journalist.

Only minimally grateful, Callender demanded that he be appointed postmaster of Richmond, Virginia, as a reward for his services to the Republican cause. When President Jefferson balked, Callender retaliated with a series of vicious articles in his Richmond newspaper. He revealed that Jefferson had paid him substantial sums to support his slanderous labors and quoted the president's words of approval. Next, the inflamed scribe accused Jefferson of fathering several children by Sally Hemings, the young mulatto who had escorted Maria Jefferson to Paris. It was "well known" among Jefferson's neighbors that he had kept Sally "as his concubine" for many years, Callender declared. One of their children was a boy of about twelve named "Tom," with red hair and a striking resemblance to Jefferson. Supposedly, Tom had been conceived in Paris, when Sally escorted Maria Jefferson across the Atlantic to join her father. Everyone in the vicinity of Monticello knew about Sally. So did James Madison, when he urged Americans to vote for Jefferson because of his "virtue."[3]

Federalist editors leaped on the Sally story and gleefully reprinted it in their newspapers throughout the nation. As the Federalists saw it, they were retaliating against the Jeffersonian Republican editors who had revived the British slanders about Washington's supposed sexual sins in the Revolution and exposed Alexander Hamilton's affair with Maria Reynolds. The *Boston Gazette* published a song about Sally, supposedly written by the "Sage of Monticello" to the tune of "Yankee Doodle." Callender reprinted it in his Virginia paper, making the accusation difficult for Jefferson's friends and family to ignore:

> *Of all the damsels on the green*
> *On mountain or in valley*
> *A lass so luscious ne'er was seen*
> *As Monticellian Sally*
>
> *(chorus) Yankee Doodle, who's the noodle?*
> *What wife was half so handy?*
> *To breed a flock of slaves for stock*
> *A blackamoor's the dandy*
>
> *When pressed by load of state affairs*
> *I seek to sport and dally*
> *The sweetest solace of my cares*
> *Is in the lap of Sally.*[4]

Next, Callender revealed that Jefferson's former friend John Walker, now a fierce Federalist, accused him of trying to seduce his wife, Betsey, thirty years ago. Although Walker had known the story for over a decade, he claimed to be outraged and threatened to challenge Jefferson to a duel. As Walker talked and wrote about it, the story took on even more lurid dimensions. He claimed that Jefferson had pursued Betsey as late as 1779, when he was a married man living in supposed contentment with Martha Wayles. Jefferson's friend Thomas Paine attempted to defend him against the ten-year extension of the story. "We have heard of a ten year siege of Troy," Paine wrote. "But who ever heard of a ten year siege to seduce?" Both stories became national sensations.[5]

II

Until this explosion, there was scarcely a mention of Sally Hemings in Jefferson's letters or the records he kept at Monticello in his Farm Book and Account Book. Sally had begun having children in 1795 and by 1802 had given birth to two girls and a boy. There was no record of a child who would approximate the age of the boy Callender described; he would have been born not long after Jefferson, his daughters, and James and Sally Hemings returned from France. The journalist claimed that Sally had as many as thirty other lovers beside President Jefferson. She was "a slut as common as the pavement."[6]

In accordance with their agreement, Jefferson had freed Sally's older brother, James Hemings, after he had trained his younger brother, Peter, to become Monticello's cook. It took James four years to complete this task. One reason may have been James's fondness for alcohol, a habit he apparently acquired in Paris. James used the money Jefferson paid him during these years to return to France. But he found revolutionary Paris a strange and unsettling place, and soon sailed back to America. He paid a visit to Monticello and was cordially welcomed by Jefferson. James talked grandly of perhaps going to Spain to find work there. Jefferson noted in a letter that he seemed to have gotten control of his drinking, a hopeful sign.

Alas, it was only a temporary reform. In the fall of 1801, penniless and depressed by his addiction to alcohol, James Hemings committed suicide in Philadelphia. Jefferson was deeply distressed by this tragic news. James had been one of his most capable and devoted servants for many years. He had served as his coachman and butler before becoming a chef. When Jefferson became president, he had offered James the post of chef at the executive mansion in Washington, D.C. But Hemings, perhaps reluctant to resume the master–servant relationship, had turned him down.

By this time it had become clear that Jefferson considered all the children born to Betty Hemings and John Wayles entitled to various degrees of freedom. They were permitted to travel around Virginia, to marry men and women of their choice, and to make arrangements with employers as far away as Richmond. They kept all the money they made. Jefferson apparently believed they were ultimately entitled to complete freedom, as third-generation mulattoes. Also important was their blood relationship to Martha Wayles; he continued to express her special concern for them by this privileged treatment.

In 1794, Jefferson freed Robert Hemings, James's older brother, who had been trained as a barber. Robert had married an enslaved woman in Fredericksburg. He persuaded her master to pay Jefferson to free Robert, who in turn promised to repay him and purchase his wife's freedom. Jefferson was not pleased by Robert's abrupt demand for freedom, which left him without a barber. But he gave Robert his certificate of manumission. The couple lived in Richmond for the rest of their lives.

In April 1792, Jefferson had given Sally's older sister, Mary, a form of freedom. She had been hired as a servant by Colonel Thomas Bell, a Charlottesville storekeeper. They apparently became lovers and soon had two children. Mary had previously had children by an unnamed father at Monticello. Jefferson kept these children under his control but agreed to sell Mary and her Bell children to the colonel. Thereafter they lived as man and wife, and no one in Charlottesville said a censorious word. That is not entirely surprising. Like her sisters and brothers, Mary's skin was probably white. When Bell died in 1800, he freed Mary and their children and made them his beneficiaries.[7]

III

President Jefferson was staggered by James Callender's assaults on his personal character. Friends such as James Madison came to his defense with scathing denials and denunciations of Callender. Madison dismissed the story as "incredible." But Jefferson made no attempt to answer the charges publicly. Only in private letters to close friends and political allies, he admitted he had, while single, "offered love to a handsome lady." This was the only charge that was true, he insisted, implicitly denying Callender's story about Sally Hemings. Eventually he negotiated a private admission of guilt with John Walker that avoided further altercation and a duel.[8]

A troubled Jefferson asked Martha and Maria to join him in the executive mansion in Washington. Opinions vary in respect to his motives. Friends thought he was trying to protect them from the ugly gossip swirling through Virginia in the wake of Callender's assault. His political foes sneered that he was trying to portray himself as a man who had retained the devotion of his two daughters and was therefore innocent of Callender's charges.

Martha and Maria left their husbands and children in Virginia and joined the president in response to his summons. For six weeks, they participated in a stream of formal dinners with congressmen and senators. Jefferson insisted on paying all their expenses, including new dresses and bonnets for Maria. One Federalist guest reported that the two daughters appeared to be "well-accomplished young women . . . very delicate and tolerably handsome." Martha enjoyed the experience, but Maria found it tiresome to make conversation with so many strangers. She was even more troubled by how much their visit had cost Jefferson. She wrote him a touching note after she returned home, hinting rather strongly that she wished he had a wife: "How much do I think of you at the hours which we have been accustomed to be with you alone and how much pain it gives me to think of the . . . solitary manner in which you sleep upstairs. Adieu much beloved of fathers . . . You are the first and dearest to my heart."[9]

IV

Callender's assault came as President Jefferson was undergoing terrific political stress as president. He had taken office hoping to reverse a decade of hostility between America and France. But his vision of the French as America's natural allies vanished in a cascade of reports and rumors that France's new leader, Napoleon Bonaparte, planned to send an army to New Orleans to create a rival nation in the Louisiana Territory, a vast swath of the continent between the Mississippi and the Rocky Mountains. France had given it to Spain in 1763 to compensate for Spanish losses in the Seven Years' War with Britain, but Napoleon had pressured Spain into secretly ceding it back to France. Another French army invaded St. Dominique (present-day Haiti and the Dominican Republic), determined to regain control of this wealthy sugar island from rebellious slaves who had seized it with the encouragement of the Federalists.

In collusion with the French, the Spanish closed the port of New Orleans, cutting off the western states' export trade. The Federalists, led by Alexander Hamilton, called for war, and numerous Jefferson supporters in the West joined the cry. Jefferson sent James Monroe to Paris as a special envoy to try to resolve the crisis. The president discovered, to his and everyone else's amazement, that Napoleon was prepared to sell New Orleans and the entire territory of Louisiana to the United States. The French army in St. Dominique had been decimated by yellow fever and other diseases, and Bonaparte had decided to cut his losses and abandon his scheme to revive the French empire in America. Although there was nothing in the Constitution that permitted the acquisition of more territory, Jefferson accepted the offer, doubling the size of the United States.[10]

The purchase of the Louisiana Territory made Jefferson immensely popular—and defused the Federalists' attacks on his personal life. He was reelected in 1804, winning four out of every five votes. The Federalist party was reduced to a hapless minority. Callender, the one man in the nation who might have continued to attack Jefferson about "dusky Sally," conveniently drowned in the James River in three feet of water, a few months after he had been beaten over the head by the federal district attorney for Virginia.[11] But no one seemed to notice or care. Jefferson's fame soared to unparalleled heights. His followers compared him to George Washington and found him superior because he had acquired "an empire for liberty" without firing a shot or losing a single soldier—and without raising taxes.[12]

V

In the midst of this improbable ascension from the depths of disgrace to the heights of fame, Jefferson's personal life received a devastating blow. His beloved younger daughter, Maria, had enjoyed a happy marriage with her cousin, Jack Eppes. He adored her with a fervor that more than matched her father's devotion to Martha Wayles. Jack won election to Congress, where he supported his father-in-law with wit and eloquence. Jefferson liked him so much that he invited him to live at the "palace," as the presidential residence was often called in its early years.

Unfortunately, Maria had inherited not only her mother's beauty and temperament but also her fragile physique. Her first baby, a daughter, was born prematurely and lived less than a month. Maria spent the next two years suffering from a variety of illnesses, including a breast infection and excruciating back pains thst reduced her to invalidism. Her next baby, a boy whom she named Francis after her father-in-law, survived but was a frail child, subject to alarming convulsions that made the family fear he was an epileptic.

Pregnant again, Maria gave birth to a girl on February 15, 1804. Before the birth she described herself as "depressed and low in spirits." Afterward she was so ill that she was unable to nurse the child. She again developed a breast abscess and suffered from constant nausea, which made it almost impossible for her to digest food. Her anguished father, hearing these reports, rushed from Washington the moment that Congress adjourned. In a letter he sent by express, Jefferson urged Jack Eppes to take Maria to Monticello. Jefferson had convinced himself that the house would work some kind of magic on her—a sign of how distracted he was. The equally distraught Eppes ordered his slaves to make a litter and carry Maria to the top of the mountain.

In a hasty letter to James Madison, Jefferson reported that Maria was going to recover, thanks to being "favorably affected by my being with her." Alas, Maria's will to live dwindled away, in a heartbreaking imitation of her mother's decline. On April 13, 1804, his sixty-first birthday, Jefferson reported to James Madison that Maria "rather weakens." She continued to have a "small and constant" fever and found it impossible to keep any food in her stomach. On April 17, 1804, Jefferson wrote in his account book: "This morning between 8 and 9 o'clock my dear daughter Maria Eppes died."[13]

Jefferson was almost as prostrated as he had been after Martha's death. It took him two months to express his feelings to anyone, and then it was a cry of almost unbearable anguish: "Others may lose of their abundance but I, of my want, have lost even the half of all I had. My evening prospects now hang on the slender thread of a single life," he told John Page. Jefferson was referring to Martha Jefferson Randolph. It was difficult for anyone who did not know him intimately to grasp the centrality of his family to Jefferson's vision of happiness.[14]

VI

Another woman who had once been close to Jefferson heard about Maria's death and wrote him a letter. It took Abigail Adams almost a month to overcome the bitterness her soured friendship with the president had left in her heart. She began by telling him that if he were "no other than the private inhabitant of Monticello," she would have written to him immediately. Only when "the powerful feelings of my heart burst through" her restraint did she feel compelled to shed "tears of sorrow . . . over your beloved and deserving daughter." She realized, thinking of her son Charles, that they had the loss of a beloved child in common. She knew the pain a parent feels when chords of affection are "snapped asunder." She had "tasted the bitter cup" and she could only hope Jefferson would learn to accept it as a decree of "over-ruling providence."

Deeply moved, Jefferson sent the letter to Maria's husband, Jack Eppes, who pronounced it "the generous effusions of an excellent heart." His son-in-law advised Jefferson to answer it expressing only "the sentiments of your heart." He urged him to avoid any mention of ex-president John Adams. Unfortunately, Jefferson replied to Abigail before he received this good advice. He began by declaring he would never forget her kindness to Maria in London. He added that her letter gave him a chance to express his regret for the "circumstances" that seemed to have "draw(n) a line of separation between us."

At first Abigail responded with assurances that she felt the same way. Encouraged, Jefferson proceeded to get into how John Adams had been "personally unkind" by appointing a raft of Federalist judges on his last night as president. Portia rushed to defend her dearest friend and soon she was condemning Jefferson for hiring "the wretch," James Thomson Callender, to defame John with "the basest libel, the lowest and vilest slander which malice could invent." When Jefferson tried to put his connection to Callender in a better light, Abigail exploded: "The serpent you cherished and warmed [has] bit the hand that cherished him and gave you sufficient specimens of his talents, his gratitude, his justice and his truth."

A discouraged Jefferson, after more futile letters, finally replied, "Perhaps I trespassed too far on your attention." Abigail, probably feeling renewed sympathy for his loss of Maria, told him that in a tribute to their lost friendship, "I would forgive, as I hope to be forgiven." She wished him success in administering the government "with a just and impartial hand." Abigail waited weeks to tell John Adams about this correspondence. After reading the letters, he wisely chose to say nothing about them.[15]

VII

Fortunately for Jefferson's peace of mind, Martha Jefferson Randolph was a remarkably healthy young woman. She gave birth to twelve children in the course of her marriage, and eleven lived to maturity. But Martha's—and Jefferson's—confidence in a happy private life slowly eroded as Thomas Mann Randolph revealed an emotional instability that ran like a dark thread through his family's history. He suffered from crushing depressions, and he slowly acquired a grievance against his wife. Martha had persuaded him to buy the Edgehill plantation, only four miles from Monticello. Whenever Jefferson returned to his hilltop mansion, Martha joined him with her children, leaving Randolph little choice but to follow them.

More and more, Randolph felt he was in competition with the great

Thomas Jefferson for his wife's affections, and was an inevitable loser. At one point he wrote Jefferson a bitter letter, saying he felt like "the proverbial silly bird" who could not "feel at ease among swans." He accused Martha of looking down on him and undervaluing his talents. At another point, he made plans to sell Edgehill and move to Mississippi to raise cotton, as many other Virginians were doing. His constantly growing family meant he was slipping into debt. The soil in Virginia was depleted and the market for its crops was depressed because of the abundant harvests from new farms in Kentucky, Tennessee, and other western states. With Martha's help, a distraught Jefferson talked him out of this reach for independence.[16]

In 1803, Randolph ran for Congress without consulting the president and defeated an old Jefferson friend by a handful of votes, turning the loser into a potential enemy. Jefferson did his best to repair the political damage and invited Randolph to live in the presidential palace with him and Jack Eppes. Randolph accepted, but in 1806 he decided that the president preferred Jack's conversation to his and picked a quarrel with his brother-in-law. Randolph wrote an incoherent letter to the president and moved into a boarding house at the other end of Pennsylvania Avenue, populated almost entirely by Federalists.

Engulfed in gloom, Randolph ate most of his meals in his room and seldom spoke to anyone. The frantic president assured him that he loved him "as I would a son" and begged him to return to the executive mansion. Randolph refused, and one friend warned Jefferson that he might commit suicide. Randolph became ill with a virulent fever that brought him to the brink of death. Jefferson sent a doctor and a series of friends who virtually camped at his bedside until Randolph recovered.[17]

This behavior was the beginning of a long sad history of personal and marital unhappiness. Randolph quit Congress but became little more than a supernumerary in Martha Jefferson's emotional life. One of the most telling signs of his sense of inferiority was his policy of permitting Jefferson to name his children. Jefferson did not ask for the privilege. Once, when he delayed, obviously hoping Randolph would name a new baby, Martha begged her father to produce a name so the child could be baptized. Of all the children, only one—their first daughter, Anne Cary—had a Randolph family name. Randolph took equally little interest in their education and development.[18]

VIII

Each of the Randolph children became part of Jefferson's family. While he was president, he carried on a delightful correspondence with the older ones. He debated with the oldest girl, Anne, about whether she should change her name to Anastasia. When Ellen, the second oldest, began reading romantic poetry, she signed her letters Eleanora, which Jefferson warmly approved. Anne was his favorite gardener. He sent her flowering peas from Arkansas, found by Meriwether Lewis and his partner, William Clark, in their famous exploration of the continent in the wake of the Louisiana Purchase. When someone sent the president rare Algerian chickens, these, too, were shipped to Anne to be raised at Monticello.

Ellen Randolph had strong intellectual interests. When she found herself puzzled by a question such as "What is the Seventh Art?" she forwarded it to the president. Jefferson replied that he thought gardening was almost as important as poetry. The shrewd grandfather suspected that Anne and Ellen exchanged their letters from him and Ellen might be ready to claim she was the one he liked best. Ellen submitted an impressive reading list to her grandfather; it included "Grecian history" in which she was "very much interested" and Plutarch's *Lives* in French. Jefferson's praise was lavish.[19]

From the start of their correspondence, the president insisted the young ladies had to answer every letter he wrote to them and he felt obligated to do the same. At one point, he claimed that Ellen was five letters behind in "her account" and threatened to "send the sheriff after you."[20] Sometimes the nation's chief executive had to deal with urgent political problems, such as former vice president Aaron Burr's attempt to separate the western states from the union in 1806. Jefferson apologized to Ellen for falling behind in his "epistolary account."

Later, when Grandpapa fell behind again and admitted it, a delighted Ellen triumphantly responded, "Your fear of being bankrupt is well founded." Jefferson wondered in an answering letter whether this meant Ellen had more "industry or less to do than myself." Ten-year-old Ellen gravely replied that she had made a real effort to spend as little time in "idleness" as possible that winter but she was inclined to suspect the president had "a great deal more to do than I have."

Equally delightful was an exchange the president had with his four-year-old granddaughter, Mary Randolph. He told Ellen to thank Mary for her letter, which was an indecipherable scrawl. "But tell her it is written in a cipher of which I have not the key. She must, therefore, tell it all to me when I come home." Ellen replied, "Mary says she would tell you what was in her letter if she knew herself."

As Jefferson neared the end of his second term, he urged Martha and her children to move to Monticello permanently, and she rapturously agreed. She assured him that her "first and most important object" would be "chearing your old age by every endearment of filial tenderness." She could barely wait to see him "seated by your own fireside surrounded by your grandchildren contending for the pleasure of waiting upon you." Her husband, Thomas Mann Randolph, was ominously absent from this vision of future happiness.[21]

IX

Most of Martha's children were girls who became worshippers of their grandfather. "From him seemed to flow all the pleasures of my life," Ellen later wrote. "When I was about fifteen years old, I began to think of a watch, but knew the state of my father's finances promised no such indulgence." One day a packet addressed to Jefferson arrived from Philadelphia. He opened it and presented Ellen with "an elegant lady's watch with chain and seals." Similar presents arrived for her as she grew older: "my first handsome writing desk, my first leghorn hat, my first silk dress."[22]

The other granddaughters received similar gifts. When Jefferson overheard ten-year-old Cornelia say, with a sob in her voice, "I never had a silk dress in my life," a splendid garment arrived from nearby Charlottesville the next day. To make sure there were no more tears, a pair of lovely dresses for Cornelia's two younger sisters was in the package. Mary Randolph heard that a neighbor was moving west and wanted to sell a guitar. But the price was far beyond the reach of her father's wallet. One morning when she came down to breakfast, there was the guitar in her chair. Grandpa Jefferson said it was hers, if she solemnly promised to learn to play it.[23]

X

Jefferson persisted in this generosity to his grandchildren in the face of ever-mounting money worries. His debts were partly a result of his expensive lifestyle and partly caused by the long recession into which Virginia sank in the years after he left the presidency. He stubbornly maintained the free-spending habits of his youth and middle age, above all the tradition of southern hospitality. As his postpresidential fame continued to grow, visitors thronged to Monticello. So did relatives and close friends. It took thirty-seven house servants to keep Monticello running. No expenses were spared to provide the visitors with sumptuous meals, while their horses consumed staggering amounts of expensive feed. Meanwhile the prices Jefferson and other Virginia farmers could obtain for their crops remained low.

For a while Jefferson tried other ways to raise money. Perhaps his best-known experiment was his nailery. He launched it in the 1790s to make nails for rebuilding Monticello. It was hard, hot work, toiling with molten metal at temperatures between 600 and 700 degrees centigrade. About a dozen slave boys between the ages of ten and sixteen produced ten thousand nails a day. Jefferson rewarded the hardest workers with money and clothing, and sometimes disciplined those who hated the work and ran away. Not a few people have criticized him for forcing boys to labor so hard.

For a while Jefferson sold his surplus nails at a brisk and profitable pace. But by the time he left the presidency, cheaper English-made nails were on the market, and the nailery became a losing proposition. Similar bad luck dogged a flour mill that Jefferson tried to build on the nearby Rivanna River. It was destroyed in a storm and abandoned for want of funds to rebuild it.[24]

Jefferson worsened his financial burden by cosigning a large loan for one of his most devoted political followers. The man died bankrupt, and the entire sum was added to Jefferson's debt, which soon totaled over $100,000—more than two million dollars in modern money.[25] In his last years, Jefferson made desperate efforts to pay his more and more impatient creditors. He tried to sell some of his land but found no takers. A financial panic in 1819 had sent land prices plummeting everywhere. With cheaper lands available in the West, Virginia farms no longer seemed a good investment.

In desperation, Jefferson petitioned the legislature to permit him to raise money through a lottery. He hoped to make enough to pay his debts and leave a surplus for Martha and her children. The legislators at first demurred, claiming lotteries were immoral. But Jefferson's friends finally persuaded them to approve the venture. Alas, ticket sales in debt-ridden

Virginia were disappointing. Attempts by his grandson Thomas Jefferson Randolph to sell tickets in other states also failed.

Jefferson's family rallied around him. Maria's son, Francis Eppes, returned property Jefferson had deeded to him as part of his mother's inheritance. "You have been to me ever, an affectionate and tender father, and you will find me ever, a loving and devoted son," he wrote. But Francis and other relatives were engulfed by the economic collapse that overwhelmed so many Virginians.

XI

One of the saddest victims of the collapse was Thomas Mann Randolph. The ex-congressman sank into ever deeper debt. Randolph worsened things by plunging into inexplicable mood swings at crucial moments. He would harvest a bumper crop of wheat, then leave it in his barns or send his overseers to Richmond to sell it too late to catch the top of the market. Neighbors remarked that "no man made better crops than Colonel Randolph and no one sold his crops for worse prices."[26]

In 1819, Randolph ran for governor of Virginia and won, but his performance in office was awful. He quarreled with everyone—the legislature, his council, even the board of the University of Virginia. He tried to expand the powers of the governor, claiming he disdained to be a mere "signing clerk," and failed disastrously. He finally retreated to Monticello, where Martha prepared one of the "skylight" bedrooms in the dome room for him. The peace and quiet enabled him to get a grip on his ravaged nerves for a little while. But the years following this respite saw a final slide into financial bankruptcy and the total collapse of his self-esteem. His marriage to Martha also deteriorated; his black moods and outbursts of bad temper finally forced her to tell him she would no longer share a bedroom with him.

The emotional and financial agonies of his daughter and son-in-law added weight to Jefferson's own mountain of debt in the final year of his life. When he tried to console Randolph by offering to deed all his property to him, Randolph went berserk and accused Jefferson of being indifferent and coldhearted. He stormed out of Monticello and became a hermit in the only piece of property his creditors had left to him, a five-room cottage in North Milton, several miles away.[27]

XII

Threaded through these Job-like woes was the tragic story of Jefferson's oldest granddaughter, Anne. She married a man named Charles Bankhead, whose solution to Virginia's economic woes was alcohol. He abused and beat Anne, even when her horrified mother was present, and at one point stabbed her brother, Thomas Jefferson Randolph, when they exchanged insults on the street in Charlottesville. Anne finally fled back to Monticello and in the first months of 1826, died while her grandfather wept beside her bed.

Meanwhile, Jefferson's debts and the mounting impatience of his creditors made his gesture of assistance to Thomas Mann Randolph meaningless. When the lottery failed, Jefferson took to his bed, suffering from an acute form of diarrhea. He sensed (or perhaps wished) he was dying and wrote farewell letters to James Madison and other close friends. In the letter to Madison, he revealed his concern for his future fame. "You have been a pillar of support through life. Take care of me when dead," he wrote, "and be assured I will leave with you my last affections."[28]

On March 16, 1826, Jefferson made his will. In it he gave freedom to five slaves. Two were Madison and Eston Hemings, sons of Sally Hemings, who had been trained as carpenters. He requested that the legislature permit them to remain in the state. Freed slaves were required by law to leave Virginia, lest they use their freedom to incite rebellion. This enabled the two young men to continue to serve as assistants to Monticello's aging chief carpenter, John Hemings, who was also freed. Earlier in the 1820s, Jefferson had permitted Sally's two older children, Harriet and Beverly, to leave Monticello. The two other men freed in the will were also members of Elizabeth Hemings's family. But there was no mention of Sally Hemings.

Jefferson slipped slowly downward, his strength ebbing. His mind remained amazingly clear and firm. He corresponded with President John Quincy Adams about treaties of commerce that he had helped negotiate decades ago. In another letter he recalled in vivid detail his memories of Benedict Arnold's Virginia raid during his ill-fated governorship. He sent these recollections to Henry Lee, a son of the cavalry hero "Light Horse" Harry Lee, who was revising his father's memoirs of the Revolution. Another letter went to Ellen Randolph, who had married a New Englander, Joseph Coolidge. He also wrote witty and charming comments about suitors arriving at Monticello in pursuit of the younger granddaughters.

As the fiftieth anniversary of the Declaration of Independence approached, a renewed wave of admiration for Jefferson and his by now legendary document swept the nation. In a letter to James Madison, Jefferson said the Declaration was "the fundamental act of union in these states." Perpetuating its principles was "a holy purpose."[29] The mayor of Washington invited him to be the leading figure in a great celebration on July 4, 1826. Jefferson was too ill to travel, but he sent a memorable statement of the Declaration's meaning not merely for their own era but for all time.

XIII

On July 2, Jefferson invited his family to his bedside and said farewell to each of them individually. He told Martha he had left a gift for her in a dresser drawer. He urged each grandchild to "pursue virtue, be true and truthful." Eight-year-old George Wythe Randolph looked bewildered. Jefferson smiled gently at him. "George does not know what all this means," he said.

Perhaps in answer to a request by Martha, Jefferson said he would not object to meeting with the Reverend Frederick Hatch, pastor of the Episcopal church in Charlottesville. But the priest should understand they would only talk as neighbors. That was his gentle way of telling Martha that he had no fear of approaching death, nor did he feel he had committed moral failures—sins—for which he had to seek absolution.

Several times Jefferson told his grandson Thomas Jefferson Randolph, who was constantly at his bedside, that he hoped he would live until the Fourth of July. On July third he seemed to be drifting down into the darkness. His private secretary, Nicholas Trist, who had married his granddaughter Virginia, could not bear to watch his agony and told him that the Fourth had arrived. Jefferson ceased struggling for life, but he continued to breathe. About 7 p.m. he awoke and found his doctor, Robley Dunglison, beside his bed. He was puzzled by his continued presence and asked him, "Is it the Fourth?"

"It soon will be," the doctor said. Studying him, Dunglison predicted he would die in a few minutes. But Jefferson remained alive. At last the clock's hands passed midnight, and those keeping watch in the bedroom breathed a sigh of relief. To their amazement, Jefferson lived another twelve hours, dying at ten minutes before noon on the Fourth, his wish fulfilled.

John Adams had died on the same day in his home in Quincy, Massachusetts. Most people agreed with President John Quincy Adams, who wrote in his diary that it was fresh evidence that America had a special destiny in this world. Humbly, the president—and the nation—stood "in grateful and silent adoration before the Ruler of the Universe."

On the evening of the Fourth, as the church bells in Charlottesville tolled, Thomas Mann Randolph appeared at Monticello, supposedly to mourn his father-in-law. Noticing that Martha was not weeping, he began to taunt her, saying she was too coldhearted to shed a tear. He asked Dr. Dunglison to give Martha some sort of medicine that would produce evidence of grief. Thomas Jefferson Randolph lost his temper and accused his father of hating Jefferson and behaving abominably.[30]

Martha Jefferson Randolph fled this appalling scene. In her bedroom, she remembered the gift Jefferson had left for her. She opened her dresser drawer and found a poem:

A Deathbed Adieu from Th. J. to M.R.

Life's visions have vanished, its dreams are no more
Dear friend of my busom, why bathed in tears?
I go to my fathers, I welcome the shore
Which crowns all my hopes and buries my cares
Then farewell my dear, my loved daughter, adieu
The last pang of life is in parting with you!
Two seraphs await me long shrouded in death
I will bear them your love on my last parting breath.

The seraphs were Martha Wayles Jefferson and Maria Jefferson Eppes, those two exquisite women that fate had torn from Jefferson's life. The next day, he was buried beside Martha and Maria in the Monticello graveyard. On his gravestone he asked his family to place this inscription:

> HERE WAS BURIED THOMAS JEFFERSON
> AUTHOR OF THE
> DECLARATION
> OF
> AMERICAN INDEPENDENCE
> OF THE
> STATUTE OF VIRGINIA
> FOR
> RELIGIOUS FREEDOM
> AND FATHER OF THE
> UNIVERSITY OF VIRGINIA

Ignoring his numerous high offices, from governor to president, Jefferson chose to reiterate his commitment to freedom. Above him on the mountain he left another epitaph, Monticello, his vision of the purpose of this freedom, a place where head and heart, architecture and art and science joined hands in the pursuit of happiness. For much of the next two hundred and fifty years, generations of Americans accepted this vision— and the man who created it—as the epitome of all that was good and fine in America.

XIV

Six months after Jefferson's death, an auction took place at Monticello. It was advertised in the *Charlottesville Central Gazette* as the sale of "the whole of the residue of the personal estate of Thomas Jefferson, dec, consisting of 130 VALUABLE NEGROES, stock, crop &C Household and Kitchen furniture." The slaves, claimed the ad, were the "most valuable for their number ever offered at one time in the state of Virginia." There were also "valuable historical and portrait paintings" including a bust of Jefferson, and the polygraph, the copying instrument he used when he wrote letters, plus "various other articles curious and useful to men of business and private families."[31]

Martha Jefferson was not present at this ordeal. She had fled to Boston with her two youngest children to live with Ellen Randolph Coolidge. For five days, her unmarried children and the executor of the estate, Thomas Jefferson Randolph, watched as the mansion was virtually stripped bare of furnishings. Even more painful was the sale of the slaves. Many members of the Hemings extended family—sons and daughters and grandchildren of Elizabeth Hemings by fathers other than John Wayles—were sold to strangers. Years later, Thomas Jefferson Randolph remembered his anguish. "I had known all of them from childhood and had strong attachments to many," he said. "I was powerless to relieve them."[32] Fifty-three-year-old Sally Hemings was not among the sold. Martha Jefferson later freed her, using a device known as giving a slave his or her "time." Technically she remained a slave and was not required to leave the state like other freed slaves. She moved to Charlottesville, where she lived with her sons, Madison and Eston, until her death in 1835.

The sale did not come close to paying Thomas Jefferson's debts. Nor did the sale of the mansion itself, a few years later, for a pathetic $7,000. But Thomas Jefferson Randolph grimly vowed that he would repay his grandfather's debts "to the last copper" if it took the rest of his life. The effort required another twenty years of backbreaking toil on the plantation he had inherited from his bankrupt father, Edgehill, and the sacrifice of all comforts and luxuries. The big, burly grandson made this personal sacrifice to redeem the good name of the grandfather he loved.

Martha Jefferson Randolph returned from Massachusetts and did everything in her power to assist her son. At one point, she and her unmarried daughters helped him copy and edit the first collection of Jefferson's writings. They launched a school for young women that flourished for many years. Martha permitted friends to persuade the legislatures of South Carolina and Louisiana to send her gifts of $10,000, which she passed on to her son. In 1836 Martha died suddenly, apparently of a stroke, at the age of sixty-four. She was buried in Monticello's graveyard beside the father she had never ceased to love.[33]

IF JEFFERSON IS WRONG,
IS AMERICA WRONG?

*O*n November 1, 1998, the British science magazine *Nature* announced the imminent publication of an article titled "JEFFERSON FATHERED SLAVE'S LAST CHILD." The text stated that tests conducted by pathologist Dr. Eugene M. Foster revealed that the DNA of a descendant of Sally Hemings's youngest child, Eston Hemings, matched the DNA of a descendant of Thomas Jefferson's uncle, Field Jefferson. "The simplest and most probable explanations for our molecular findings," Dr. Foster wrote, "are that Thomas Jefferson was the father of Eston Hemings Jefferson . . ."[1]

A media explosion tore across America and the entire world. In an article in the *Washington Post,* David Murray, head of the Statistical Assessment Service in Washington, D.C., counted 295 editorial and news citations, 8 pieces in newsweeklies, and 31 broadcast transcripts. National Public Radio announced, "The proof is finally in . . . DNA testing has ended [the] debate." The *Des Moines Register* proclaimed Jefferson was an "adulterer on Mount Rushmore." The *New York Times* quoted a Jefferson scholar who said, "If people had accepted this story, he never would have become an American icon . . . The personification of America can't live 38 years with a black woman."[2]

II

Behind this uproar lay 250 years of debating Thomas Jefferson's reputation. His fame had continued to expand in the decades before the Civil War, but his public image became more complicated as Americans began to argue about slavery. Both the enemies and the defenders of the "peculiar institution" found support in his life. His writings were full of denunciations of slavery. But he also revealed grave doubts about African-Americans' intellectual abilities and was a strong advocate of states' rights in the ongoing argument about the power of the federal government. Jefferson feared a bloody race war if slavery were abolished instantly, as growing numbers of its northern critics, soon called abolitionists, demanded. In 1820, he lamented that southerners had "the wolf by the ears" and could not let him go. He saw in the growing disagreement "the [death] knell of the union."[3]

Some abolitionists revived James Thomson Callender's 1802 accusation about Jefferson's relationship with Sally Hemings. They saw it as proof of the moral degeneracy of slavery. "The best blood of Virginia flows in the veins of slaves, even the blood of Jefferson," they declared. Soon a story was circulating in anti-slavery circles that one of Sally Hemings's daughters had been sold in a New Orleans slave market for $1,000. According to Dr. Levi Gaylord of Sodus, New York, it was "attested to by a southern gentleman" who had witnessed the ghastly event. Next a novel, *Clotel, or the President's Daughter,* written by a fugitive slave and first published in France, created another sensation. The book opened with a slave auction in which Sally Hemings and two of her daughters were sold to the highest bidders.[4]

Visiting British writers such as Charles Dickens and Frances Trollope eagerly participated in the uproar. They told their readers about "hospitable orgies" at Monticello and claimed that Jefferson was the father of "unnumbered generations of slaves." They portrayed him sitting at his dinner table, waited on by a half dozen of his own black children. The British used these supposed facts to ridicule the idea that all men were created equal. Jefferson's sins became a weapon to blunt America's worldwide appeal as a universal democracy.[5]

III

In 1861, the Civil War erupted, killing 620,000 young Americans. Thomas Jefferson's reputation collapsed. He became linked in many northern minds with "The Slave Power" that they blamed for bringing this catastrophe upon the nation. In the South he was equally execrated because he was considered responsible for the antislavery crusade that had led to the region's defeat and desolation. For a while, it looked as if he were destined to be a dismissed and derided founding father.

Jefferson's fame was reborn in 1871 with the popularity of a book by his great-granddaughter, Sarah Nicholas Randolph, *The Domestic Life of Thomas Jefferson*. Randolph's introduction declared the seemingly modest purpose of the book: "I do not . . . write of Jefferson either as of the great man or as of the statesman. My object is only to give a faithful picture of him as he was in private life—to show that he was, as I have been taught to think of him by those who knew and loved him best, a beautiful domestic character."

The book barely mentioned politics. That contentious world was viewed, if at all, as an intrusion on the family's happiness. Randolph's goal was to give readers an appreciation of "the warmth of his [Jefferson's] affections, the elevation of his character, and the scrupulous fidelity with which he discharged the duties of every relation in his life." No public man's character "had been more foully assailed than Jefferson's," she continued. "And none so fully exposed to the public gaze, nor more fully vindicated."[6]

The Domestic Life was warmly reviewed in newspapers and magazines in the North and South. *The Nation,* a magazine hitherto given to damning Jefferson, praised it extravagantly. One of his iciest New England critics admitted that the man Randolph portrayed with so many convincing quotations from his letters was "entirely amiable and charming" and deserved to be "more mildly judged" than he had been in recent years.[7]

In 1874, James Parton's *Life of Thomas Jefferson* did even more to revive Jefferson's reputation. Parton insisted that the essence of Jefferson's character was love: "In every other quality and grace of human nature he has often been equaled, sometimes been excelled, but where has there ever been a lover so tender, so warm, so constant, as he? Love was his life. . . . He knew no satisfying joy, at any period of his life, except through his affections."

Parton also projected an image of Jefferson as a hero of American culture. At thirty-two he could "calculate an eclipse, survey an estate, tie an artery, plan an edifice, try a cause, break a horse, dance a minuet and play a violin." Totally carried away, Parton declared: "If Jefferson was wrong, America is wrong. If Jefferson was right, America is right." The biographer convinced himself—and hundreds of thousands of readers—that Thomas Jefferson and America were virtually one indissoluble entity.[8]

IV

In 1873, *The Pike County Republican* published a story about Thomas Jefferson that struck a very different note. The editor of this small Ohio newspaper was fifty-three-year-old Samuel F. Wetmore, an abolitionist who had been born in Ohio and had worked on several newspapers in the Midwest before launching the *Republican* in 1868. His friend and future son-in-law, Wells S. Jones, a Union brigadier general in the Civil War, owned 1,700 acres in Pike County and had announced his intention to make this slice of southern Ohio a Republican bastion. That proved to be a difficult task. Jones lost a run for the state senate in 1867, largely because he proposed giving African-Americans the right to vote in Ohio. The defeat led him to persuade—and probably finance—Wetmore to start the *Pike County Republican*. Helpful patronage came from Washington, D.C., where Republicans were in power under President UIysses S. Grant. They made Wetmore the postmaster in Waverly as well as a U.S. marshal.

By 1873, Wetmore was a worried man. The Republican Party was in trouble, both in Ohio and in the nation. President Grant had won the war as a consummate general, but he left a lot to be desired as a political leader. Washington, D.C., swirled with rumors of scandals in his administration. Worse, Grant had signed into law a bill raising the pay of the members of Congress and the Supreme Court. Angry Democrats and not a few worried Republicans denounced the "salary grab" as little more than theft. The Democrats of Ohio were especially vociferous.

This was the atmosphere in which Samuel F. Wetmore announced a series of articles about the ex-slaves who were now living in Pike County. He called the series "Life Among the Lowly." For readers in 1873, the title had an instant familiarity. The phrase was the subtitle of Harriet Beecher Stowe's bestselling novel, *Uncle Tom's Cabin*. Wetmore's great-grandmother had been a Stowe. He was also friendly with the family of Dr. Levi Gaylord, the man who had first published the fictitious story about a daughter of Thomas Jefferson being sold in New Orleans.[9]

Wetmore's choice to launch the "Lowly" series was Madison Hemings, Sally Hemings's son. The newsman began by admitting that his subject had not experienced many of the physical cruelties of slavery. But "we must say the system was cruel at best. To keep such a man in the condition of a slave, however well treated in other respects, was a sin of very deep dye . . . If he had been educated and given a chance in the world he would have shone out as a star of very great magnitude. But he was kept under, by his own father, an ex-president of the United States, and a man who penned the immortal declaration of independence which fully acknowledges the rights and equality of the human race!"

Wetmore described Madison as five feet ten inches tall, "sparely made, with sandy complexion and a mild grey eye." These details "accord[ed] very nearly with the description given of Thomas Jefferson, except that he was six feet one and a half inch in height." Thereafter, the story was told in Madison's words, ghostwritten by Wetmore. The heart of the brief narrative was his description of Sally Hemings's experience in Paris as nine-year-old Maria Jefferson's companion and nurse:

> *Their stay (my mother's and Maria's) was about eighteen months. But during that time my mother became Mr. Jefferson's concubine, and when he was called back home she was* enceinte *by him. He desired to bring my mother back to Virginia with him but she demurred. She was just beginning to understand the French language well, and in France she was free, while if she returned to Virginia she would be re-enslaved. So she refused to return with him. To induce her to do so he promised her extraordinary privileges, and made a solemn pledge that her children should be freed at the age of twenty-one years. In consequence of his promise, on which she implicitly relied, she returned with him to Virginia. Soon after their arrival, she gave birth to a child, of whom Thomas Jefferson was the father. It lived but a short time. She gave birth to four others and Jefferson was the father of all of them. Their names were Beverly, Harriet, Madison (myself) and Eston—three sons and one daughter. We all became free agreeably to the treaty entered into by our parents before we were born . . .*

Madison went on to tell how he was "named . . . by the wife of James Madison, who was afterward President of the United States. Mrs. Madison happened to be at Monticello at the time of my birth, and begged the privilege of naming me, promising my mother a fine present for the honor. She consented and Mrs. Madison dubbed me by the name I now acknowledge, but like many promises of white folks to the slaves she never gave my mother anything."

Madison said he learned about Jefferson's great fame only after he died. "About his own home he was the quietest of men. He was hardly ever known to get angry." He was "uniformly kind to all about him." But he "was not in the habit of showing partiality or fatherly affection to us [slave] children. We were the only children of his by a slave woman." Toward his white grandchildren, however, he was "very affectionate."[10]

Madison's story was brought to the attention of James Parton, who was publishing installments of his forthcoming biography of Jefferson in the *Atlantic Monthly*. In July 1873, Parton discussed the story of "Dusky Sally" and stated politely that Madison Hemings was "misinformed." His real father was a "near relation" of Mr. Jefferson, "who need not be named." Parton had been told by Henry S. Randall, author of a biography of Jefferson published in 1858, that the father of Sally's children was Peter Carr, the son of Dabney Carr, Jefferson's brother-in-law. Peter and his younger brother Samuel had been raised at Monticello. Randall's source was Thomas Jefferson Randolph, Jefferson's grandson.[11]

John A. Jones, the editor of Pike County's Democratic newspaper, *The Waverly Watchman*, also dismissed Madison Hemings's story, with a minimum of politeness: "Hemings, or rather Wetmore, gives a very truthful account of the public and private life of the Jefferson family; but this no doubt, was condensed from one of the numerous lives of Jefferson which can be found in any well regulated family library . . . There are at least fifty Negroes in this county who lay claim to illustrious parentage . . . They are not to be blamed for making these assertions. It sounds much better for the mother to tell her offspring that 'master' is their father . . ."[12]

V

Eight months later, Wetmore published another installment of his "Life Among the Lowly" series—an interview with Israel Jefferson, also an ex-slave from Monticello. By this time, the political sky was darkening for the Republicans. In September, the American economy had collapsed and the Panic of 1873 had plunged the nation into a severe depression. Ohio voters elected William "Foghorn" Allen as governor, the first Democrat in twenty years, and gave the party a majority in the state legislature. Not a single Republican was elected from Pike County. Wetmore's newspaper crusade was a dismal failure.

Israel told Wetmore that after Jefferson's death, he was sold to a neighbor. Later he married a free mulatto woman who inspired him to purchase his freedom from his new master, to be paid over several years. In this process Israel had to take a last name, and chose Jefferson because "it would give me more dignity to be called after so eminent a man."

Israel recalled how he participated in the "exciting events attending the preparations of Mr. Jefferson and other members of his family on their removal to Washington DC" in 1800, when he was elected president. Four years later, Israel started working as a waiter at Jefferson's table and claimed that thereafter "the private life of Mr. Jefferson was very familiar to me." For fourteen years, he had "made the fire in his bedroom and his private chamber, cleaned his office, dusted his books, run . . . errands and attended him at home." He often escorted important visitors into Jefferson's chamber. Israel said that "Sally Hemmings" (Wetmore's spelling) was "employed as [Jefferson's] chamber maid." He [Jefferson] was "on the most intimate terms with her. . . . in fact, she was his concubine." Based on his "intimacy with both parties," Israel confirmed that Madison Hemmings [*sic*] was "the natural son of Mr. Jefferson, the author of the Declaration of Independence, and that his brothers Beverly, Eston and sister Harriet are of the same parentage."[13]

VI

Sometime in 1874, Thomas Jefferson Randolph received in the mail a copy of Israel Jefferson's recollections in the *Pike County Republican*. The founder's grandson was still living on his farm, Edgehill, four miles from Monticello. Randolph wrote a six-page letter responding to Israel's claims. According to Thomas Jefferson's Farm Book, a record, Randolph noted acerbically, "in Mr. J's handwriting," Israel was born on December 28, 1800. Mr. Jefferson left for Washington on December 1, 1800. Israel was describing as his earliest recollection something that happened twenty-eight days before he was born.

As for Israel becoming a waiter at Jefferson's table in 1804 (when he was four years old), Randolph acidly reported that from 1801 to 1809 Israel and his entire family were on the list of slaves leased to a Mr. Craven, who had a farm some distance from Monticello. Again he was quoting from the Farm Book, "in Mr. J's writing." A record of February 10, 1810, places Israel among Jefferson's farm workers, not at the Monticello house. When he went to work in the house (at an unspecified future date) he labored as a "scullion" in the kitchen. It was unlikely that he would know anything about the private life of Mr. Jefferson. "Israel was never employed in any post of trust or confidence about the house at Monticello."

Another reason why Randolph rejected Israel's confirmation of Madison Hemings's story was the living arrangements inside Monticello: "Mr. Jefferson and his daughter with her large family occupied the same wing of the building. The private access to their apartments was contiguous." There was no possibility of Mr. Jefferson conducting a clandestine love affair with Sally Hemings with any hope of secrecy. That was why "every member of this family repelled with indignation this calumny."

Turning to the motive for Wetmore's version of Israel's and Madison Hemings's stories, Randolph asked, "Can it be other than the necessity which the[se] writers [northern abolitionists] feel to pander to that morbid hatred of the southern white man which devours with obscene malignity every calumny or absurdity which can blacken or degrade his character?"

Thomas Jefferson Randolph never mailed this letter. Further study of the *Pike County Republican* probably convinced him it would be a waste of a stamp. The letter was found decades later in the files of the University of Virginia library.[14]

VII

Madison Hemings's Wetmore-ghosted story and Israel Jefferson's dubious confirmation of it vanished from the public mind. The Democratic Party, struggling to escape the stigma of favoring slavery and failing to give more than half-hearted support to the Civil War, turned to Thomas Jefferson as their savior. William Jennings Bryan, the orator who virtually took over the party in 1896 and was nominated for president three times, constantly invoked his name. In the 1920s, Franklin D. Roosevelt became convinced that he and Jefferson shared a political destiny. FDR participated vigorously in a fundraising campaign to purchase Monticello and make it a national shrine. The Thomas Jefferson Memorial Foundation took possession of Monticello on July 5, 1926, the day after the national celebration of the 150th anniversary of the Declaration of Independence. Roosevelt joined the foundation's board and remained a member to the end of his life.[15]

In 1929, the stock market crashed and FDR emerged as the leader of the Democratic Party and the spokesman of Jefferson's ideals.[16] He ordered

White House aides to make sure that a wreath was laid on Jefferson's grave at Monticello every year during his presidency. He constantly quoted Jefferson in his speeches. In 1938, the mint issued the Jefferson nickel and the U.S. Post Office issued the Jefferson three-cent stamp. On April 13, 1943, the two hundredth anniversary of the man from Monticello's birth, President Roosevelt dedicated the Thomas Jefferson Memorial on the edge of the Tidal Basin in Washington, D.C. World War II made the occasion doubly meaningful. "Today," FDR said, "in the midst of a great war for freedom, we dedicate a shrine to freedom." It was, he declared "a debt long overdue."

Jefferson took his place in the ultimate American pantheon, within sight of George Washington's soaring monument and Abraham Lincoln's brooding seated statue. It began to look more and more like James Parton was right. Jefferson and America were one and the same glorious spiritual entity.

VIII

Thirty years later, Samuel F. Wetmore's ghostwritten account of Madison Hemings's recollections experienced a rebirth. The historical currents that had levitated Thomas Jefferson's reputation for more than a century underwent a drastic reversal in the aftermath of Vietnam and Watergate. For many people, America became a flawed superpower and Jefferson, the symbol of her greatness, no longer merited unquestioning respect. A series of biographers, novelists, and movie producers accepted Madison's story as true and portrayed Jefferson as Sally Hemings's lover, while a chorus of historians insisted they were wrong. Readers interested in this evolution will find a detailed account of it in the Appendix.

In 1997, historian Joseph Ellis summed up the prevailing view of the scholarly community: "Short of digging up Jefferson and doing DNA testing on him and Hemings descendants," they had come as close to the truth as the available evidence allowed. The stage was set for the media explosion.

IX

At a Charlottesville, Virginia, dinner party in 1996, wealthy Winifred Bennett asked Dr. Eugene Foster whether DNA could be used to resolve the uncertainty surrounding Thomas Jefferson and Sally Hemings. Foster had recently retired after many years as a professor of pathology at Tufts University School of Medicine. He knew that scientists had made large strides in the science of genetics. One of the breakthroughs was the identification of individual Y chromosomes in male DNA. Over generations, these tiny entities develop distinctive mutations, which become the genetic hallmarks of a particular family. If an individual's Y chromosomes matched those of another individual, the chances were good that they shared a common ancestor.[17]

The procedure resembles seeking an exact match from DNA found in blood or other body substances to decide paternity lawsuits and convict criminals, especially sex offenders. In such cases the matches or mismatches are virtually unchallengeable in court. The odds in favor of certainty are well over a million to one. But DNA identification of an ancestor through Y chromosomes does not come close to such exactitude. The most it can deliver is a probability. This crucial point was ignored in the media explosion.

Dr. Foster expressed an interest in obtaining DNA samples of Jefferson and Hemings descendants. Mrs. Bennett said she would pay the costs. Since Thomas Jefferson had no sons, it was necessary to find other male Jeffersons with the family's chromosomes. Foster contacted Herbert Barger, a leading member of the Jefferson Family Association, who supplied him with phone numbers for seven descendants of Field Jefferson, Thomas Jefferson's uncle. One of them agreed to give samples of his blood.

Next came the task of finding male descendants of Sally Hemings. Barger suggested contacting the Thomas Woodson Family Association, which had 1,400 members who claimed descent from Jefferson through their ancestor, the young slave identified by James Thomson Callender in his 1802 exposé as resembling Jefferson so closely, sarcastic neighbors called him "Master Tom." (The Woodson name came from a later owner.) Barger also found descendants of Peter Carr and his brother Samuel; the latter had been named by a Jefferson granddaughter as Sally's lover. Next, Barger helped Foster find a descendant of Eston Hemings. Unfortunately, there were no known male descendants of Madison Hemings.

Dr. Foster flew to England with the blood samples and had them analyzed by British DNA specialists. He summarized the results in his brief article in *Nature*. But Foster did not write the article's headline: JEFFERSON FATHERED SLAVE'S LAST CHILD. Those words, chosen by *Nature*'s editor, triggered the media explosion.

Few people bothered to evaluate the significance of two additional conclusions from the DNA tests. The Y chromosomes of five descendants of Thomas Woodson failed to match the Jefferson DNA. Did this mean "Master Tom" had been fathered by someone else? The DNA of the descendants of those prime suspects, the Carr brothers, also failed to match the Jefferson DNA. At first glance, this failure seemed to refute the assertions of Thomas Jefferson Randolph and other members of the Jefferson family. But this conclusion would turn out to need further evaluation.[18]

X

On November 9, 1998, four days after the *Nature* article appeared in print, Dr. Foster published a letter in the *New York Times*. Earnestly, with a hint of muted indignation, he stated that "the genetic findings my collaborators and I reported in the scientific journal *Nature* do not prove that Thomas Jefferson was the father of one of Sally Hemings's children. We never made that claim." Apparently, Dr. Foster was laying heretofore invisible stress on the word "prove." He insisted that he had repeatedly said before the tests began that they would "not prove anything conclusively." All he ever hoped to do was provide some "objective evidence that would bear on the controversy." He and his fellow researchers "had not changed our position."[19]

Three days before Dr. Foster's letter appeared, Thomas B. Moore, a lawyer with a wide background in medical litigation, wrote an even more critical letter to the *New York Times*. On the basis of the evidence Foster presented, Moore declared, "no court of law would hold that Thomas Jefferson had a child by Sally Hemings." The most Foster's evidence could prove, Moore maintained, was that "sometime over the last 300 years or so, a descendant of Jefferson's grandfather had a relationship that produced a male child who is an ancestor of one of the living and tested male descendants of Sally Hemings." This could have happened "in the 17th, 18th, 19th or 20th centuries." Foster began his letter to the *Times* by agreeing with Moore. No one seemed to realize it, but James Thomson Callender's 1802 story had entered the ambiguous wonderland of statistical probability.[20]

XI

At a November 1, 1998, press conference, Daniel P. Jordan, president of The Thomas Jefferson Memorial Foundation, said that he had seen Dr. Foster's article only "forty-eight hours ago." The foundation would need "more time to evaluate it carefully." Fourteen months later, on January 26, 2000, the foundation issued a report by a research committee that declared Thomas Jefferson was the father, not only of Eston Hemings but of all of Sally Hemings's children. They based their conclusion on Madison Hemings's story and on the fact that Jefferson was at Monticello when Sally conceived each of her children. The committee bolstered their conclusion with a statistical study by staff archaeologist Fraser D. Neiman that concluded the probability of Jefferson's guilt was a near certainty—99 percent.

The committee admitted that "many aspects of this likely relationship remain unclear." The nature of the relationship, the longevity of Sally's first child, and the identity of Thomas Woodson were among the mysteries. Finally, "the implications of the relationship between Sally Hemings and Thomas Jefferson should be explored to enrich the understanding and interpretation of Jefferson and the entire Monticello community." One of these implications soon became apparent. President Daniel P. Jordan announced that the word "Memorial" was being dropped from the foundation's title. Apparently, the foundation no longer thought Thomas Jefferson was worthy of being "memorialized" by them—and presumably by the American people.[21]

XII

Behind the scenes, an angry confrontation was taking place at Monticello. One member of the research committee, Dr. White McKenzie Wallenborn, retired professor of clinical medicine at the University of Virginia, had written a minority report, disagreeing with the majority conclusion. Dr. Wallenborn's opinion was not mentioned in the press release announcing Jefferson's guilt. This omission led to some heated exchanges between Wallenborn and Daniel Jordan. An obviously reluctant Jordan finally released the report in April 2000, two months after the majority report.

With it came a fierce rebuttal from Lucia C. Stanton, Monticello's senior research historian.

The lone dissenter admitted there was "significant historical evidence" that Jefferson could be the father of Eston Hemings." But Dr. Wallenborn argued there was "significant historical evidence of equal stature" that indicates Jefferson was not Eston's father, and was also not the father of Sally Hemings's other children. Wallenborn maintained that the Carr brothers were by no means eliminated by the DNA test on Eston Hemings's descendant. One of them could still have been Sally's lover and fathered some or all of her other children. He noted that the Carrs had been identified as Sally's lovers not only by Jefferson family members but by Monticello overseer Edmund Bacon. As for the statistical study, Wallenborn dismissed it because it lacked information on Sally Hemings's whereabouts at the times of her conceptions and where Randolph Jefferson, Thomas Jefferson's younger brother, and other males with Jefferson DNA were at these times.

Lucia C. Stanton refuted Wallenborn's arguments in a style that radiated contempt. Her four-page statement was organized under headings that dismissed each argument before it was discussed: **1. Jefferson denied the relationship (and by implication, Jefferson would not lie). 2. Edmund Bacon denied the relationship (and by implication, Bacon would not lie). 3. Thomas Jefferson Randolph denied the relationship (and by implication, Thomas Jefferson Randolph would not lie).** Stanton is a respected scholar who has written a fascinating book, *Free Some Day,* about the lives of Monticello's slaves. The tone of her rebuttal is a good example of the overheated atmosphere that pervaded Monticello at this time.[22]

XIII

Later in 2000, CBS Television ran a four-hour miniseries starring Carmen Ejogo as Sally Hemings, Diahann Carroll as Sally's mother, Elizabeth, and Mario Van Peebles as Sally's brother, James Hemings. The film was written and co-executive produced by former actress Tina Andrews. "It's a love story," insisted Ms. Andrews in an interview. "The fact that they were together 40 years and remained so despite extraordinary circumstances makes me want to believe that there was some tenderness and emotion involved." Ms. Andrews and her co-producers cited the Thomas Jefferson Foundation's findings as the basis for their drama.[23]

In the same year PBS Television's investigative show *Frontline* produced a documentary, "Jefferson's Blood," that explored the controversy. Although there were occasional comments that the DNA findings were "not definitive," most of the participants assumed Jefferson's guilt. "Blood tests all but confirmed" went one statement. "DNA subjected this great man to a fall" was another remark.

At one point, a descendant of the Woodson family dismissed the DNA findings, which disproved Jefferson's role in his ancestor's birth. He insisted his family's oral history was true—that Jefferson had fathered "Master Tom" in Paris with Sally Hemings. He added that today Jefferson would be convicted of the rape of a child. The show distributed print interviews with principal witnesses, such as Dr. Foster, who reversed himself and said, "It would be possible, but highly, highly, highly highly improbable" that Jefferson was not the father of Eston Hemings. His words reflected the importance of Fraser Neiman's follow-up statistical study in confirming Jefferson's paternity.[24]

XIV

The Thomas Jefferson Heritage Society was created by Jefferson family descendants and others who disagreed with the Thomas Jefferson Foundation's conclusions. In 2000, the TJHS played a leading role in convening a Scholars Commission of thirteen historians, many of them authors of books on Jefferson. After a year of study and fifteen hours of face-to-face discussions, they concluded that while reasonable people could differ on the question, they found no convincing evidence of Jefferson's paternity, either of Eston Hemings or of Sally Hemings's other children.

The Scholars Commission hoped to garner major publicity for their five-hundred-page report. It was released to the public at the National Press Club in Washington, D.C., on April 12, 2001, the eve of Jefferson's birthday. They were more than a little disappointed. Another meeting a few blocks away won far more media attention. Hemings family descendants and some Jefferson family members who sided with them met at the White House with President George W. Bush.

Backers of the Scholars Commission angrily maintained that the meeting was arranged by members of the Thomas Jefferson Foundation board. President Bush knew nothing about the Scholars Commission press conference and saw no reason why he should not welcome the Hemings descendants and their Jeffersonian friends. The episode suggested the dispute about Sally Hemings was becoming a publicity war, aimed at controlling public opinion. The truth seemed almost—but not quite—irrelevant.

XV

What seemed to journalists and historians probabilities strong enough to be called certainties in 1998–2000 have slowly been eroded by doubts. After reviewing the report of the Scholars Commission, *American Heritage* magazine concluded: "whatever one's views, it is hard to deny that honorable people can and do disagree about Jefferson and Hemings . . . It's important for the public to realize that the purported Jefferson-Hemings liaison remains a disputed possibility, not an established fact."[25]

On February 24, 2003, the Thomas Jefferson Foundation revised their statement about Sally Hemings. They admitted that the evidence for a relationship between her and Jefferson was "not definitive" and "the complete story may never be known." The Foundation encouraged visitors to Monticello and their website "to make up their own minds as to the true nature of the relationship, based on what evidence does exist." This was close to a reversal of their previous statement: *The Thomas Jefferson Foundation stands by its original findings—that the weight of evidence suggests that Jefferson probably was the father of Eston Hemings and perhaps the father of all of Sally Hemings' children.* The foundation's new stance was—and still is—remarkably close to the one urged by Dr. White McKenzie Wallenborn.[26]

The mystery of who fathered "Master Tom" Woodson has had a growing impact on the believability of the pro-paternity argument. In an article published not long after the DNA tests, Michele Cooley-Quille, a Thomas Woodson descendant, described in impressive detail the history of her family, which includes distinguished people in every generation. Ms. Cooley-Quille is a clinical psychologist. Dismissing the DNA conclusions, she asked, "From what should the tapestry of history be woven? Hairy threads of DNA? Stories told? Or words written?"[27]

This writer discussed the Woodson conundrum with Dr. Kenneth Kidd, a Yale Medical School geneticist, who said it was possible that a male with different Y chromosomes had intruded into the Woodson family line at some point in its history and the Woodson volunteers from whom DNA samples were taken had descended from him. Dr. Kidd cited a well-known genetic motto, "the father is always uncertain." The dictum adds weight to Thomas Moore's contention (seconded by Dr. Foster) that the Jefferson DNA of Eston Hemings's descendant could have come from anyone who had acquired the same Y chromosomes in the decades before Jefferson's death or in the two and a half centuries since his demise.

While one sympathizes with the Woodsons' desire to believe their oral tradition, the answer to Cooley-Quille's large question about the tapestry of history would seem to be complexity. History is written from scientific, written, and oral data. The key criterion for achieving certainty is evidence that is verifiable. Here oral history falls short, especially when it is confused with oral tradition.

Oral history is collected by trained interviewers and is an important part of today's historical profession. But it has recognized limitations. The human memory is a very unreliable recording instrument. Oral tradition has far more serious limitations. Its unreliability is inevitable as it travels down the generations. It remains a valuable part of historical memory. But in a court of law it would be banned as hearsay evidence. That makes it hard, if not impossible, to see what role oral tradition can play in proving Thomas Jefferson's paternity.

Recently, Harvard professor Henry Louis Gates, one of the nation's leading African-American historians, published an article in which he described his family's oral tradition that they were the descendants of former slave Jane Gates and her owner, Samuel Brady. Professor Gates set out to "prove or disprove" the story. He found white descendants of Samuel Brady who gave him blood samples for DNA testing. He compared the DNA results with DNA-tested blood from his black relatives—and was amazed to discover "the tests established without a doubt that Brady was not the father of Jane Gates' children." One of his relatives dismissed his findings. "I've been a Brady eighty-nine years and I'm still a Brady," she told him.[28] Though Mr. Gates does not do so, his relative could cite Dr. Kidd's motto, "the father is always unknown," and argue that an interloper in the family line has disrupted the descent of Samuel Brady's DNA. The story testifies to the unreliability of both oral tradition and DNA evidence.

XVI

In 2008, two biostatisticians, William Blackwelder and David Douglas, found grievous fault with archaeologist Fraser Neiman's statistical study that declared Jefferson's guilt a 99-percent certainty. Neiman used a sampling method called Monte Carlo, which is often used by businessmen to evaluate investments. Another version is used by insurance companies. But experts warn that Monte Carlo has serious flaws. People put too high a probability on outcomes produced by the method. A whole industry called AIE, Applied Information Economics, has been developed to train Monte Carlo practitioners to develop more realistic probabilities.

Blackwelder and Douglas published their critique on a website to invite further discussion. The man who asked them to undertake this task is Steven T. Corneliussen, a science writer who works with physicists in Virginia. Blackwelder is a biostatistical consultant at the National Institutes of Health; Douglas is a physicist and senior scientist at the Thomas Jefferson National Accelerator Facility in Newport News. Statistics and probability theory and computer simulations are Douglas's specialty. All three men were troubled by what they saw as a serious misuse of science in Neiman's study.

Blackwelder condemned Neiman's conclusion, that "doubt about Jefferson's paternity can no longer be reasonably sustained." The veteran biostatistician called this "a gross misinterpretation" of the study. Douglas's criticism was equally harsh. He found that Neiman miscounted the probable conception "windows" for Sally Heming's pregnancies. (The term refers to the interval between the end of a menstrual cycle and the start of another one during which a woman may be fertile.) In four of her six pregnancies, Sally could have conceived while Jefferson was absent from Monticello. Douglas even found a distinct possibility that Jefferson was absent from Monticello at the time of Eston Hemings's conception—the only child to which DNA has linked a Jefferson. With Blackwelder's full agreement, Douglas concluded the probability of Jefferson's presence at all six conceptions was less than 50 percent. Douglas also faulted Neiman's unscientific presentation, which omitted crucial details that would enable other statisticians to replicate the study.[29]

XVII

These surges of uncertainty have led to increasing doubts about the reliability of Samuel F. Wetmore's ghosted narrative of Madison Hemings's life in the *Pike County Republican*. It seems only fair to apply the same standard of proof to Wetmore's journalism that Jefferson paternity advocates have applied to the testimony of Thomas Jefferson Randolph and others who have denied Jefferson's fatherhood. They claim these people's affection and loyalty to Thomas Jefferson prompted them to lie about his relationship with Sally Hemings.

What was Wetmore's motivation in claiming Jefferson's paternity? The answer: hatred and contempt for Thomas Jefferson and the Democratic Party. Viewed in this light, the Wetmore-Hemings story begins to look more and more like a recycling of James Thomson Callender's vindictive 1802 assault, with Wetmore in control.

Recent research has added strength to this suspicion. In his opening sentences, Wetmore claims that Madison Hemings was five feet ten and one half inches tall, giving him a strong resemblance to Jefferson. In the Virginia census of 1833, Madison was measured as five feet seven.[30] Why did Madison Hemings tolerate this distortion? It seems likely that Wetmore had convinced him it was important to "improve" his story in various ways to make it more appealing to readers. If Hemings were willing to agree to let Wetmore misstate his size, would he not be equally ready to say that his mother had only one lover, Thomas Jefferson? This would correct the cruel accusation that James Thomson Callender had flung at Sally: she was "a slut as common as the pavement." It would make an innocent Sally another victim of Thomas Jefferson, the uncaring slave owner.

Madison Hemings was the only one of Sally Hemings's children who never passed for white. In census after census, he was listed as a Negro. That makes a reader dubious about his "sandy complexion," which supposedly added to his resemblance to Jefferson. Madison's brother Beverly and sister Harriet left Monticello with Jefferson's permission in the early 1820s and vanished into the white world. His brother Eston moved from Ohio to Wisconsin and passed there. The nasty remark Wetmore has Madison make about "white persons" for Dolley Madison's failure to give his mother a gift after his birth suggests he had few if any warm feelings for whites, and especially for Thomas Jefferson, the man who had enslaved him. These feelings were probably exacerbated by Madison's experience in Pike County, Ohio. The largely Democratic citizens of Waverly, the county seat, refused to permit blacks to live within the town limits.

Wetmore's presence as the narrator is visible in other ways. At one point, Hemings tells how familiar he was with Martha Jefferson Randolph's children. He claims they taught him to read—and he reels off the names of the eleven who lived to adulthood: *Ann, Thomas Jefferson, Ellen, Cornelia, Virginia, Mary, James, Benjamin Franklin, Lewis Madison, Septemia, and George Wythe.* Madison had not seen any of these people for forty-seven years. This is surely Wetmore the ghostwriter at work, with *The Domestic Life of Thomas Jefferson* or some other biography of Jefferson on the desk beside his manuscript.

In the same category is Madison's astonishing knowledge of Jefferson's early life. "Thomas Jefferson, the author of the Declaration of Independence, was educated at William and Mary College, which had its seat at Williamsburg. He afterwards studied law with Geo. Wythe and practiced law at the bar of the general courts of the Colony. He was afterwards elected a member of the provincial legislature from Albemarle county." Jefferson was sixty-two years old when Madison was born. This passage is almost certainly Samuel F. Wetmore copying word for word from a popular biography.

Equally dubious is Hemings's claim that he never knew how famous Jefferson was until after his death. Even granting that Jefferson left the presidency when Madison was a toddler, wouldn't the teenage Hemings notice and wonder why hundreds of visitors came to Monticello each year to pay homage to Jefferson's fame? Again, this is Wetmore the ghostwriter at work. He is trying to make readers sorry for Hemings, whose famous "father" paid so little attention to him. Once more we see that contempt for an unfeeling Thomas Jefferson is the desired outcome of Madison's story.

Another dubious statement is Hemings's claim that Jefferson was extraordinarily healthy: "Till within three weeks of his death he was hale and hearty and at the age of 82 years walked erect and with a stately tread. I am now 68, and I well remember that he was a much smarter man, physically, at that age than I am." When Jefferson was sixty-eight, Madison was six years old. That puts this recollection in the same class as Israel Jefferson's story about waiting on Jefferson's table at the age of four. Throughout his later life, Jefferson suffered from crippling attacks of rheumatism, which he frequently mentioned in his letters. In 1794, 1797, 1802, 1806, 1811, 1813, and 1819 agonizing pain in his back, hips, and thighs often kept him from walking.[31]

Several historians have pointed out with not a little sarcasm that records make it clear Dolley Madison was not at Sally Hemings's bedside when Madison was born, as Wetmore-Hemings claim in their narrative. Madison was born on January 19, 1805. On this date Dolley was in Washington, D.C., with her husband, James Madison, Jefferson's secretary of state. Paternity proponents have come up with a theoretical answer—Dolley made a documented visit to Monticello in the fall, and that was when she promised Sally a gift if she named the child after her husband. Underlying the entire story is the assumption that the deeply religious Dolley and her husband were aware of—and approved of—Jefferson's relationship with Sally. There is not a shred of proof for this assertion. The only documented evidence is a James Madison statement we have already seen: he said Callender's accusation was "incredible."[32]

Further doubts about the reliability of Madison Hemings's narrative arise when we discover how many details can be traced to Callender's original account, published three years before Madison was born. Madison traced his mother's birth back to Elizabeth Hemings's relationship with John Wayles. Callender misspelled the name as "Wales" and so did Wetmore-Hemings. Wayles is called a "Welchman", in both accounts, when his background was English. These errors are of no great import, but they tell us who was in control of the story. Clearly, it was not Madison Hemings; it was Wetmore writing with copies of Callender's articles or quotations from them on his desk. This is another reason to suspect that Madison's story is Callender with the window dressing of a first-person narrative.

Newspaper ethics in the nineteenth century did not put a high value on accuracy. "Faking" a story (embellishing it or inventing it wholesale) was accepted journalistic practice. Indifference to facts was virtually universal, as the example of the newspaper reporting on George Washington's supposed father-son relationship to Thomas Posey make dolorously clear. When we consider Israel Jefferson's story and its veritable tissue of lies (documented by Thomas Jefferson Randolph), the evidence strongly suggests Samuel F. Wetmore was a practitioner of the shoddy art of faking the truth.[33]

Two years after he wrote the Madison Hemings article, Wetmore was fired as postmaster of Waverly for stealing $155 from his accounts. He resigned as editor of the *Pike County Republican* and vanished from the local scene. This was not the first time Wetmore revealed a dishonest streak. In 1871, a man sued him for failing to repay a debt of $362—about $5,000 in today's money. Toward the end of the Civil War, on March 31, 1865, when Wetmore was forty-four, he joined the army and received a $100 signing bonus, then complained of "rheumatism" and was mustered out thirty-nine days later. He never repaid the balance he owed on his $150 clothing allowance or his $100 munitions allowance. A few months before Wetmore disappeared, he was sued in the Waverly court on behalf of an infant, Adaline Rose. Unfortunately, the archival records of this lawsuit have been lost. But it has some of the earmarks of a paternity suit. Could the man who wrote Madison's Hemings's story be guilty of the same indiscretion for which he pilloried Thomas Jefferson? Such ironies are not uncommon in history.[34]

Wetmore's brother Josiah took over the newspaper and issued a statement that betrayed not a little agitation. At one point, he claimed Samuel was "severely ill." At another point he admitted that Samuel had "given rise to scandal, by withdrawing without consultation." The new editor added that Samuel had sent a message "from a distant city, hinting at a continued journey," which suggested "a prolonged absence." So great was the turmoil inside the Wetmore family, no one seemed to notice how incongruous it was to claim someone was severely ill and then report he had left his wife and three children to flee to a distant city.

Samuel Wetmore's absence turned out to be permanent. No one in Waverly ever heard from him again. Although the *Pike County Republican* frantically eulogized him as "a man who never used tobacco or any other narcotic in any form," it seems likely that an erratic character had come to a bad end.[35]

XVIII

The difference between the short, tan-skinned Madison and his tall younger brother, Eston, who had reddish hair and a striking resemblance to the Jeffersons, suggests that the two men had different fathers. That casts further doubt on the Wetmore-Hemings assertion that Sally Hemings never had sexual relations with anyone except Jefferson. Further fueling this doubt are the recollections of a French visitor, Comte de Volney, who spent three weeks at Monticello in 1796. He was amazed by the atmosphere of sexual freedom. "Women and girls . . . do not have any censure of manners," he wrote in his journal, "living freely with the white workmen of the country or hired Europeans, Germans, Irishmen and others . . ." Another French visitor made similar observations around the same time.[36]

The most important person in Sally Hemings's life almost certainly was her mother, Elizabeth Hemings. She had six children by John Wayles and eight more by other fathers, some white, some black, after she came to Monticello. This makes it seem likely—or at least plausible—that Sally, too, had several lovers. The relaxed sexual atmosphere at Monticello also reduces the significance of various slave children, from Callender's "Master Tom" (Woodson) to Eston Hemings, resembling Thomas Jefferson. The Thomas Jefferson Foundation made this one of the chief points in their 2000 conviction of Jefferson—now withdrawn. If Peter Carr fathered some of Sally's early children, they might well resemble Jefferson. He or his brother Samuel might also have fathered children by other Hemings women. As we shall soon see, there is another Jefferson relative who also might have enjoyed Monticello's relaxed sexual mores.[37]

XIX

Dr. Walllenborn had a point when he called for a reexamination of the role of Peter Carr in Sally's life. The two men who named him as Sally's lover, Thomas Jefferson Randolph and overseer Edmund Bacon, admittedly had motives to shade or deny the truth. But they did not testify at the same time or in the same place. On the contrary, their accounts are separated by many years and several hundred miles. Randolph spoke at his Edgehill farm, virtually in the shadow of Monticello, in the mid-1850s; Bacon talked in Kentucky after the Civil War had begun. There is no evidence that either was aware of what the other man had said. This lends a modicum of credibility to their words.

One historian has proposed a scenario that would explain why Sally might have told her son Madison that Jefferson was his father. Peter Carr married a Baltimore heiress in 1797, after Sally had conceived but not yet given birth to her second child. If Sally were his mistress, the situation may have been charged with explosive jealousy. Bitterness and anger may have led her to turn to other lovers and toward the end of her life to say Thomas Jefferson was their father, as an act of revenge against Peter Carr. This scenario makes Sally a woman with a broken heart—a victim not only of the monstrous injustices of the slave system but the duplicity of a faithless lover.

Peter Carr's heart, too, may have been damaged. Thomas Jefferson had regarded him as the son he never had and expended a great deal of time and money to educate him, hoping he would become a national leader. But his political career faltered and expired early, and he bumbled through the rest of his life like a man in a daze. Thomas Jefferson Randolph reported hearing Carr express his shame over his affair with Sally and the embarrassment it had caused Jefferson. This underscores the possibility that he felt lifelong regret.[38]

XX

In recent years there has been a growing inclination among historians to take Randolph Jefferson seriously as a potential lover of Sally Hemings. He lived twenty miles from Monticello and often visited his famous brother. Randolph was twelve years younger than Thomas Jefferson, and this rather large age gap meant there was not much intimacy between them. But they remained friendly, and Jefferson was always ready to help his brother out of financial and personal difficulties. Randolph inherited 2,200 acres of prime farmland on the south side of the James River and enough slaves to make a comfortable living. But he was a poor businessman, often in debt.

Isaac Jefferson, a Monticello ex-slave interviewed in 1843, described Randolph as "a mighty simple man [who] used to come out among the black people and play the fiddle and dance half the night." Monticello's slaves called him "Uncle Randolph"—a glimpse of how friendly and down-to-earth they found him. Randolph seems to have been such a frequent visitor to Monticello that his appearance there was no cause for special comment. In a letter to her father, Martha Jefferson Randolph remarked on "Uncle Randolph" being "in the house" and giving a "dram" to a sick slave, which made him feel better. It is unlikely that each of his visits was noted in any formal way. This subtracts not a little from the claim by the Jefferson Foundation Research Committee that Randolph could be dismissed as a paternity candidate because there is no "documented" evidence of his being at Monticello when Sally Hemings conceived.

Randolph was too fond of drams for his own good. At one point, Thomas Jefferson urgently advised him to get his drinking under control. It is not hard to envision Randolph—and Sally—participating in the late-night revels Comte de Volney described. When Eston Hemings was born, Randolph was fifty-one years old and had been a widower for a decade. A letter from Thomas Jefferson has survived, inviting Randolph to Monticello during the "window" of time when Sally conceived Eston.

Shortly after Eston's birth, Randolph married again. His new wife was considered "a controlling woman." Someone with a sharper tongue called her a "jade of genuine bottom." She seems to have been more than capable of ending Randolph's inclinations for Sally and any other woman toward whom his eye wandered. Sally had no more children after Eston. For generations, Eston's children and grandchildren described themselves as descended from a Jefferson uncle. Only after Fawn Brodie, the first Jefferson biographer to assert his guilt, talked to later descendants in the 1970s did they begin to claim Thomas Jefferson was their ancestor.[39]

XXI

A sexual relationship between Thomas Jefferson and Sally Hemings will always remain a possibility. But is it a probability? The writer of history must factor into the puzzle so many contrary realities, from Samuel F. Wetmore the abolitionist propagandist to explanations of how Eston Hemings's descendants might have acquired Jefferson DNA to eyewitness claims that other men were Sally's lovers. Not to be dismissed is Thomas Jefferson Randolph's documented refutation of the Wetmore-ghosted Israel Jefferson's story. That prompts a skeptic to suspect a similar indifference to the truth in Wetmore's Madison Hemings story, which has been proclaimed by some pro-paternity advocates as a kind of gospel truth. At least as important is Thomas Jefferson's acute sensitivity to slurs on his reputation and his denial of Callender's accusation to close friends. In this light, the word "probability" retreats from the certainty that the pro-paternity advocates claim for it.

Hardest of all the pro-paternity claims to believe is the assertion that the relationship between Jefferson and Sally Hemings lasted thirty-eight years. Thirty-eight years of furtive sex in a house swarming with visitors and grandchildren? A respected historian of Monticello has suggested that the entire controversy should be considered a historical Rorschach test that tells us more about the person who believes—or doubts—than it reveals about Jefferson.[40]

One thing seems clear: the American public remains emotionally involved with the story. In the fall of 2007, this writer discussed the current atmosphere at Monticello with a young historian on the University of Virginia faculty. He had been asked to serve as a guide at Monticello for several weeks. The foundation's latest policy, he was told, was not to mention Sally Hemings. He obeyed this dictum, but at the end of virtually every tour he conducted, someone asked him. "Is it true about Jefferson and Sally?"

The controversy has unquestionably played a part in the reevaluation of Jefferson's role in America's founding. A new generation of historians has faulted him for his failure to take a stronger stand against the continuation of slavery. A focus on his duplicity in his political dealings with George Washington, Alexander Hamilton, and John Adams has made it hard to see him as a spotless icon of American idealism. Doubt has even been cast on his role as the author of the Declaration of Independence.[41]

Jefferson's fame will nevertheless remain large. But he is no longer a demigod who looms above the other founding fathers as a unique symbol of America. In a 2001 poll, when Americans were asked to rate the greatest president, Jefferson received only 1 percent of the vote. In recent polls he has done better. A 2009 C-Span survey ranked him number seven behind Harry S. Truman and John F. Kennedy.[42]

This moderation of Jefferson's fame is not such a bad thing. Whether Thomas Jefferson is right or wrong—whether he is, in the words of historian Peter Onuf, "a proxy for America"—should not be, and never should have been, crucial in Americans' political vision of themselves. None of the founding fathers, not even George Washington, need or deserve that sort of sanctification to retain their importance in our national memory.

Beyond these large political thoughts, we should not allow differences about Sally Hemings to obscure the other women in Jefferson's life. Hundreds of vivid letters tell us how much he loved those two beautiful tragedy-haunted "seraphs," Martha Wayles Jefferson and Maria Jefferson Eppes, whom he hoped to greet beyond death's darkness—and the tall, earnest daughter who devoted her life to him, Martha Jefferson Randolph.

BOOK SIX

James Madison

A SHY GENIUS MAKES A CONQUEST

In December 1779, in response to a plea from George Washington that Virginia send her "ablest men" to the Continental Congress, the state legislature, with Governor Thomas Jefferson's warm approval, nominated James Madison. It was not quite the tribute that some biographers have tried to make it. As the year 1780 began, Congress's prestige had fallen so low that it was difficult to find anyone willing to waste his time in Philadelphia. In the year 1779, Virginia had named sixteen men as delegates. Seven resigned, four failed to serve or went home, and four did not show up until the following spring.

The war for independence had become interminable, and Congress had made a colossal mess out of the country's finances. With $230 million in circulation, Continental currency was turning into waste paper. Military and civilian morale plunged with the dwindling power of the dollar. "Congress" had become a word tinged with contempt. Some people may have thought that the short, thin, morbidly shy new delegate was the best Congress could expect for its mediocre ranks.

One delegate described the twenty-nine-year-old Madison as "just from college"—a graphic indication of the first impression he made. The wife of another Virginia delegate, Martha Bland, described him as a "gloomy stiff creature" with "nothing engaging . . . in his manners—the most unsociable creature in existence." But even this nasty critic—her husband, Theodorick Bland was a windbag who became Madison's political enemy—eventually admitted that Madison was "clever in Congress."[1]

James Madison's brilliant mind, and his readiness to work hard and think even harder, gradually made him a leader in this feckless legislature. For the next four years, Madison grappled with the problems of the faltering war effort, rancorous quarrels between political factions inside Congress, and the struggle to unify thirteen often recalcitrant states. Washington's triumph at Yorktown in 1781 put independence within America's grasp, but it seemed to only exacerbate the problem of creating a nation.

At Thomas Jefferson's suggestion, Madison lived in a boardinghouse on the corner of Fifth and Market streets. The landlady's daughter, Mrs. Eliza Trist, was a charming woman described by one historian as the house's "presiding angel." Jefferson urged Madison to "cultivate her affection." It soon became apparent that she and Jefferson were partners in trying to help Madison cope with his shyness. Mrs. Trist developed a protective attitude toward him. When there was talk of electing him governor of Virginia, she wrote to Jefferson, strongly advising against it. "He has a soul replete with gentleness, humanity and every social virtue," she said. But his "amiable" disposition would never be able to tolerate the abuse that went with the governorship. "It will hurt his feelings and injure his health, take my word."[2]

Among the congressmen in the boardinghouse was William Floyd of New York, who had brought his wife and three children with him. The youngest of these offspring was thirteen-year-old Kitty, who was extremely pretty, vivacious, and talented at the piano. By 1782, when Madison was thirty-one and Kitty was fifteen, the congressman was in love with her. Jefferson joined forces with Mrs. Trist to encourage the match. In fact, everyone in the boardinghouse seemed to concur in urging Miss Kitty to say yes. The elders did not seem to realize that this was a poor tactic when dealing with someone from a different generation.

In a scene that can readily be imagined as agonizing, Madison finally asked Kitty whether she was willing and her answer was yes. By this time, we are in 1783 and Kitty was sixteen. Although the age gap might sound wide to modern ears, Madison was not robbing the cradle. Marriage at sixteen or seventeen was not unusual for a woman in this era. Martha Jefferson was seventeen when she married Thomas Mann Randolph Jr. Madison was elated by Kitty's response and rushed the news to Jefferson. The wedding would have to wait until the end of the congressional year. Madison expressed his gratitude for Jefferson's interest in his quest. It confirmed feelings of friendship that Madison "reciprocated."[3]

In the spring of 1783, Congressman Floyd took Kitty and the rest of his family home to Long Island. Madison rode with them to New Brunswick, New Jersey, a distance of sixty miles—strong evidence of how deeply his feelings were engaged. Embroiled most of the time in the struggle to rescue the bankrupt United States from chaos, the congressman did not realize he had a rival. Nineteen-year-old William Clarkson, a medical student and son of a prominent Philadelphia doctor, spent a great deal of time in the Trist boardinghouse, much of it leaning on the harpsichord admiring Kitty's playing.

There is a tradition in the Floyd family that someone around Kitty's age in the Trist ménage secretly encouraged her to look with favor on the smitten Clarkson. Back on Long Island, beyond the reach of the pro-Madison pressure group, Kitty thought things over and decided her heart was voting for Clarkson. Her embarrassed parents told her to write Madison a letter. She obeyed, displaying very little grace. Soon he was telling Jefferson about the "profession of indifference" he had received from Kitty and the "disappointment" of his plans "by one of those incidents to which such affairs are liable."[4]

Fifty years later, when James Madison was editing his papers, the wound was still painful enough to prompt him to scratch out most of the letter. Biographers have been able to decipher only a few phrases, such as having hoped for a "more propitious fate." Jefferson did his best to console his friend. He confessed that "no event has been more contrary to my expectations." If Kitty's decision were final, Jefferson philosophized that "the world presents the same resources for happiness and you possess many within yourself." To relieve his friend's pain, he urged "firmness of mind and unremitting occupations."[5]

Consoling though it was to have such a sympathetic friend, Madison's humiliation must have been intense. In Congress he dealt each day with the most momentous imaginable issues—how to raise enough money to prevent the American army from overthrowing the government, how to persuade the various states to cede their conflicting claims to western territory to Congress, and how to improve the unworkable Articles of Confederation under which the nation was supposed to govern itself. The French ambassador considered him a man of weight and decision. But he had been bested in a contest for the affection of a beautiful young woman by a lowly medical student.

II

James Madison was born in 1751—making him nineteen years younger than George Washington and eight years younger than Thomas Jefferson. In 1764, when John Adams married Abigail Smith, Madison was just emerging from boyhood on his father's four-thousand-acre plantation in the wooded hill country of Orange County, Virginia. Like the Washingtons, the Madisons traced their ancestry back to an Englishman who arrived in Virginia in the middle of the seventeenth century. They had prospered tilling the rich, reddish soil of the Virginia Piedmont.

Madison's boyhood was uneventful and happy. His mother, Nelly Conway, gave birth to eleven more children after James's arrival, seven of whom survived to adulthood. He had brothers and sisters to play with, and there were other large families on neighboring plantations. His parents were affectionate and devoted to their brood. James's only distress was his frail physique. He never grew beyond five feet six or exceeded one hundred pounds, and his health was poor. He suffered from a severe digestive complaint, cholera morbus, which forced him to live on gruel much of the time. This made him something of an anomaly in a society that prized masculinity and physical prowess.

As the oldest son, James Madison Jr. was strongly influenced by his father, who was the largest landowner and leading citizen of Orange County. As his family multiplied, James Madison built a spacious mansion, Montpelier (the Mount of the Pilgrim), to house them. Deprived of his own father at the age of nine, he was a deeply paternal man who instilled in his namesake a strong sense of public responsibility.

James Sr. worried about James's poor health. He decided to send his son to the College of New Jersey in distant Princeton rather than to William and Mary, where most Virginians went. Both Madisons feared the endemic fevers of the Virginia lowlands. Neither realized the decision would turn out to be one of those transformative moments that would alter the course of James Madison Jr.'s life.

Before he left for Princeton, James had acquired a first-class education from some of the best teachers in Virginia. He read Latin, Greek, and French and was well versed in English literature. Arriving in Princeton in the fall of 1769, he easily passed examinations that enabled him to skip his freshman year. In massive Nassau Hall, the college's main—and only—building, he plunged into a regimen that called for hard study and serious thought. The president of the college was a recently arrived Scottish minister, the Reverend John Witherspoon, a tall, stern, beak-nosed man of God who was determined to make the school the best in America.

Behind the rigorous facade, however, college boys remained college boys. Dormitory life was full of practical jokes and rampant teasing of newcomers, forbidden midnight feasts smuggled from nearby taverns, and ingenious assaults on tutors and anyone else who suffered from a surfeit of self-importance. Only the imperious Witherspoon was exempted from such high-jinks.

James Madison seems to have participated in these indoor sports with zest. He was a ribald rhymester who penned raucous attacks on members of the Cliosophic Society, a mostly New England group who were the eternal rivals of the Whig Society, to which James belonged:

> Great Allen founder of the crew
> If right I guess must keep a stew
> The lecherous rascal there will find
> A place just suited to his mind
> May whore and pimp and drink and swear
> Nor more the garb of Christians wear
> And free Nassau from such a pest
> A dunce a fool an ass at best.

Moses Allen, the target of this jape, did not take it seriously, any more than Samuel Spring, the poet laureate of the Clios, resented Madison describing him paying a visit to Clio in her private room and emerging a eunuch, "my voice to render more melodious." It was all in fun, and Madison's participation in it reveals a side of him that his public personality long concealed. He made dozens of friends at Princeton, most of them from other colonies, many of them young men of talent, such as poet Philip Freneau and Hugh Henry Brackenridge, America's first novelist.

III

On the serious side, Madison decided to cram three years into two and save his father money. Virginia was mired in recession during his college years. Part of it was caused by droughts and other varieties of bad weather, part by the boycotts of English commerce that the Americans imposed in their mounting quarrel with Parliament. In several letters to his father, Madison apologized for spending too much money. Apparently, the father, with two other sons to educate and four daughters who would need dowries, expected his oldest son to help him deal with severe financial stress.

Telescoping his college years was a tribute to Madison's powers of concentration, but it was a serious mistake. James added to his woes by staying at Princeton for another year of intense study. When he returned to Virginia in 1772, he was exhausted. He began suffering epileptic-like seizures that caused him to lose consciousness. No one has ever satisfactorily diagnosed this illness. It might be simplest to describe it as a nervous collapse from overwork. For the next three years, Madison did little but stay home and read and correspond with college friends such as William Bradford, scion of a wealthy Philadelphia printing family.

In one letter, he told Bradford he was "too dull and infirm" to expect anything extraordinary in his future. In fact, he had all but resigned himself to a short, unhappy life. Sounding like a septuagenarian, he envied Bradford's "health, youth, fire and genius." Bradford replied by scoffing at his hypochondria and expectations of imminent demise. He remarked that Madison's worries about his health might have an opposite effect—guarantee him a long life. It was an offhand prophecy that would prove to be uncannily on the mark. Madison would outlive most of his contemporaries, including Bradford.[6]

Two more years found Madison healthier but still undecided about his future. He made a pass at studying law but found Old Coke's leaden prose intolerable. The ministry, another obvious choice for a Princeton graduate, was equally unappealing. Early in 1774, Madison was well enough to embark on a trip to Philadelphia to visit Bradford and other college friends. He saw the street demonstrations in support of the Bostonians who had recently defied royal authority by throwing 9,659 pounds of English tea into Boston harbor.

Energy began surging through Madison's frail frame. At Princeton he had been indoctrinated in a Presbyterian hostility to arbitrary power. Back in Virginia, he begged Bradford to send him the latest political news. When the First Continental Congress met in September 1774, Madison bemoaned taking his trip earlier in the year. He yearned to be with Bradford, watching the delegates grapple with the natural rights of Americans versus the legal rights of Parliament. The art and science of government fascinated Madison. As one historian has put it, he found his vocation in the American Revolution.

IV

When the shooting war began, Madison was elected colonel of the Orange County militia. It was largely a tribute to his father's local prestige. It may also have been a fatherly attempt to lure James out of the library. His poor health never permitted him to serve a day as a soldier, and he soon resigned the appointment. His younger brother William joined the army to uphold the family honor.

Another paternal power play had a more positive effect. Madison was elected to the Virginia Convention, the extralegal legislature that had replaced the colonial House of Burgesses. In April 1776, at the age of twenty-five, James decided to risk the miasmas of Williamsburg and took his seat among this body of politicians.

In this first venture into public life, Madison said next to nothing throughout the proceedings as Virginia adopted a constitution and a bill of rights. As one of the youngest delegates, his silence was understandable, but it foreshadowed a style that flowed from his diminutive size and unprepossessing appearance. Madison was the polar opposite of the flamboyant leader of Virginia's revolution, Patrick Henry, whose defiant shout, "Give me Liberty or Give me Death!" had become one of the mottos of the Revolution.

Henry was a man's man in every sense of the word, ready to flay an opponent with words or a whip and quick to defend his honor with a gentleman's ultimate recourse, a pistol. Henry must have filled James Madison with a rueful envy—and not a little despair. Behind his rhetorical facade, the great man was an ignoramus. Madison's brainpower exceeded Henry's by at least ten to one. But the younger man's thin, reedy voice, his parchment-like skin, and his diffident manner virtually guaranteed that no one would ever listen to him.

In the fall of 1776, Madison participated in a minor way in a debate over religious freedom in the Virginia legislature. His Princeton education made him strongly sympathetic to the Baptists and Presbyterians in Virginia, who were often persecuted and sometimes jailed by zealots supporting the established Anglican Church. The leader of the assault on the established church was Thomas Jefferson, but not even his prestige as the author of the Declaration of Independence did him much good against the angry majority who sided with the religious status quo. Madison was more a spectator than a participant in the debates, and he later recalled that Jefferson paid little or no attention to him, because of "the disparities between us."

That phrase reveals a great deal about Madison's self-image at this time. The tall, lanky Jefferson, with his genial, outgoing manner and gift for the smashing phrase, was another icon that James Madison could never become. The comparison was as painful to Madison as the contrast to Henry. The younger man admired Jefferson's wide-ranging knowledge of the law, philosophy, and literature. Here was someone Madison would do almost anything to have as a friend. But there seemed to be no hope of such a relationship ever developing.

V

Back home in Orange County, Madison received a rude shock. He ran for reelection to the Virginia Convention and lost. Still a college idealist, he had disdained to offer the voters what they had become accustomed to getting from political candidates in Virginia—unlimited access to a liquor barrel. He had decided booze was "inconsistent with the purity of moral and republican principles." The voters thought he was a cheapskate or had gotten too big for his breeches.

Once more Madison retreated to his father's library and continued to read deeply on the art and science of government. He might have stayed there for decades were it not for the good offices of his father's friends in high places. In 1778, the legislature elected Madison to Governor Patrick Henry's council, an eight-man body that was supposed to advise the chief executive on matters of politics and policy.

Madison moved to Williamsburg once more and took a room with his second cousin, yet another James Madison, who was president of William

& Mary College. He was soon embroiled full time in the problems of taxation and finance, army recruitment, Indian affairs, and the myriad other matters that fighting a war and running a government dumped on Governor Henry's desk. The Virginia constitution, fearful of creating a tyrant, gave the governor and his councilors virtually equal power, creating what Madison later sarcastically called "eight governors and a councilor." Ninety percent of the power remained with the legislature, prompting Madison to describe the job as "a grave of useful talents."

In 1779, Thomas Jefferson was elected governor, and Madison's attitude toward the councilor's job changed dramatically. Virginia faced a looming challenge as the British shifted the focus of the war south, and everyone realized that Henry had spent his three years as governor talking big and doing nothing to prepare the state for serious warfare. There were woeful shortages of everything from guns to supplies. We have already watched a dismayed Governor Jefferson discover the weakness of the state's militia law, which all but made cowardice a virtue.

In this crisis atmosphere, Jefferson found himself listening far more often to the twenty-eight-year-old Madison than to any of his other councilors. The soft-voiced little man had an uncanny ability to cut through details and fasten on the heart of a problem—and suggest a realistic solution. The beginning of a friendship that would powerfully influence the history of the United States took shape during these hectic days in Richmond. It grew in depth and intensity during Madison's years in Congress. Without Jefferson's sympathy and support, Madison's humiliating debacle with Kitty Floyd might have sent him back to Virginia a recluse for the rest of his life.

VI

Somewhere deep in his unconscious or perhaps in his conscious mind, James Madison seems to have resolved not to seek romance again until he was a man of importance. For the next decade, he devoted himself exclusively to the future of the United States of America. He took Jefferson's advice about unremitting occupations, altering it in only a single respect— he narrowed his occupation to one: his role as a public man.

Americans from 1784 down to our prosperous and powerful present generation have been the beneficiaries of Kitty Floyd's decision to jilt

James Madison. One can only speculate what the ex-congressman might have done if he had carried her off to Virginia and began enjoying the pleasures of married life. Instead, he focused his powerful intellect on solving the fundamental problem confronting the nation—how to persuade the thirteen quarrelsome, semi-independent states to cede sufficient power to a central government to preserve the federal union.

When Thomas Jefferson went off to France to replace Benjamin Franklin as ambassador, Madison turned to the ultimate American hero, George Washington. Instinctively, Madison seemed to reach out to more commanding figures to help him achieve his goals. He had already won Washington's respect and attention by constantly supporting the army and other national interests in Congress. At the end of the war, Washington had won Madison's respect by peaceably resigning his commission as commander in chief of the army, even though Congress had sent his soldiers home unpaid and embittered.

The two men began a momentous correspondence, in which the ex-general made his vision of America's future plain: an "indissoluble" union of states under a single head was vital to the nation's survival. How to achieve this goal was beyond Washington's capacity—but not Madison's. While he began politicking for a convention to overhaul the ramshackle Articles of Confederation, he went to work on the masterly rearrangement of powers that became the United States Constitution.

Even here, when the document that was his brainchild was presented to the historic convention in Philadelphia in 1787, Madison remained in the background. Edmund Randolph, the splendidly handsome, fulsomely oratorical governor of Virginia, introduced the "Virginia Plan" to the convention. But Madison played a leading role in the debates that followed. Delegate William Pierce of Georgia said that thanks to his "spirit of industry and application . . . he always comes forward as the best informed man of any point in debate." He was a unique combination of "the profound politician and the scholar."[7]

Historically speaking, the convention was Madison's finest hour, his rendezvous with fame and greatness. But from a personal point of view, the most satisfying moment came the following year, when he took on Patrick Henry at the Virginia ratifying convention and persuaded the delegates to accept the constitution by a whisker-thin majority. Speaking in his low soft voice, with reams of notes concealed in his hat, Madison bested Henry's anti-federal fireworks with calm, unwavering logic. It did not hurt to have an invisible backer, George Washington, whose enormous prestige reinforced this reasoned persuasion. But the triumph, as far as the political world of Virginia was concerned, belonged to Madison.

VII

Elected a congressman from Virginia—the resentful Henry blocked his appointment to the Senate—Madison won passage of the Bill of Rights and became the most powerful voice in the new House of Representatives. By this time, at least among his fellow politicians, his fame almost equaled the eminence of his two heroes, Jefferson and Washington. Moving in this aura, James Madison felt ready to resume his search for a wife.

While attending Congress in New York, the nation's first capital, he met an attractive widow, Henrietta Maria Colden, who had married into the family of Cadwallader Colden, a prominent prewar New York politician. Although he had remained neutral, many members of his family chose the king's side. Mrs. Colden's husband had become a British officer and died not long after they retreated to London with the rest of the British army. Mrs. Colden had returned to New York with two sons to try to regain some of the family's property, which had been either confiscated or neglected during the war.

Henrietta Colden was one of the few women who had a membership in her own name in the New York Society Library—one of the first semipublic libraries in the nation. The books she took out, according to the records of this venerable institution, were impressive and undoubtedly explain one reason why Madison pursued her. She read the Roman historians, Tacitus, Suetonius, and Julius Caesar, as well as the historian of Rome, Edward Gibbon, the French savant Jean-Jacques Rousseau, and similar weighty authors. According to one admirer, she combined her brainpower with "feminine graces," which prompted some to refer to her as "the celebrated Mrs. Colden."[8]

Nothing came of Madison's interest in her. No one knows whether she rebuffed the diminutive congressman or he had second thoughts about marrying into a family that was strongly tinged with Toryism. It is more than a little likely that Mrs. Colden, who was Scottish, shared her late husband's sympathies. But the progression from sixteen-year-old Kitty Floyd to this elegant cosmopolitan lady was unquestionably a sign of Madison's new sense of himself as a man who had achieved fame.[9]

VIII

Politics now transferred Madison to Philadelphia, the next national capital. He had numerous friends there from his years as a confederation congressman and he was soon enjoying the lively social life of the City of Brotherly Love. Politically he sided with his friend and fellow Virginian, Thomas Jefferson, and acquired more fame by opposing Hamilton's Bank of the United States and President Washington's policy of neutrality in the mounting conflict between England and revolutionary France. But marriage remained very much on his mind.

At dinners and receptions, he often met Dolley Payne Todd, wife of a young Quaker lawyer and one of the most lively, attractive women in the capital. He probably bowed to her as often in the street; they lived only three blocks apart. In 1793, her husband and one of their sons died in a yellow fever epidemic. The twenty-five-year-old Dolley became one of the most sought-after widows in the city. According to one somewhat legendary story, a veritable corps of would-be husbands used to station themselves at the head of her street and wait for her to appear.

One of her suitors was suave Senator Aaron Burr of New York, whose wife was dying of cancer back home. Burr already had a reputation as an irresistible lady-killer. He lived in the nearby boardinghouse run by Dolley's mother and in 1794 helped Dolley draw up a will leaving all her property to her surviving son, Payne Todd, and making Burr the executor—and Payne's guardian. Rumors swirled that Burr was going to propose the moment he heard the hourly expected news of his wife's death.

Madison decided on a preemptive strike with a neat political twist. He offered to back Burr as the next American ambassador to France, and he persuaded his friend and former ambassador James Monroe to support him. The ambitious Burr was grateful. Paris was a diplomatic post that was virtually guaranteed to get a man's name in the newspapers. Madison took the proposal to President Washington, who thunderously rejected the idea. For reasons never completely understood, he loathed Aaron Burr.

Having tried to do this large favor for his chief rival, Madison asked Burr if he would introduce him to Dolley Payne Todd. How could the senator say no? Dolley's reaction makes plain how high Madison had risen on fame's ladder. She rushed a note to her best friend, Eliza Collins: "Thou must come to me. Aaron Burr says the great little Madison has asked to be brought to see me this evening." Dolley was not alone in using this phrase to describe Madison—but for her it seems to have had a romantic ring.[10]

Madison came and was enthralled. Buxom, dark-haired, and bubbling with high spirits, Dolley was used to being the center of attention. Like Madison, she was the oldest child of a large family. Philadelphia friends were struck by her beauty and cheerful disposition from the moment she arrived in their city at the age of fifteen. Her Quaker father had freed his slaves and moved north to launch a business career. One man recalled how her "soft blue eyes" and "engaging smile" had raised the mercury in numerous "thermometers of the heart to fever heat."

Further cementing Madison's attraction, Dolley had been born in North Carolina and raised in Virginia, and had numerous relatives there. A fellow Virginian, Congressman Richard Bland Lee, was engaged to Eliza Collins, and Dolley's younger sister Lucy was about to marry George Steptoe Washington, nephew of the president.

Madison launched a whirlwind courtship. He enlisted Dolley's cousin Catherine Coles, the wife of Congressman Isaac Coles of Virginia, to write her a teasing letter, telling her how much she had mesmerized the great little man: "To begin, he thinks so much of you in the day that he has lost his tongue, at night he dreams of you and starts in his sleep calling on you to relieve his flame for he burns to such an excess that he will be shortly consumed and he hopes that your heart will be callous to every other swain but himself."[11]

For good measure, Catherine Coles added, "He has consented to every thing that I have wrote about him with sparkling eyes." For even better measure, Madison enlisted none other than Martha Washington, who reportedly wrote Dolley a warm letter, urging her to marry her ardent forty-three-year-old suitor.[12]

Dolley played the reluctant game for a while. But in the summer of 1794, while visiting her sister Lucy in Virginia, she wrote Madison a letter, saying yes. Madison replied that he had received her "precious favor" and hoped she could "conceive the joy it gave me." The once jilted suitor anxiously added, "I hope you will never have another deliberation on that subject. If the sentiments of my heart can guarantee those of yours, they assure me they can never be cause for it."[13]

The couple were married on September 15, 1794, at Harewood, the Virginia plantation of George Steptoe Washington. Before the afternoon ceremony began, Dolley found time to write a letter to Eliza Collins, who had just married Congressman Lee. The bride's remarks about James Madison were notably unromantic. She mentioned her "respect" for him and thought their marriage would give her "everything that is soothing and grateful." What she meant was clear in the next sentence: "My little Payne will have a generous & tender protector." She was thinking of her two-year-old son, Payne Todd, who needed a father.

The word "love" went unmentioned in this intimate semi-confession. She signed the letter "Dolley Payne Todd." That evening, after the ceremony and the wedding dinner, she added beneath the previous signature: "Dolley Madison! Alas!"[14]

PARTNERS IN FAME

*D*olley Payne Todd wrote another letter on her wedding day that suggests it was a good thing James Madison campaigned vigorously for her hand. The letter was to her lawyer, William Wilkins, who had helped her settle her late husband's complicated estate. He was not pleased by the news of her marriage. He gave his *very* reluctant approval and confessed that he was "not insensible to your charms." Another hint of his feelings was the way he called her "Julia," apparently a private name. This was a device lovers used—as we have seen in the correspondence of John and Abigail Adams.

Wilkins added words that may further explain Dolley's "alas." He warned her that "the eyes of the world" were on her and "your enemies have already opened their mouths." Wilkins probably meant Dolley's Todd in-laws, who had already quarreled with her over her late husband's estate. They were likely to be unhappy about her new husband obtaining control of her modest inheritance and would probably inform their fellow Quakers that she had married an Anglican. That meant Dolley would be "read out" of their congregation. Aside from the romantic feelings she had aroused in Wilkins, Dolley's marriage to Madison represented a radical break with her Quaker past.

Dolley may not have married for love. But as the newlyweds visited friends on other plantations in Virginia, she soon realized that for her husband, there was no other reason. Beneath the rational logical persona James Madison presented to his fellow politicians, he was a romantic, a man whose heart spoke to him as often as his head. For him, marriage was a step that could be authorized only by his heart. The intensity of Madison's feelings swiftly awoke a similar response in Dolley. Whatever ambivalence she felt about their marriage vanished forever.

Madison's heart also explained his loyalty to Thomas Jefferson and his hostility to Alexander Hamilton's attempt to transform America into an industrial mirror image of Great Britain.[1] Madison unhesitatingly shared his inner political self with Dolley. A warm letter of congratulations from Jefferson no doubt helped unite politics and personal affection. Jefferson himself was adept in that department. His letter included a plea not to retire from politics. "This must not be," he wrote. He hoped Mrs. Madison would "keep you where you are for your own satisfaction and the public good." How could any woman resist such a challenge?[2]

II

During the next three years in Philadelphia, Dolley experienced the excitement of being a political insider. She saw first hand the bruising partisan warfare of the 1790s, and participated in it as James Madison's wife. She observed the toll that the insults and accusations of his opponents sometimes took on her husband's fragile health—and also realized that he and his fellow politicians enjoyed such risks as well as the other less than wonderful effects of the pursuit and use of power in the name of a cause. As a fellow Virginian, Dolley had no difficulty identifying with the Jeffersonian Republicans' hostility to Hamilton and his commercial ways.

Philadelphia was a lively city, especially for political insiders. There was an almost perpetual round of balls and dinners. At the center of the action were a number of wealthy women who were determined to find a role for their sex in the new republic. They embraced the idea that this could be done by urging men to live up to the ideals of republicanism. It was a very American twist on the role that aristocratic French women had created for themselves in Paris. Dolley was a frequent and always welcome guest at their parties and dinners, and had a unique opportunity to study their methods and estimate their success. Perhaps the most memorable social event staged by these women was the 1795 Washington's Birthday Ball given by the city's dancing assembly. It attracted 450 members of the city's political and social elite.[3]

James and Dolley also enjoyed an ultimate compliment that very few

Philadelphians received: an invitation to dine with the Washingtons "in a family way"—at a private meal with several other couples rather than at the far larger weekly official dinners. Even though Madison opposed many of the president's policies, Washington still regarded him with affection for his contributions to the creation of the Constitution. Martha demonstrated her fondness for Dolley by giving her a lovely cream pitcher from a set given to the president by a French nobleman.[4]

Dolley's teenage sister, Anna Payne, lived with the Madisons and was as attractive as Dolley. Suitors thronged their parlor day and night. The young ladies and young matrons like Dolley all wore the latest French fashions, which revealed not a little of their figures. Abigail Adams was shocked by certain young women, notably wealthy Anne Bingham's daughter, Marie, who wore dresses you "might literally see through." The men thought differently, of course. Even New Englander Harrison Gray Otis adored Miss Bingham's costume, which enabled him to see her legs "for five minutes together."[5]

Dolley's friends took a dim view of Abigail Adams. One of the most outspoken was Sally McKean, daughter of a powerful Pennsylvania politician. Sally referred to Abigail as "that old what shall I call her—with her hawk's eyes." She described one of Abigail's Smith nieces as "not young and confounded ugly," and told how she and Abigail had recently departed for Boston, "where I suppose they want to have a little fuss made with them for dear knows they have had none made here."[6]

Dolley and her sister Anna, perhaps underscoring their divorce from Quakerism, never uttered a critical word about the French styles Abigail deplored. Madison seems to have had no objections to viewing bosoms and legs galore in his house on Spruce Street and elsewhere. One Federalist politician, noting how marriage had made Madison "more open and conversant than I ever saw him before," wondered if Dolley could take credit for relieving him of the bachelor "bile" that had made him such a combative political opponent.[7]

III

When John Adams defeated Thomas Jefferson for the presidency in 1796, James Madison decided to retire from the fray for a few years. Jefferson had become vice president and returned to Philadelphia as their party's chief spokesman. The Federalists seemed likely to be in power for years to come. Madison, Dolley, and her sister Anna retreated to Montpelier for the next three years. James Madison Sr. had begun to decline into old age, and his oldest son took charge of their large extended family.

During these first Montpelier years, the intimate side of the Madison marriage flowered. Together they enlarged and redecorated the mansion, using furniture and art shipped to them by James Monroe, who was in Paris as America's ambassador. Dolley learned how to preside at large parties and dinners in a style befitting a southern hostess.

There was only one disappointment in these years of tranquil happiness. In spite of evident ardor on both sides, Dolley did not become pregnant. Neither partner ever publicly expressed disappointment about this nonevent. In addition to Dolley's son Payne, they had so many nieces and nephews in the families of their siblings that there was never any sense of deprivation. But it must have caused an occasional pang in their otherwise all-but-perfect union.

IV

One of Jefferson's first official acts when he became president in 1801 was his appointment of James Madison as his secretary of state. To underscore Madison's importance, the president invited him and Dolley to live at the executive mansion. The future White House was a vast unfinished semi-barn in which Jefferson and his secretary and a few servants rattled around, the president said, "like mice in a church." A member of the departing Adams administration called the house "a large naked ugly looking building." The rest of the so-called Federal City was in a similar unfinished state. In the words of one wit, the place was mostly houses with no streets and streets with no houses.[8]

The Jeffersonian Republicans liked it that way. They saw the oozing swamps and muddy roads and generally primitive landscape as the ideal site from which to govern a nation on pure democratic principles—an atmosphere that could never be achieved in the two previous capitals, New York and Philadelphia. Those cities were full of wealthy merchants and artful lawyers ready and eager to corrupt and ultimately dominate the political process.

For the moment, Washington, D.C., was a city—and a society—that was little more than an embryo, waiting for leaders to nurture and guide it. Not a few people had grave doubts about the future of this idealistic vision. One exasperated legislator, living in a boardinghouse with twenty or thirty fellow politicians, muttered that they reminded him of a tribe of monks. All they did was legislate by day and argue with each other by night. No one brought wives or children to this semi-wilderness.[9]

. In the executive mansion, the widower president seemed to have left women out of his formula for political perfection. He entertained lavishly, drawing on a wine cellar stocked with the expertise acquired during his sojourn in Paris as America's ambassador, but his guests were invariably all men. This was not entirely accidental. As we have seen, Jefferson had acquired a distinct hostility to the way French women participated in France's politics, with their crowded salons and their readiness to bestow sexual favors on men in power.

Into this social vacuum came thirty-three-year-old Dolley Madison, wife of the second most important man in Washington. (Vice President Aaron Burr was a widower and had had a falling-out with Jefferson.) Dolley began by charming President Jefferson as she charmed everyone. On the rare occasions when he invited women to one of his dinner parties, he asked Dolley to join him and act as the hostess. But neither Dolley nor her husband was inclined to accept Jefferson's invitation to become his permanent guests. They soon moved to a comfortable three-story brick house on F Street, two blocks east of the Executive Mansion.

V

While her husband and Jefferson grappled with the turbulent politics of a Europe in which Napoleon Bonaparte became a primary player and a Republican Party that began splitting into quarrelsome factions, Dolley put herself in charge of creating a civilized Washington. Day after day, she braved the atrocious roads in her elegant green carriage, paying calls on the few women who had accompanied their husbands to the capital, and on the relative handful of diplomats who had come from Britain and France and a few other countries, sometimes bringing their wives and children.

Dolley also paid cheerful attention to the numerous local families who had moved from Virginia and Maryland, hoping to share the Federal City's promised prosperity. With her sister Anna in residence, often joined by her sister Lucy, who was always ready to escape the rural society of her husband's plantation, Harewood, Dolley began giving lively dinner parties at which the number of women roughly equaled the number of men.

Dolley must have known she was doing something that Thomas Jefferson did not entirely approve. But the president may have realized it was a job that needed doing. Early in his first term, a group of Federal City ladies began fretting because there were no "levees"—the large receptions hosted by presidents Washington and Adams. Jefferson opened the executive mansion's doors to the public only twice a year, on July fourth and January first. The ladies decided to force the president's hand. One day they arrived at the White House in their party clothes, hair coiffed and jewelry glittering, hoping to embarrass him into giving them a levee.

Jefferson had just returned from a horseback ride, and was covered with dust and grime from the capital's primitive roads. But he did not lose his cool. He pretended that each of the ladies had come separately, and by wonderful coincidence they had all arrived at the same time. After fifteen minutes of forced good humor, the ladies departed in a very disgruntled mood. This experience may have made the president more tolerant of Dolley's parties.

She also took advantage of President Jefferson's dependence on James Madison as his chief adviser and most trusted political confederate. The president was not going to provoke a quarrel with the secretary of state by criticizing his beloved wife. Even when Dolley began giving a New Year's party that competed with the president's reception at the executive mansion, Jefferson never said a negative word. Everyone trudged dutifully to the mansion to pay their respects—and then headed for F Street, where there was a party they would enjoy.

Dolley's dinners were not the small affairs that Jefferson preferred because they gave him an opportunity to press his ideas and political plans on his guests. She liked big parties because they enabled people to relax. Senator John Quincy Adams noted in his diary that there was "a company of about seventy persons of both sexes" at one dinner. Nevertheless, he found a chance to have a conversation about politics with James Madison, which began Adams's exit from the moribund Federalist party.

Madison's shyness made him awkward and reserved when he met people individually or spoke at public events. But seated at his own dinner table, with Dolley weaving good humor into the conversation, he relaxed and became almost as charming as his spouse. At one dinner party,

Champagne was poured with a lavish hand. Madison drank his share and observed somewhat ruefully that tomorrow would almost certainly begin with a headache. It was hard to judge when one exceeded his limit.

An impish smile played across Madison's face as he observed that tomorrow was Sunday. Why not conduct an experiment and find out exactly how much Champagne it took to induce a hangover? The victims would have the next day to recover. Soon, Champagne was being lugged into the dining room by the case. No one got drunk, but one diner recalled that the conversation grew more and more animated and humorous remarks flew in all directions. If anyone kept a record of how many heads were throbbing on Sunday, it has vanished.[10]

VI

Within eighteen months of her arrival in the Federal City, Dolley Madison had established her house as the social center of Washington. She was clearly violating President Jefferson's dictum against women in politics, but she got away with it by ingeniously blending friendship and hospitality with political concerns until most outside observers were hard put to separate them. Dolley had a rare ability to choose her women friends wisely. Among the most important was Margaret Bayard Smith, the wife of Samuel Harrison Smith, editor of Washington's only newspaper, *The National Intelligencer*. Smith had been invited to Washington by Jefferson to serve as his semi-official spokesman.

Margaret Bayard Smith was a talented writer and a shrewd woman. She liked Dolley from the moment they met. Dolley's lively good humor and "affable and agreeable manners" won her wholehearted affection. In a letter to her sister, Mrs. Smith expressed amazement at the warmth of her feelings in so short a time. She felt almost as fond of Dolley's sister Anna. "It is impossible for an acquaintance with them to be different," she wrote.[11]

Another important woman friend was Anna Maria Thornton, wife of Dr. William Thornton, the architect who had won the competition to design the capitol. He was a man for all seasons, a talented inventor and businessman. The Thorntons and the Madisons were next-door neighbors, and Anna Maria had a personality almost as lively as Dolley's. A third vital woman friend was Marcia Burns Van Ness, the wealthiest woman in Washington. She had inherited $1.5 million from her landowner father before marrying John Peter Van Ness of New York. The money and Marcia's charming personality made the Van Nesses the Federal City's social leaders before Dolley arrived on the scene, and they were easily persuaded to join forces with her.[12]

Among the innovations Dolley introduced at her dinners and late-evening teas was gambling at cards. The favorite game was loo, a version of euchre, in which players bet on their ability to win tricks. The stakes were low, but the fun was high. Ladies, when they lost, squealed in the most piquant way that they had been "looed." Dolley was an enthusiastic player. During this diversion, she frequently paused to inhale some snuff, a smokeless form of tobacco to which she soon became addicted. Like tobacco users before and since, she urged all her friends never to become fond of the habit—but found it impossible to stop using it. Not a few people thought it added to Dolley's image as a woman of the world.

This public personality may have helped Dolley achieve some of her most important political-social successes in these early Washington years. Far more than the ascetic Republican Thomas Jefferson or the shy, reserved James Madison, she was crucial to making the diplomats from foreign nations feel welcome in the primitive capital city. Here, Dolley had an inside track; her Philadelphia friend, acerbic, beautiful Sally McKean, had married the handsome young Spanish minister Carlos Fernando Martinez de Yrujo. She introduced Dolley to many of the wives of other ministers, notably the French minister's spouse, Marie-Angelique de Turreau, who had a wicked sense of humor. Dolley told her sister Anna that in her company "I crack my sides laughing."[13]

A dividend of this friendship was Dolley's acquisition of the French language. Marie-Angelique was a clever and encouraging teacher. She also undertook to instruct Dolley in dressing with Parisian panache. The closeness of their relationship made Dolley supersensitive to the way red-faced, mustachioed General Louis Marie Turreau treated his pretty wife. More than once, when she disagreed with him in public, he struck her. It should be added that she was not exactly a shrinking violet; she once hit him in the head with a flatiron.[14]

Turreau was blatantly unfaithful to his wife, regularly riding through Washington in his gilded carriage to the house of a woman "of easy virtue." At other times, he insisted on bringing prostitutes into their home.

In 1805, when Dolley was in Philadelphia undergoing surgery for an ulcerated knee, General Turreau and three male friends attempted to visit her in her bedroom. She declined to entertain them, implying in a letter to Madison that she was worried about her reputation. It was also an undoubted pleasure to tell the swaggering wife-beater to go away.[15]

Dolley proved to be a valuable asset to both her husband and President Jefferson when the purchase of the Louisiana Territory from France led to a serious quarrel with Spain. Secretary of State Madison insisted the western part of Spanish-owned Florida was part of the historic transfer. Minister Yrujo stormed into Madison's office at the State Department and screamed insults in his face. Madison declared him persona non grata, and Yrujo and his wife retreated to Philadelphia. But Dolley's close friendship with the former Sally McKean enabled the government to maintain at least a semblance of friendly relations. Dolley told her sister that she still felt "a tenderness" for the Yrujos, "regardless of circumstances."[16]

VII

Dolley became even more important when President Jefferson decided to apply his ideals of Republican simplicity to dealing with the new British minister, Anthony Merry, and his ultra-dignified wife, Elizabeth. At an official dinner in the executive mansion, Jefferson announced that the guests would be seated at the table without the usual attention to honor and importance. "Pell-mell" was his name for this new etiquette, which enabled the president to ignore Mrs. Merry and lead Dolley to the place of honor beside him at the table. She knew this was a bad move and whispered urgently, "Take Mrs. Merry."

The president ignored her. Secretary of State Madison extended his arm to Mrs. Merry, but she was obviously outraged and insulted. Her flustered husband was left standing at the door without a woman to escort. When he attempted to sit down beside Sally Yrujo, a congressman who took the president's pell-mell rule too literally pushed him aside, and the minister was left to wander to a chair near the bottom of the table.

The infuriated Merrys were convinced that Jefferson was expressing his disrespect for both them and their country. They refused all further invitations from the president. But they decided they could and would accept an invitation from the Madisons after Dolley called on Mrs. Merry and did her best to make amends to the formidable lady. This was no easy task, because Dolley could not admit that Jefferson was in the wrong.

At first, things went no better at the Madisons. Mrs. Merry dismissed Dolley's dinner as a mere "harvest home" supper—peasant fare. Dolley kept her temper and calmly replied that it was the American custom to prefer "abundance to elegance," evidence of the "superabundance and prosperity of our country." She was aware of the "elegance of European taste" but chose to dine "in the more liberal fashion of Virginia." Mrs. Merry was temporarily reduced to silence. The French military attaché, hearing of Dolley's reply, wrote home that "Mrs. Madison has become one of America's most valuable assets."[17]

The contretemps between the Merrys and the president got into the newspapers. The Federalists sided with the British minister, and in the ugly style of the day, some of their reporters began spreading nasty slanders about Dolley and her sister Anna. They claimed that Dolley was Jefferson's secret mistress with Madison's covert approval because he was impotent—as his failure to produce a child supposedly proved. Soon other papers were suggesting that Madison and Jefferson "pimped" Dolley and Anna to win the goodwill of visiting foreign officials. The president bemoaned the way "the brunt of the battle" was falling on "the secretary's ladies," but he declined to call off his ridiculous and unnecessary social war.[18]

Dolley continued to woo Mrs. Merry. She persuaded her and Mr. Merry that dinners at the Madisons could be regarded as private affairs, so there was no need to invoke rules of precedence or worry about national honor being impugned. She sent her small gifts, such as a bottle of perfume whose scent Mrs. Merry admired. Soon, Dolley was describing their relationship as "unusually intimate," though the term applied only to the current moment. She never knew when the large, combative lady would get angry "at persons as well as circumstances."

Dolley was more than a little surprised—and pleased—when Mrs. Merry, hearing she was ill, appeared at the Madison's F Street house offering to be her nurse and spent three hours with her. Behind this feminine bridge-building lay some important political conversations between Merry and Madison, which helped repair some of the damage the president had inflicted with his pell-mell etiquette.[19]

Dolley Madison learned a great deal from this attempt to intrude politics on social occasions in such a literal way. Although she never publicly revealed her opinion of President Jefferson's experiment, in years to come she made it clear by her actions and style that she considered it an unfortunate blunder. Her tact was a tribute to her political shrewdness—and her generous heart.

Dolley also learned much from watching Mrs. Merry in action. Too often the ambassador's wife almost relished the conflict and the attention it won for her in the public spotlight. She was much too quick to speak for herself as well as her husband, which enabled President Jefferson and his supporters to christen her a virago unworthy of a shred of sympathy. Dolley concluded that a woman who waded into the contentious side of politics aroused the always lurking hostility between the sexes and won no friends for her side of the argument.

At this point in her journey to fame, Dolley was demonstrating her talents as a politician, but she still hesitated to apply that term to herself. She was under the influence of President Jefferson's opinion that women—especially American women—should stay out of politics. While she was being treated for her ulcerated knee in Philadelphia, she wrote a revealing letter to her husband, who had remained in Washington. "You know," she began, "I am not much of a politician but I am extremely anxious to hear (as far as you may think proper) what is going forward in the cabinet." She knew that Madison did not want his wife to be "an active partisan," and she assured him there was not "the slightest danger" of such a thing. She remained conscious of her "want of talents" and her wariness about expressing opinions "always imperfectly understood" by her sex.[20]

VIII

After the triumph of the Louisiana Purchase, President Jefferson's second term was almost bound to be an anticlimax. It soon became something much more unpleasant—one of the least successful four years in the history of the American presidency. Relations between both France and Great Britain deteriorated steadily as the two superpowers battled for global supremacy. They blockaded each other's ports and forbade all neutral trade. The British were especially obnoxious, repeatedly kidnapping American sailors from ships at sea under the pretext that they were deserters from the royal navy. President Jefferson, having reduced the army and the navy to skeleton forces in his passion for minimum taxes, was looking weak and feckless. His secretary of state proposed taking a leaf from the history of the American Revolution and declaring an embargo on all commerce between America and Europe. Madison confidently assured the president it would starve England into submission.

These boycotts, as the revolutionaries had called them, were very effective in the 1760s and 1770s. But the Madison-Jefferson embargo was a national disaster. The loss of American commodities such as wheat and cotton had only a minimal impact on the two superpowers, but it devastated the American economy. Exports declined 80 percent from 1807 to 1808. One disgusted critic compared it to "cutting a man's throat to cure a nosebleed." New England, where commerce was a way of life, was soon in semi-revolt, condoning and even encouraging wholesale smuggling in blatant violation of the law.

A dismayed and baffled President Jefferson grew more and more discouraged. In his final year in office, he virtually abdicated, handing over most of his executive responsibility to his secretary of state. He even began shipping the furniture he had brought to the Executive Mansion back to Monticello. In this atmosphere of disillusion and disarray, James Madison became a candidate for president. He had Jefferson's backing, but that was not worth much. Few presidents have been more unpopular in their final year in office.

The secretary of the navy, Robert Smith, accused Jefferson of launching the embargo against the advice of the majority of his cabinet. Senator William Plumer of New Hampshire, another hostile Republican, wrote: "Madison has acquired a complete ascendancy over him." Complicating matters was the prevailing code that a candidate could not campaign openly for president. Confessing a desire for power stirred fears of executive tyranny in too many minds, especially among Republicans. Soon two other candidates were in the race: Jefferson's aging vice president, George Clinton of New York, and James Monroe, who had almost as much claim to being Jefferson's chief disciple as Madison.[21]

There was little doubt that James Madison needed help. He found his rescuer in his own household. By now there were few more astute observers of the political scene than Dolley Madison. "Public business was perhaps never thicker," she wrote cheerfully to her aunt. Dolley was not even slightly intimidated. Political nominating conventions were far in the future. Candidates were chosen by congressional caucuses of both parties. This added heft to Dolley's social skills. She brushed off claims from Monroe's backers that Madison was a Federalist in disguise. As for Vice President Clinton, he was suffering from New York's long-running jealousy of Virginia's power. She did not say these things publicly, of course. But the VIPs who thronged her dinner parties did not hesitate to voice them.[22]

Monroe's chief backer was Congressman John Randolph of Virginia. He was a veritable walking, talking compound of all the neuroses long associated with his family. He disliked women in general, but Dolley's lush beauty and her revealing gowns stirred raging antagonism in his dour soul. He began telling Monroe and anyone else who would listen that Dolley was promiscuous and using her favors to promote Madison's presidency.[23]

A local Federalist newspaper ran a pseudo-ad for a book that supposedly told all about a powerful, impotent man with an oversexed wife. Soon other anonymous stories were sprouting, even naming some of Dolley's supposed lovers. The Madisons took this mudslinging seriously enough to refute one story by inviting one of Dolley's rumored flames to a small family dinner at their F Street house. The key to dealing with such slanders, Dolley told a friend, was to "listen without emotion" when they were repeated in your hearing, knowing that "they were framed but to play on your sensibility." This was a lesson thin-skinned Abigail Adams never learned. It is still good advice for anyone and everyone in politics.[24]

Other papers with a tilt to the Republican radicals claimed Dolley was a secret Federalist and British supporter. They investigated her first husband's death and concocted a story of her abandoning him as he writhed in the final throes of yellow fever. Again and again they assayed Madison as "cold"—a synonym for impotence—and wondered how he could be expected to lead the country when he lacked the strength to satisfy his wife.

Dolley began blaming some of these assaults on Monroe personally, because he remained silent, never saying a word in defense of Madison, his supposed close friend. At one dinner party, she made several uncharacteristically cutting remarks about Monroe and his wife. Her disapproval had an impact in political Washington, and Monroe soon faded as a candidate. Meanwhile, Dolley's tireless entertaining was a tactic that the widower vice president, George Clinton, soon saw as an insurmountable advantage. He, too, retreated from any public confrontation, and Madison faced only one serious opponent in the general election, Charles Cotesworth Pinckney of South Carolina, the Federalist whom Alexander Hamilton had tried to make president instead of John Adams in 1800.[25]

Pinckney's chief issue was the hated embargo, but Congress repealed it on Jefferson's last day in office. Madison was elected in what passed for a landslide, with Pinckney winning almost no support outside New England. When he grudgingly conceded, he added a nasty but historically significant embellishment. He said he had lost to "Mr. and Mrs. Madison. I might have had a better chance if I faced Mr. Madison alone."[26]

IX

Dolley Madison entered the executive mansion or president's "palace" with a great many people watching her. The wife of a New York congressman who had backed George Clinton noted that she had grown more dignified. She seldom played loo or wore revealing French dresses. But on inauguration day, her F Street parlor was jammed with visitors all "now worshipping the Rising Sun." The comparison was more apt than the writer realized. Dolley had a plan already worked out, aimed at making the president's house the social center of Washington, D.C.[27]

She began her campaign with an inaugural ball at Long's Hotel that attracted more than four hundred mesmerized guests. Dolley wore a velvet gown with a train so long, it cried out for several young pages or ladies-in-waiting to deal with it. Her friend Margaret Bayard Smith implied as much. "She looked like a queen," she wrote. Even more eye-catching was Dolley's gleaming white satin turban trimmed with bird-of-paradise feathers. Shrewdly, she limited her jewelry to a pearl necklace and earrings and a few bracelets. The effect was a striking combination of royal elegance and American simplicity. When the dancing began, and the master of ceremonies offered to lead her to the floor, she replied, "I don't dance." Again, everyone was charmed by this calm adherence to her Quaker roots.[28]

In the glow of this performance, the president, worn out by the long inaugural day, was scarcely noticed. Dolley proceeded to deal with this problem, too. At dinner, she sat herself between the British and French ambassadors and soon had them smiling and chatting. Gone was President Jefferson's pell-mell etiquette. The representatives of the warring great powers led the way to the dinner table. General Turreau, still the French ambassador, escorted Dolley; she concealed her dislike of him with the smile of a master diplomat. Behind him, the Briton who had replaced Ambassador Merry escorted Dolley's sister Anna. Ex-president Jefferson, also a guest, gave not so much as a hint of disapproval.[29]

People swarmed onto the dance floor to get a closer look at Dolley. Soon the room was so crowded that some ladies grew faint. An alarmed male guest broke the upper panes of several windows to let in more air. Everyone went home talking about Mrs. Madison. In *The National Intelligencer,* her friend Margaret Bayard Smith came close to exhausting her supply of admiring verbs and adjectives. To no one's surprise, the paper christened Dolley "The Presidentress."[30]

This only emboldened Dolley to push ahead in her campaign to make herself and other women a vital part of James Madison's presidency. Ignoring blasts of vituperation against the president from Congressman Randolph, she led groups of women to the visitors' gallery to watch both houses of Congress in action. Occasionally she led similar groups in visits to the Supreme Court. As for Randolph, she decided to treat him as a public amusement. After one of his performances, she asked a visitor if he had heard about it. "It was as good as a play," she said. Soon people were marveling at the way women were playing a part in American politics that was "not known elsewhere."[31]

The heart of Dolley's plan became visible when President Madison asked Congress for $24,000 (about 400,000 modern dollars) to renovate the executive mansion and buy much-needed furniture, china, and other civilizing necessities. Aside from essential work such as shoring up the roof, widower Jefferson had done little or nothing to finish the house during his eight years. Congress acquiesced, and Dolley went to work with architect Benjamin Latrobe. As soon as possible, she wanted a large drawing room and a small parlor for entertaining. Also on the list was a state dining room. Dolley had decorated all three of these rooms in her head before Latrobe went to work. Although some luxury items were unavailable because of the aftereffects of the embargo, the architect managed to find acceptable substitutes.[32]

First to be finished was what dazzled guests called "Mrs. Madison's Parlor" (the Red Room in the modern White House). The dominant feature was the sunflower-yellow damask silk draperies that adorned the windows. High-backed sofas and chairs had the same lush color, as did a damask fireboard in front of the mantel. Another eye-catcher was Gilbert Stuart's regal portrait of Dolley.

Her first reception in the room, on May 31, 1809, swiftly became the talk of Washington. Military music filled the air, and a buffet offered ice cream, punch, cookies, and fruit. A smiling Dolley, in another spectacular gown and turban, dominated the room. Mrs. Madison's Wednesday "drawing rooms" quickly became a destination for everyone in Washington and many beyond the city's borders.

Writer Washington Irving described his eagerness to attend during a visit to the capital. "I swore by all the Gods I should be there," he said, when he learned Dolley was having a reception on the day he arrived in town. Wangling an invitation, Irving found himself in "the blazing splendor of Mrs. Madison's drawing room," where he met "a crowded collection of great and little men, of ugly old women and beautiful young ones" and in ten minutes was "hand in glove with half the people in the assemblage." He found Dolley to be "a fine portly buxom dame who has a smile and a pleasant word for everybody." As for her sisters, Anna and Lucy, they were "like the two merry wives of Windsor."[33]

On New Year's Day, 1810, Dolley and the president held their first reception in the much larger Oval Room, which had remained an unused wasteland during Jefferson's administration. This time the impact of Dolley's decorations was nothing less than palatial. Great gold lamps lined the entrance, and a huge mirror gleamed above the mantle. The walls were papered in rich cream, and the woodwork shadowed in blue and gray. The floor-to-ceiling windows were adorned with red silk velvet curtains, which were matched by the red cushioned furniture, with thirty-six "Grecian" chairs. "The President's house is a perfect palace," gasped one visitor.[34]

The state dining room, which opened off the Oval Room, was even more palatial. The ceilings were three times the height of the rooms in an average house. A gigantic sideboard occupied an entire wall. At the far end of the room hung a life-sized portrait of George Washington by Gilbert Stuart. Public approval of Dolley's interior decorating was virtually universal. Members of both political parties competed for an invitation to the mansion that everyone began calling "The White House." A Baltimore newspaper warmly approved the title. It was, they opined, "the people's name."[35]

As guests by the hundreds swarmed into the White House and swirled around Dolley, some people wondered whether she was almost too successful. She enjoyed herself hugely, but her sixty-year-old husband did not seem to have such a good time. One 1809 visitor described Madison as "a very small thin pale-visaged man of rather a sour, reserved and forbidding countenance." He seemed "incapable of smiling" but talked agreeably to all comers. By 1810, another guest thought Madison looked as if he were "bending under the weight and cares of his office." Whereas Dolley remained "a robust and hearty lady."[36]

X

Madison's personality and leadership style were not well suited to an executive role. He was at his best in Congress or on a committee, where his weak voice and mild manner did not matter so much because the logic and depth of his arguments were so persuasive. A president leads in a very different way. Compounding Madison's problems was the disintegration of the Republican Party into a half dozen factions, each with its own ambitious leader. As a result, the United States drifted irresolutely through the political turbulence that was tearing the world apart as the war between Great Britain and Napoleonic France rumbled toward a climax.

Madison had no illusions about Napoleon, who played a cat-and-mouse game with America, agreeable one week, nasty the next. But his focus on dominating Europe gave him little chance to harm the United States. Britain's high-handed attitude toward American ships at sea was another matter. Their arrogance slowly convinced Madison that only a war would settle America's relationship with the mother country. He apparently discussed this growing conviction with Dolley. In December 1811 she wrote to her sister Anna, now married to a Maine congressman, "I believe there will be a war as M sees no end to our perplexities without it."[37]

Unfortunately, President Madison could not convince some members of his cabinet or key members of Congress to prepare the country for a serious conflict. He let newly arrived western politicians such as Henry Clay of Kentucky do the orating in Congress. Dubbed The War Hawks, they assured everyone that the British, embroiled with Napoleon, were pushovers and would never be able to defend thinly populated Canada. New England's politicians and the states that followed their lead, such as New

York and New Jersey, remained stubbornly opposed to the war. Dismaying proof of Madison's failure to rally support in Congress was the Senate's approval of a declaration of war by a mere six votes. The margin in the House of Representatives was not much better: 79–49.

Nevertheless, Madison signed the war resolution in June 1812. The Federalist *Alexandria Gazette* promptly accused him of persuading many congressmen to vote for the declaration with invitations to Dolley's Wednesday drawing rooms and splendid dinners. The paper called her parties "extravagant imitations of a royal court" and claimed Americans were being taught to bow and curtsy before the president and his wife and otherwise "play the parasite."[38]

No one paid much attention to these squawks of protest. Throughout Madison's first term, Dolley's parties and dinners had grown more lavish and splendid. She hired a French chef who served duck and venison cooked in an elaborate style seldom seen in American kitchens. Dolley regularly sat at the head of the table and took charge of the conversation, freeing the president from the task, which he did not enjoy or handle well. This enabled him to relax and indulge in genial small talk with nearby guests, who often came away charmed.

Far from being aristocratic affairs, Dolley's parties were democratic with a small "d." Anyone could come, once they had been introduced to Mrs. Madison or the president. George Washington and John Adams held "levees" at which guests remained stationary, waiting for their host to greet them and exchange a few words. Dolley encouraged her guests to feel free to move around all three of her redecorated rooms, chatting with friends and with her or the president if they were so disposed. At dinners, not a few people marveled at the way she sat diplomats such as the Russian minister beside a local tradesman and his wife.

XI

The first test of the popularity of the war with Britain was Madison's campaign for reelection. His opponent was DeWitt Clinton, nephew of vice president George Clinton, who had died in 1811. The candidate was a good politician and a popular former mayor of New York City with a strong Republican following in his native state. He was backed by the Federalists, who remained a force in New England. Dolley struggled to maintain her public neutrality, but she was heard to refer to Clinton as "that fellow" when he paid a visit to Washington.

The Federalists cried petticoat politics and tried to convince people that Madison lacked the forcefulness to lead the country in a war. In the original thirteen states, Clinton came within one electoral vote of beating Madison. But in the new states of the West, the War Hawks were dominant, and they gave the president a comfortable margin of victory.

Clinton did not claim that he would have won if President Madison lacked Dolley. The words would have been superfluous. Everyone who read a newspaper knew that Dolley was an essential part of the Madison administration. But no one anticipated that the war would make her a national heroine.

HOW TO SAVE A COUNTRY

*T*he war did not go well. In the preceding years, Madison had been unable to stop his penny-pinching secretary of the treasury, Albert Gallatin, from blocking congressional resolutions to expand the country's armed forces. The Americans began the conflict with no regular army worth mentioning. Their navy consisted of a handful of frigates and a fleet of pathetic gunboats, each armed with a single cannon, which President Jefferson had designed as defenders of America's ports. In 1811, Congress had voted to abolish Alexander Hamilton's Bank of the United States, making it almost impossible for the government to raise money. Worst of all, the British defeated Napoleon and the United States found itself fighting the most powerful army and navy in the world, alone.

By 1813, Secretary of the Treasury Gallatin was telling the president, "We have hardly enough money to last to the end of the month."[1] Along the Canadian border, American armies stumbled into ruinous defeats. A huge British naval squadron blockaded the American coast. The only good news came from victories over lone British men-of-war by warships of the tiny American navy. In Congress, New Englanders sneered at "Mr. Madison's War," and the governor of Massachusetts refused to permit any of the state's militiamen to join the attack on Canada. Madison fell ill and the aged vice president, Elbridge Gerry, also grew so feeble that Congress began arguing about who would be the next president if they both died.

Dolley Madison's White House receptions and dinners became the only place in the nation where hope and determination continued to flourish. Soon she herself became a symbol of America's refusal to be daunted by British power. Although she was born a Quaker, Dolley maintained she had always believed in fighting back "when assailed." Her instinct for defiant drama came to the fore at an 1812 ball given by naval officers celebrating Congress's decision to expand the American navy. Everyone's spirits rose when news of an American victory over the British frigate *Macedonian* off the Canary Islands reached the White House. A few minutes later, a young lieutenant arrived at the ball carrying the flag of the defeated ship. Senior naval officers paraded it around the floor and laid it at Dolley's feet.[2]

At her social events, Dolley struggled, in the words of one observer, "to destroy rancorous feelings, then so bitter between Federalists and Republicans." Affability and good manners remained her watchwords, and members of Congress, weary with flinging curses at each other during the day, were willing to relax and even discuss compromise and conciliation in the evening. Their wives and daughters were almost all allies of Mrs. Madison. By day Dolley was a tireless visitor, leaving her calling cards all over the city. Before the war, most of her parties attracted about three hundred people. Now attendance climbed to five hundred, and young people began calling them "squeezes."

There were times when Dolley felt the pressure of presiding over these crowded rooms. She confessed to one friend, "My head is dizzy!" But she maintained what another observer called her "remorseless equanimity," even when the war news was bad.[3] Critics heaped scorn on the president, calling him "Little Jemmy" and reviving the smear that he was impotent, making it a symbol of his failures as commander in chief. But Dolley seemed immune to such slanders. President Madison might look as if he had one foot in the grave, but Dolley remained blooming and seemingly tireless. More and more people began bestowing a new title on her: the First Lady. Dolley had created a semipublic office as well as a unique role for women in the American government.

By this time, the relationship between James and Dolley had moved several light years beyond the diffidence with which she brought up politics in her letters to him in 1805. They had both abandoned the idea that a woman should not and could not think about the thorny subject. As early as the first summer of his presidency (1809) Madison had been forced to rush back to Washington from a vacation at Montpelier, leaving Dolley behind.

In a note he wrote just after he returned to the White House, he told her that he intended to bring her up to date on "intelligence" just received from France in his next letter. Meanwhile he sent her the morning paper, which had a story on the subject. In a letter two days later, he sent "the foreign news in the inclosed papers" and discussed a recent speech by the British prime minister. There was little doubt that Dolley had become the president's political partner in every imaginable sense of the word.[4]

II

Dolley's charms and political acumen had their limits. The British were relentless in their determination to reduce Americans to obedient colonists once more. Checked by an American naval victory on Lake Erie and the defeat of their Indian allies in the West, they concentrated their assault on the coastline from Florida to Delaware Bay. Again and again their landing parties swarmed ashore to pillage homes, rape women, and burn public and private property. The commander of these operations was a strutting red-faced admiral named Sir George Cockburn. As arrogant as he was ruthless, he sent word to Mrs. Madison that he soon expected to "make his bow" in her drawing room as the ruler of a captured Washington, D.C.

Dolley did her best to reassure her friends. She told them it was impossible for a British army to get within twenty miles of Washington. But many people began moving wives and children and furniture out of the city. The drumbeat of news about British landings elsewhere intensified local criticism of the president. Some people claimed that Dolley herself was planning to flee Washington and if Madison attempted to follow her, they would make sure that he and the city would "fall" together. At one point Dolley exploded in a letter to a friend: "I am not the least alarmed at these things but entirely disgusted & determined to stay with him."[5]

On August 17, 1814, a large British fleet anchored at the mouth of the Patuxent River, only thirty-five miles from Washington. Aboard were four thousand veteran troops under the command of a tough professional soldier, General Robert Ross. Soon they were ashore without a shot being fired at them and they began a slow, cautious advance on Washington. There was not a single trained American soldier in the vicinity to oppose them. All President Madison could do was call out thousands of militia. The commander of these jittery amateurs was Brigadier General William Winder, an aged veteran of the Revolution. Madison had appointed him early in the war for only one reason: his brother was the governor of Maryland.[6]

When Winder's incompetence became glaringly obvious, friends urged Dolley to flee the city. Thousands of Washingtonians were crowding the roads. Dolley demurred: "I am determined to stay with my husband," she said. She welcomed Madison's decision to station one hundred militiamen under the command of a regular army colonel on the White House lawn. Not only was it a gesture of protection on his part; it was also a declaration that he and Dolley were going to stand their ground. She applauded when the president joined the six thousand militiamen who were marching to confront the British in Maryland. She was sure his presence would stiffen their resolve to protect the capital.

Doing her share to display defiance, Dolley decided to give a dinner party on August twenty-third, after the president had ridden off with the army. All her guests either ignored her invitations or sent hasty regrets. *The National Intelligencer* had reported that the British had received six thousand reinforcements, panicking the city. Again and again, Dolley ascended to the White House roof to scan the horizon with a spyglass, hoping to see evidence of an American victory. Meanwhile, Madison sent her two scribbled messages. The first assured her that the British would easily be defeated; the second warned her to be ready to flee on a moment's notice.

If the worst transpired, Madison had told her to save the cabinet papers and all the other public documents she could fit in her carriage. Late in the day, Dolley began a letter to her sister Lucy, describing her situation. Her carriage was loaded with trunks full of public papers. "All my friends and acquaintances have gone," she wrote. The army colonel and his hundred-man guard had also fled. But Dolley refused to budge. "I am determined not to go until I hear Mr. Madison is safe," she continued in her letter. She wanted to appear beside him "as I hear of much hostility toward him . . . disaffection stalks around us." She felt her presence might deter enemies who were ready to harm the president.

At dawn the next day, after a mostly sleepless night, Dolley was back on the White House roof with her spyglass. Resuming her letter to Lucy, she told her that she had spent the morning "turning my spy glass in every direction and watching with unwearied anxiety, hoping to discern the approach of my dear husband and his friends." Instead, all she saw was "groups of military wandering in all directions, as if there were a lack of arms or of spirit to fight for their own firesides!" She was seeing the egregious disintegration of the American army that was supposed to be confronting the British at nearby Bladensburg, Maryland.

Soon the boom of cannon rattled the White House windows. The battle remained beyond the range of Dolley's spyglass. She was spared the sight of the American militia fleeing at their first glimpse of the charging British infantry. President Madison was swept away in the rout, along with General Winder and everyone else with some authority. In the White House, Dolley stood her ground. She had found a wagon, and packed it with the red silk velvet draperies of the Oval Room and the silver service and the blue and gold Lowestoft china she had purchased for the state dining room.

Resuming her letter to Lucy, Dolley wrote: "Would you believe it, my sister? We have had a battle or skirmish . . . and I am still here within sound of the cannon!" Gamely, she ordered the table set for a dinner for the president and his staff, and insisted that the cook and his assistant begin preparing it. "Two messengers covered with dust" arrived from the battle-field, urging her to flee. Still she refused, determined to wait for Madison. She ordered the dinner to be served. She told the servants that if she were a man, she would post a cannon in every window of the White House and fight to the bitter end.

The arrival of a close friend, Major Charles Carroll of the prominent Maryland family, finally changed Dolley's mind. He told her it was time to go and she glumly acquiesced. As they headed for the door, Dolley saw the Gilbert Stuart portrait of Washington in the state dining room and declared she could not abandon it to the enemy to be mocked and desecrated. While Carroll watched, all but rending his garments with anxiety, Dolley ordered the servants to take down the painting. It was screwed to the wall, and they lacked the tools or the time to deal with the problem. Dolley told them to break the frame and extract the canvas. At that point, "two gentlemen from New York" appeared, to see if they could help. Dolley gave them the painting with orders to conceal it from the oncoming British at all costs. Finally, with amazing self-possession, she closed her letter to Lucy: "And now, dear sister, I must leave this house . . . where I shall be tomorrow, I cannot tell!"

Again on the way to the door, Dolley further endangered Mr. Carroll's sanity by spotting a copy of the Declaration of Independence in a display case and stopping to extract it and stuff it into one of her suitcases. As they reached the door, one of the president's free black servants, Jim Smith, who had accompanied him to the battlefield, rode up on a sweaty horse and shouted: "Clear out! Clear out!" The British were only a few miles away. Dolley climbed into her carriage and rode off with Carroll to a comfortable refuge at his family mansion, Belle Vue, in Maryland.

III

The British arrived a few hours later, as darkness fell. With them was Admiral Cockburn, eager to savor the results of the victory. He and General Ross issued orders to burn the Capitol and the Library of Congress and headed for the White House. They were vastly amused to find the dinner Mrs. Madison had ordered still on the table in the dining room. "Several kinds of wine in handsome glass decanters were cooling on the sideboard," one officer wrote. They sampled some of the dishes and drank a toast to "Jemmy's health."

Soldiers roamed the house, grabbing souvenirs. One man strutted around with one of President Madison's hats on his bayonet and boasted that he would parade it through the streets of London if they failed to capture "the little president." Admiral Cockburn commandeered a portrait of Dolley and a cushion from one of her chairs, which inspired him to make vulgar comments about the size of the first lady's derriere.

Tiring of their fun and games, the British got down to business. Under Admiral Cockburn's direction, 150 men smashed out the windows and piled the furniture in the center of the various rooms. Outside, fifty of the marauders seized poles with oil-soaked rags on the ends and surrounded the house. At a signal from the admiral, men with torches ignited the rags and the poles were flung through the smashed windows like fiery spears. Within minutes a huge conflagration soared into the night sky. Not far away, the Americans had set the navy yard on fire, destroying numerous ships and warehouses full of uniforms, ammunition, and other war materiel. For a while, it looked as if all Washington was ablaze.

The next day, the British continued their depredations, burning the Treasury, the State and War Department, and other public buildings. But they were distracted and not a little spooked by a freak storm that suddenly erupted, with hurricane-force winds and violent thunder and lightning. Just before this display of nature's seeming wrath, an ammunition dump on Greenleaf's Point exploded while the British were preparing to destroy it. Thirty-five men were killed and forty-five suffered horrific injuries. The shaken British commanders decided to retreat to their ships.[8]

Meanwhile, Dolley received a note from Madison, urging her to join him in Virginia. After not a little wandering they were finally reunited. The president had barely slept in days. He had been in the saddle for five and six hours at a stretch. Dolley was deeply worried about his health. But he was determined to return to Washington as soon as possible, lest he be accused of cowardly flight. He insisted on Dolley staying in Virginia until he knew the city was safe. The moment this surety was confirmed, "you cannot return too soon." The words convey not only Madison's need for her, but his awareness that Dolley was an equal and in some ways a more potent symbol of his presidency. A chance encounter with some Washington refugees amply confirmed this remarkable fact. When they saw Dolley, they cheered her.[9]

Four days later, Dolley was back in Washington with her husband. They found hospitality at the home of her sister Anna Payne Cutts, who had taken over their house on F Street. The sight of the ruined capitol—and the charred, blackened shell of the White House—was almost unbearable for Dolley. For several days she grew morose and tearful whenever she thought of them. But she soon regained her legendary self-control. She realized that the president needed her help. One friend who saw Madison at this time described him as "miserably shattered and woebegone. . . . he looks heartbroken."[10]

The president saw the ravaging of Washington as a humiliating personal defeat. He felt betrayed by the incompetent general he had appointed and by the ragtag army that had abandoned him. He blamed the soldiers' lack of courage on the deluge of insults and denunciations of "Mr. Madison's War" from New England. The sneering Yankees had demoralized the country.

IV

Dolley undertook the task of restoring the president's battered morale. He was being assailed by advice from friends and pseudo-friends urging him to move the government to a safer place. The common council of Philadelphia issued an invitation to return to their hospitality, declaring they were ready to offer buildings for both the president and Congress. Dolley fervently maintained they should stay in Washington and the president agreed with her. He issued a call for an emergency session of Congress to meet on September nineteenth. Meanwhile, Dolley persuaded the Federalist owner of a handsome brick dwelling on New York Avenue and 18th Street known as the Octagon House to let the Madisons use it as an official residence. She opened the social season there with a reception on September twenty-first that was so crowded, the proponents of deserting Washington were nonplused and not a little annoyed.[11]

To her surprise and delight, Dolley found unexpected support elsewhere in the country. The numerous newspaper accounts of her festivities had made the White House a popular national symbol. People reacted with outrage when they heard that the British had burned the mansion. Next came a groundswell of admiration as newspapers reported Dolley's refusal to retreat and her rescue of George Washington's portrait and the copy of the Declaration of Independence. A paper with a national circulation declared, "The spirit of the nation is roused."

Madison's opponent for the presidency in 1812, DeWitt Clinton, said there was only one issue worth discussing now. Would the Americans fight back? His answer was a resounding *yes*. A revived President Madison issued a proclamation on September first, "exhorting all the good people" of the United States "to unite in their hearts and hands . . . to chastise and expel the invader."[12]

The citizens and soldiers of Baltimore soon demonstrated they were listening to these bold words. The British fleet assaulted the busy port in mid-September, hoping to batter its guardian fort into submission and force the city to pay a huge ransom. The men manning the guns in Fort McHenry fiercely resisted a nightlong bombardment. Francis Scott Key, an American aboard the British flagship, was sure that the fort would surrender. He had gone aboard the ship at the request of President Madison to try to negotiate the release of a doctor, William Beanes, seized by a British landing party. When Key saw the flag still flying at sunrise, he scribbled a poem that began, *Oh say can you see by the dawn's early light, what so proudly we hailed at the twilight's last gleaming?* Within a few days, the words of "The Star Spangled Banner" were put to the music of a popular drinking song and were being sung all over Baltimore.

Good military news reached Washington from more distant fronts. An American fleet on Lake Champlain won a surprise victory over a British armada escorting an invading army. The discouraged redcoats fought a half-hearted battle with American forces and retreated to Canada, "leaving their sick and wounded behind," the victorious American general reported. In Louisiana, an American army commanded by General Andrew Jackson seized Pensacola and Mobile, depriving another invading British army of a place to disembark. President Madison hailed these victories in a message to Congress.

But the unhappy politicians, meeting in the cramped quarters of Blodgett's Hotel, which they shared with the Post Office and Patent Office, were unimpressed. The House of Representatives voted 79–37 to consider abandoning Washington. Madison staunchly resisted the idea. Dolley summoned all her social resources to persuade the congressmen to change their minds. Octagon House dinners and receptions became mini-versions of her White House galas. For the next four months, the congressmen and senators debated while Dolley and her allies worked on them. Finally, both houses of Congress voted to stay in Washington and approved funds to rebuild the Capitol and White House. The president worked equally hard to achieve this victory, using patronage and all the other resources of his office.

V

The Madisons' worries were by no means over. In December 1814, the Massachusetts legislature called a conference of the five New England states to meet in Hartford, Connecticut. Vermont and New Hampshire refused to attend, but Connecticut and Rhode Island sent delegates. Rumors swept the nation that the Yankees were going to secede. At the very least, they were likely to demand a semi-independence that could become the death knell of the union. They were determined to demand a number of vital powers from the president and Congress, such as the right to declare an embargo and to order state militia into the national army. Another probable ultimatum was leaked to the press by one of the delegates: President Madison's resignation.[13]

Meanwhile, the British army attacking New Orleans had landed and clashed with General Jackson's troops. If they captured the Queen City, as many people already called it, they would control the Mississippi River Valley. In Hartford, the disunion convention adjourned and sent three delegates to Washington to confront the president with their demands. On the other side of the Atlantic, American envoys, headed by Albert Gallatin, Madison's secretary of the treasury, were negotiating with the British about possible peace terms. The reports they sent to Madison were glum. The British were making outrageous demands, aimed at reducing the United States to subservience. "The prospect of peace appears to get darker and darker," Dolley wrote to a friend.

On January 14, 1815, a profoundly worried Dolley wrote to Hannah Gallatin, the treasury secretary's wife: "The fate of N Orleans will be known today—on which so much depends."[14] She was wrong. The rest of January trickled away with no news from New Orleans. Meanwhile, the delegates from the Hartford Convention reached Washington, and the Madisons were relieved to learn the Federalists were not recommending secession. But they wanted amendments to the Constitution restricting the president's power and vowed to call another convention in June if the war continued. There was little doubt that this second session would recommend secession.

Federalists and other pessimists predicted New Orleans would be lost, and some people called for Madison's impeachment. On Saturday, February fourth, a messenger reached Washington with a letter from General Jackson. The president opened it with shaking hands and read the story of an astonishing total victory. Jackson and his men had routed the British veterans, killing and wounding almost 2,500 of them with a loss of only seven Americans. New Orleans—and the Mississippi River—would remain American territory. The news swept through Washington, and as night fell, thousands of cheering celebrators marched along the city's streets carrying candles and torches. Dolley placed candles in every window of Octagon House. In the tumult, the delegates from the Hartford Convention slunk out of town, never to be heard from again.[15]

Ten days later, even more astonishing news arrived from Europe. On February fourteenth, Henry Carroll, secretary to the peace delegation, reached Washington. He had landed in New York the previous day, and rumors of peace preceded him. Some people cheered him and others watched breathlessly as he rushed to Octagon House. A buoyant Dolley urged all her friends to come to a reception that evening. When they arrived, they were told that Carroll had brought a draft of a treaty of peace and the president was upstairs in his study, discussing it with his cabinet.

The house was jammed with congressmen and senators from both parties. A reporter from *The National Intelligencer* marveled at the way these political enemies were congratulating each other, thanks to the warmth of Dolley's smile and everyone's rising hopes that the war was over. "No one . . . who beheld the radiance of joy which lighted up her countenance" could doubt, the reporter wrote, "that all uncertainty was at an end." This was a good deal less than true. The president was not thrilled by the document, which offered little more than an end to the fighting and dying. But he decided that accepting it on the heels of the glorious news from New Orleans would make Americans feel they had won a second war of independence.

Outside the room where the president was making up his mind, Dolley had shrewdly stationed her pretty cousin Sally Coles. When the door opened and Sally saw smiles on every face, she rushed to the head of the stairs and cried: "Peace, Peace."[16] Octagon House exploded with joy. People rushed to embrace and congratulate Dolley. The butler began filling every wineglass in sight. Even the servants were invited to drink, and according to one account, they took two days to recover from the celebration.

For a few days, no one found fault with anyone or anything. From a president with a popularity rating that was close to zero, James Madison ascended to the zenith of national hero. Everyone knew there were two reasons for this miraculous transformation: General Andrew Jackson—and Dolley. Soon demobilized soldiers were marching past Octagon House. Dolley stood on the steps beside her husband, accepting their salutes. The partners in politics had become partners in fame.

VI

After the brutal partisanship of their first six presidential years, the Madisons' last two years in Washington were a comparative love feast. The president's popularity continued to soar. Congress was amazingly cooperative. Dolley's parties at Octagon House soon outsqueezed the crowds she had drawn at the height of her White House entertaining. Only one thing cast a shadow on their high spirits: the behavior of Dolley's handsome son, Payne Todd. Hoping to profit from the example of John Adams, whose son, John Quincy, had benefited from his youthful diplomatic experiences, Madison had appointed Payne to a secretarial post in the peace delegation.

Alas, Payne's reaction to this exposure to European culture was closer to that of Benjamin Franklin's grandson. Like Temple, Payne was primarily interested in pursuing women. He added an even more dangerous vice: gambling. When he returned home, he owed $6,500—over $100,000 in today's dollars. Payne had borrowed the money from London bankers, using his stepfather's credit. Now he blithely assumed Madison would pay these debts. Though he was deeply disappointed in the young man, the president paid—and never said a word to Dolley, who continued to adore Payne as the incarnation of manly perfection.

In the final weeks of his term, the president seemed to acquire new vitality and dignity. He broke his rule of never attending parties outside his residence and made a surprise appearance at a Christmas ball at the French embassy, where he chatted and joked with all comers. His secretary of the navy, William Jones, told a friend he had never seen Madison so happy. And why shouldn't he be? Jones added. He had "the applause of the nation" and he would soon be on his way home, liberated from public office.

Dolley presided over her final receptions at Octagon House with a constantly beaming smile. Everyone was welcome, from ambassadors to "the under-clerks at the post office." The last party turned into a frequently tearful farewell as Dolley's women friends talked of her innumerable acts of kindness for so many people. They soon began bewailing the prospect of life in Washington without her.[17]

The citizens of Georgetown, remembering Dolley's sixteen years of hospitality, decided to give a party for her. It was one of the most elaborate galas in the social history of Washington. The house was decorated with transparencies and paintings that recalled the major events of the Madisons' lives, along with framed tributes in poetry and prose. As a farewell gift, these decorations were carefully packed and shipped to Montpelier, where the Madisons used them to decorate one of the rooms in the mansion.

VII

Once settled in Montpelier, Dolley became a busy hostess again. The house had almost as many visitors as Thomas Jefferson's Monticello. At one point she fed ninety people (all of them local gentry) at a holiday dinner. She repeatedly assured her Washington friends that she was happy and contented. But her letters soon revealed that she missed the capital and its politics and colorful social life. As early as 1818, she confessed to her sister Anna that the beauty of spring at Montpelier reminded her of "the many happy scenes I have passed [in Washington] never, I fear, to return."

To another close friend, Dolley confessed boredom: "Our amusements in this region are confined to books and rural occupations." More and more, her letters begged friends for the latest political news and gossip. When her sister Anna, busy with numerous children and a lively social schedule, failed to write to her for a few days, Dolley came close to reproaching her for neglect.[18]

Theoretically, Dolley could have gone to Washington for a week or two at any time. But her husband was so dependent on her affection and company that she refused any and all invitations. Madison's health remained fragile. He regularly succumbed to flu-like fevers and was seldom without a cold during the winter months. He recruited Dolley to help him organize his voluminous papers, a task that stretched from months into years. Dolley recognized the importance of the papers to the nation's history and labored at his side, writing letters to old friends in pursuit of documents and taking dictation from Madison when rheumatism began making it difficult for him to write. "I cannot press him to forsake a duty so important, or find it in my heart to leave him during its fulfillment," she told one friend who sent her an invitation for a visit.[19]

The one person who might have eased Dolley's sense of isolation, her son Payne, spent as little time as possible with his mother and stepfather. Instead, he filled Dolley's nights and days with worry and anxiety, disappearing for weeks at a time. Only when bills he ran up at hotels and restaurants, or unpaid loans from Madison friends arrived in the mail did they discover where Payne was spending his time—and their money. In 1829, they were appalled to learn that he was in debtor's prison in Philadelphia. In his old age, Madison estimated he had spent $40,000—more than 800,000 modern dollars—settling Payne's debts. Most of the time, Madison never mentioned this drain on their finances to Dolley, knowing how much it would upset her.[20]

Dolley's brother, John C. Payne, was another drinker and gambler who became a financial leech. After various expensive adventures, he settled on a farm near Montpelier with a wife and growing family, where he continued to drink and require constant supervision. Another unpleasant surprise was the financial collapse of Anna's husband, Richard Cutts, who went bankrupt in the Panic of 1819. Although he had a government job, Cutts's bad investments overwhelmed him. He was soon on the brink of destitution. Dolley persuaded Madison to buy their house on Lafayette Square near the White House, and the Cutts continued to live there, thus surviving the crisis without public embarrassment.[21]

VIII

For twenty years, Dolley remained at Madison's side at Montpelier. Toward the end of his life, rheumatism made him an almost helpless invalid. But his mind remained marvelously unimpaired, and politicians continued to visit him and seek his advice on the country's problems, above all the mounting conflict over slavery and states' rights. In the 1830s, a coalescence of these two ideas threatened to sunder the American union that Madison had devoted so much time and thought to creating.

Orators in South Carolina began accusing Congress of favoring northern merchants with a policy of high tariffs. They insisted they had the right to "nullify" an act of Congress if it suited them. If Congress objected, the South Carolinians were prepared to secede from the union, which they argued was merely a "confederation" that individual states could vote to leave whenever it pleased them.

As the quarrel deepened, President Andrew Jackson turned to Madison for advice. The president's private secretary, Nicholas Trist, was married to Thomas Jefferson's granddaughter Cornelia Randolph, and belonged to the Philadelphia family that had played an intimate role in Madison's life during his years as a continental congressman. Trist became a link between Madison and Jackson.

The anti-tariff zealots had an alarming number of allies in Virginia. In 1831, a group met at the Orange County Court House, only five miles from Montpelier, and issued a statement that Madison found "extraordinary" for its total ignorance of the Constitution's arrangement of political power. If Virginia joined the South Carolinians, the union might well collapse. The zealots pointed to the two sets of resolutions that Madison and Jefferson had written to protest the Alien and Sedition Acts. Madison's statement had been endorsed by the Virginia legislature, Jefferson's by Kentucky. Jefferson's language was far more radical. He had used the word "nullification." The Virginia zealots tried to convince themselves and others that Madison agreed with Jefferson. Northern orators such as Daniel Webster truculently declared Congress's power was virtually absolute. Suddenly the threat of civil war was in the political air.

Though his rheumatic fingers could barely grip a pen, Madison plunged into the controversy, with Dolley's enthusiastic help. He published a long letter in the influential *North American Review,* declaring that the union created by the ratified Constitution was indissoluble, and explaining how it worked: Congress had certain powers such as the right to tax; other powers were left to the states. It was a "mixed" government with power carefully distributed to avoid two ever-present threats, tyranny and anarchy. The zealots were courting anarchy with their reckless talk of secession over a minor disagreement about the tariff. As for Mr. Jefferson's tilt toward nullification, Madison dismissed it as part of his friend's habit of "expressing in strong and round terms the impressions of the moment." Jefferson's entire political career testified to his devotion to the union.

The zealots' reaction was neither respectful nor friendly. They dismissed Madison as senile. He fired back hard-hitting answers that clearly proved they were wrong. Soon, to his and Dolley's delight, other Virginians were castigating the "youthful arrogance" of the nullifiers, who dared to ignore the living voice of the man who had created the Constitution. President Jackson, no great shakes as a constitutional thinker, instinctively sided with Madison. When South Carolina nullified a new tariff law and all but seceded from the union in 1833, Jackson was ready to act with the confidence that he had Madison's backing and approval. The president ordered the nullifiers to retract their stand or face an invading army under his personal command. The South Carolinians collapsed and the Union was preserved. The Madison partnership had saved the country a second time.[22]

IX

To an amazing extent, Dolley retained her vigor and good looks during these Montpelier years. When her colleague in decorating the White House, architect Benjamin Latrobe, visited her, he was stunned. "It seemed to me I had parted with her only yesterday, so little has time been able to change her personal appearance," Latrobe later wrote. "Not a wrinkle, no alteration in her complexion, no difference in her walk."[23] This escape from what Latrobe called "the spoiler" of old age made Dolley's devotion to Madison all the more remarkable. She could have found excuses to escape the narrow world in which she was confined. But her love for her husband made such an idea unthinkable. "I never leave him [for] more than a few minutes at a time," she told one friend, "and have not left the enclosure around the house for the last eight months."

Madison was completely bedridden for the last six months of his life. In her letters during these final days, Dolley called him "my patient," and in one letter described how she remained constantly at his bedside, "so deep is the interest, & sympathy I feel for him." During this trying period, she gratefully accepted the nursing help of her niece, Anna Payne, who lived not far from Montpelier. At least as important was the presence of Paul Jennings, Madison's slave valet, who had served him in this capacity for sixteen years.

Annie, as Anna Payne was known, soon became devoted to both Madisons. She reported how Dolley guarded her patient against an attempt to exploit him as a national symbol, at the expense of his dignity. By late June 1836, it was evident that Madison had only a few days to live. The doctors offered to prolong his life with stimulants so he could die on the Fourth of July, like Thomas Jefferson and John Adams. Madison declined the offer and Dolley tearfully supported him.

On the morning of June 28, 1836, Paul Jennings was at his bedside as usual. He noticed that Madison had trouble swallowing his breakfast. A visiting niece asked her uncle whether anything was wrong. Madison calmly replied, "Nothing more than a change of mind, my dear. " Before anyone could speak or move, Jennings said, "His head instantly dropped, and he ceased breathing as quietly as the snuff of a candle goes out." He died, Annie Payne wrote, "in the full possession of all his noble faculties."[24]

X

One of the first letters Dolley received after her husband's death came from Martha Jefferson Randolph. On July 1, 1836, she wrote from Washington:

> *I heard yesterday, my very dear friend, of a misfortune that I believe we were both too well prepared to accept. I would if possible be with you immediately, but shall be detained here some days by circumstances over which I have no control. Friday evening, probably, I shall be at [the Orange County] Court House and if you can send your carriage for me next morning, Cornelia [her daughter] and my self will go to you, if however you should have consented to withdraw yourself for a time from scenes of so much former happiness and present sorrow, tell me frankly, my dear friend, and we will delay our visit until you return home. One line left at the Court House will inform me of your present plan and determine mine. God bless and support you dear friend, under your present affliction, prays most affectionately and unalterably*

> *M Randolph.*[25]

Other letters came from Louisa Catherine Adams, a woman who had modeled her Washington career on Dolley's and took similar satisfaction in having helped her very different husband become president. President Andrew Jackson told Dolley "my own sensibility at the loss sustained by yourself and the nation" could add little to the overwhelming evidence of "the nation's sympathy."[26]

Dolley responded wholeheartedly to this outpouring of admiration and affection for her and her husband. She welcomed Martha Jefferson Randolph and three of her daughters, as well as nieces Anna Payne and Mary Cutts, to Montpelier. Young people were exactly what she needed to raise her spirits. Soon she was confiding to one of her oldest friends, Eliza Collins Lee, how she was dealing with her sorrow. She had resolved to "be calm, and strive to live long after him—that I should proceed to fulfill the trust he reposed in me." She was referring to the publication of Madison's papers, on which they had labored for so many years.[27]

This trust was almost betrayed by Dolley's attempt to involve her son Payne in the sale of the papers. She sent him to Philadelphia and New York to sound out publishers. It is hard to imagine a worse spokesman for a project that had to be sold as a noble venture, vital to the nation's understanding of its past. One New York writer who tried to advise Payne remarked that he was "the last man in the world to compass such a business." There was not much enthusiasm for the idea in the publishing world, and Payne managed to dissipate what little there was with his high-handed demands and arrogant style. Eventually, friends with political connections intervened and Dolley sold the papers to Congress in two installments, for a total of $50,000.[28]

Alas, much of this cash was committed to paying generous bequests in Madison's will to the College of New Jersey (Princeton), the University of Virginia, and other institutions he cared about. Still more was consumed by the debts Madison had accumulated in his last years to pay for Payne's extravagances and gambling and to keep Montpelier solvent. After the Panic of 1819, Virginia sank into an almost permanent recession. The same dismaying decline had ruined Thomas Jefferson's hopes of paying off his far larger debts at Monticello.

Dolley struggled to make Montpelier profitable, but it was almost impossible. Madison had sold three-fourths of the acreage to stay solvent during his lifetime and she had to support over one hundred slaves, many of them too old to work. She was extremely reluctant to sell them without their consent. She apparently sold a few to neighbors, which meant a family would not have to be separated. For most of these years, Dolley relied on an overseer to run the plantation. She also depended on Payne, who behaved in his usual irresponsible style, selling books, paintings, some of Madison's manuscripts, and an occasional slave without asking his mother's permission.

XI

Starting in 1837, Dolley was only a summer visitor at Montpelier. That year she moved to the Cutts house in Washington. Her sister Anna had died and her children had grown to adulthood. The house was available, and Dolley effortlessly rejoined the capital's social scene. In the month of December, she made no less than sixty-five calls on old friends and new acquaintances. Everyone who mattered or wanted to matter bombarded her with invitations to teas, balls, and dinners. President Martin Van Buren took great pleasure in welcoming her to the White House. One man who sat next to her at a dinner in 1839 enjoyed her company from start to finish. "The old lady is a very hearty good-looking woman of about 75," he wrote to a friend. "Soon after we were seated we became on the most friendly terms & I paid her the same attentions I should have done to a girl of 15—which seemed to suit her fancy very well."[29]

In 1844, Dolley was forced to give up her struggle to save Montpelier. She sold the property and its slaves to Henry W. Moncure, a wealthy Richmond merchant, who had already bought 750 acres in 1842. It caused her considerable anguish. She took eight months to sign the final papers, handing over the mansion. "No one," she wrote apologetically to the patient Moncure, "can appreciate my feeling of grief and dismay at the necessity of transferring to another a beloved home."[30]

Thereafter, Dolley became a permanent Washington resident. She was still hard pressed for cash. One of her most devoted friends was Paul Jennings. Dolley had sold him to Senator Daniel Webster with the understanding that he would be able to buy his freedom. Jennings visited Dolley regularly, sometimes bringing her food and often cash. "Mrs. Madison was beloved by every body in Washington, white and colored," he said.[31]

During these final years, Dolley became more than a popular guest at balls and dinners. She was hailed as a national treasure, a woman who had taken tea with George and Martha Washington and knew personally each of the next eleven presidents and their wives. Sarah Polk, another first lady who shared the political as well as the social side of the presidency with her husband, was especially fond of Dolley. In September 1845, when the Surviving Defenders of 1814, the men who had fought the British forays against Washington and Baltimore, gathered to commemorate their efforts, they marched in a body to Dolley's house to pay their respects.[32]

Invariably, Dolley accepted these tributes as "a token of remembrance of One who has gone before us." In death as well as in life, she shared her fame with James Madison. It was in this spirit that she joined Elizabeth Hamilton in the campaign to rescue George Washington's proposed monument. The two matrons watched from the White House while an "immense audience" cheered the laying of the cornerstone.

During these five years of continual admiration, Dolley grew frail. When an old friend sent her the first rose of the summer from her garden, Dolley thanked her extravagantly and then added that she had risen that morning with the sun and "felt as if I could fly with the aid of a sweet breeze then blowing, but now at 10 o'clock I am so ready for a nap I may not sign my name to this."[33] On July 8, 1849, Dolley was unable to rise from her bed. For several days, she slipped in and out of consciousness. When she awoke, her friend Eliza Collins Lee said, she would "smile her loving smile, put her arms out to embrace those she loved, and who were near her, and gently relapse into the rest that was peace."[34]

Dolley Madison died on July 12, 1849, at the age of eighty-one. President Zachary Taylor cancelled all government business and ordered a state funeral. The crowd was stupendous. All the mourners seemed to sense they were saying farewell to a woman who had not only lived history but made it. Another seventy years would pass before women won the right to vote and became accepted members of the political community. But Dolley's example remains a landmark on this long, torturous road. She demonstrated what women can bring to the often acrimonious world of American politics—good humor and tolerance of opposing points of view. Equally important is the virtue that won her respect as well as affection: her courage. Above all, Dolley epitomized the significance of a woman's love in the lives of men who sought history's most elusive prize, fame.

Appendix:

THE EROSION OF JEFFERSON'S IMAGE
IN THE AMERICAN MIND

Samuel F. Wetmore's ghostwritten account of Madison Hemings's recollections did not entirely disappear after its 1873 publication in the *Pike County Republican*. The African-American historian Arthur Calhoun concluded the story was "probably true" in his 1917 *Social History of the American Family*. W.E.B. Du Bois, a central figure in the evolution of African-American identity, cited it as true in his 1935 book *Black Reconstruction*. In 1954, *Ebony* magazine ran a photo essay, "Thomas Jefferson's Negro Grandchildren," about a group of African-Americans who said they were descended from Jefferson. Many were descendants of Sally Hemings. Others claimed Joe Fosset, son of Mary Hemings, Sally's sister, as their ancestor and asserted Jefferson was Joe's father. The magazine abandoned its usually moderate tone as it discussed the subject. The *Ebony* writer heaped sarcasm on white historians, claiming they were well aware that Jefferson had fathered numerous "slave concubines" by Sally and other members of the Hemings family.[1]

In 1960, Jefferson scholar Merrill D. Peterson printed a summary of the Madison Hemings story in his landmark book, *The Jefferson Image in the American Mind*. He found it "more credible" than most recollections of Jefferson. But he ultimately concluded the "miscegenation legend," as he called it, was not true. Peterson cited a story about an African-American carpenter named Robert Jefferson, who died in Ohio in 1882. This Jefferson had been a house slave who belonged to a man named Christian in

Charlestown, Virginia. He was born in 1803 and claimed that his mother repeatedly told him Jefferson was his father and he "had no reason to doubt her word." Peterson decided Madison's story was refuted by too many "incredible" black claims like Robert Jefferson's and "by the overwhelming evidence" of Jefferson's personal life.[2]

In 1968, historian Winthrop Jordan wrote another landmark book, *White Over Black: American Attitudes Toward the Negro, 1550–1812.* Jordan, a descendant of abolitionists, included more than a few paragraphs about Thomas Jefferson. He dismissed the story of Sally Hemings as stained by "the utter disreputability of the source" (James Thomson Callender). But he noted that Jefferson had been at Monticello nine months before the births of each of Sally's children.[3]

In 1974, historian Douglas Adair, long head of the Institute for Early American Culture at Williamsburg, came to Jefferson's defense in a posthumous book of essays titled *Fame and the Founding Fathers.* Included was a hitherto unpublished essay, "The Jefferson Scandals." Adair had done considerable research into the private lives of Jefferson's nephews, Samuel and Peter Carr. He emphatically backed Thomas Jefferson Randolph's testimony that they were guilty. Adair found that the older brother, Peter, had enjoyed a long-running relationship with Sally Hemings, so intense that it left him incapable of a happy marriage. The testimony of an overseer, Edmund Bacon, who worked at Monticello from 1806 to 1822, seemed to confirm this assertion. Bacon said he had seen ____ (he discreetly omitted the name) coming out of Sally's cabin on "many a morning." Adair believed he was talking about Peter Carr. His brother, Samuel Carr, had been the lover of Sally's niece, Betty, the child of her half sister, Mary.

Adair condemned with special vehemence the idea that sixteen-year-old Sally Hemings had conceived her first child when she was living in Jefferson's Paris residence. That would have meant Jefferson had traveled home from France with his teenage daughters "in the tight enforced intimacy of shipboard . . . with his pregnant mulatto mistress as the fourth member of the family group." He was even more skeptical that Sally became "the overruling passion" of Jefferson's later life—that Jefferson's desire for her approached the "obsessive"—making him "oblivious to and contemptuous of the public opinion of the great world and the private judgments of his intimate family." This did not jibe with the "thin-skinned censure-allergic Virginian" that he and other historians found when they studied Jefferson's political life.

If this story were true, Adair concluded, it would require everyone "not merely to change some shadings" in the portrait of Jefferson, but to "reverse the picture of him as an honorable man, painted by both the contemporaries who knew him well and the multitude of later scholars who have studied with care every stage in his career."[4]

Outside the scholarly community, Douglas Adair's exoneration of Jefferson was scarcely noticed. Several months before its publication, historian Fawn Brodie published *Thomas Jefferson, An Intimate History*. Brodie was a psychobiographer best known for her controversial life of Joseph Smith, the founder of the Mormon church. The details she reported about Smith's sex life and other matters earned her an excommunication from the church.

Brodie's *Intimate History* accepted Madison Hemings's narrative as true. She described it as "the most important single document" relating to the story of Jefferson and Sally Hemings. Her psychosexual study of Jefferson concluded that his affair with Sally was "a serious passion that brought both parties much private happiness over a period lasting thirty-eight years."

One reviewer called the book "a psychoanalytic history of Jefferson's complex mind and motivations" and praised it as a "compelling, compassionate case history." Numerous scholars disagreed. David Herbert Donald, Charles Warren Professor of American History at Harvard, accused Brodie of trying to portray Jefferson as "a secret swinger" and suggested that the title of her book should have been *By Sex Obsessed*. He thought Brodie refused to believe that the real Jefferson was "somewhat monkish, abstemious, continent and virtually passionless."

Garry Wills, professor of history at Northwestern University and the author of several books on or related to Jefferson, was even tougher on Brodie. "Two vast things, each wondrous in itself, combine to make this book a prodigy," Wills wrote. "The author's industry—and her ignorance."[5] Wills was outdone by Julian Boyd, the editor of the Jefferson Papers, who had spent thirty years of his life studying all of Jefferson's "recorded actions." He called Brodie's Jefferson a fiction created by those who "so eagerly embrace the concept of collective guilt, who project views of the rights of women and blacks into the past."[6]

In spite of these denunciations, Brodie's book was a publishing success. She was interviewed on the *Today* show, and her version of Jefferson and Sally soon created what the media call "a buzz." The book was on the *New York Times* bestseller list for thirteen weeks and sold 80,000 copies in hardback and 270,000 in paperback. *The Los Angeles Times* named Brodie Woman of the Year.[7]

The success of Brodie's book cast a shadow on the 1976 celebration of the two hundredth anniversary of the Declaration of Independence. One scholar wrote a troubled letter to Dumas Malone, the nation's leading Jefferson scholar, bemoaning the way *Ebony* was "spreading the Hemings canard all over their bicentennial issue and blacks hereabouts are reading it gleefully." In numerous articles around this time, *Ebony* regularly cited Brodie's book as the source for their denunciations of Jefferson. They accepted Brodie's contention that it was a love story, but claimed this only "intensified his [Jefferson's] hypocrisy."

Dumas Malone, close to completing his six-volume Jefferson biography, joined the fray with an article in the *Journal of Southern History* debunking Madison Hemings's story. Malone ended his article with a quotation from the editor of the *Waverly Watchman,* Pike County's Democratic newspaper. That sharp-tongued partisan compared Madison Hemings's claim to Jefferson's paternity to "a pedigree printed on the numerous stud horse bills" each spring. "No matter how scrubby the stock," the owners invented "an exalted lineage for their property." Such offensive language infuriated blacks already suspicious of the white Jefferson scholars' establishment— and alienated white sympathizers.[8]

In 1979, Barbara Chase-Riboud published *Sally Hemings, A Novel.* The Philadelphia-born author, who is also a gifted sculptor, accepted Fawn Brodie's premise of a deeply serious, caring relationship and created a story that opened in 1830 in Albermarle County not far from Monticello. Sally Hemings was free and living with her two freed sons, Eston and Madison. From there the story is told from several points of view and moves back and forth in time, with many pages devoted to Paris, where Jefferson supposedly fell in love with Sally.

According to the fictional Sally, Martha and Maria Jefferson were fully aware of the affair. Like Sally, they were in awe of Thomas Jefferson and never dreamed of objecting to the liaison. The novel closed with Jefferson dead and Sally freed by Martha after a bitter, mutually abusive exchange.

A few months later, Sally watched while all Monticello's blacks except some of the Hemings clan were sold at auction to pay Jefferson's debts. Sally bitterly regretted her inability to buy her sister Critta's children; she felt the scene was "my condemnation to everlasting hell."

The novel was the target of negative remarks from several historians. Chase-Riboud fired back in an interview in the *New York Times:* "These men have an overwhelming investment in Thomas Jefferson, they've spent their whole lives writing about this man. I have a similar emotional investment in Sally Hemings . . . I find it extraordinary that certified historians are rebutting a novel."

Some of these scholars were soon doing much more than rebutting her book. When they learned that CBS-TV and later NBC-TV were considering a miniseries based on the novel, they launched a letter-writing campaign to persuade the network to abandon the project. Both networks dropped the idea. These Jefferson defenders did not seem to realize that they were perilously close to violating the First Amendment. Barbara Chase-Riboud, out several hundred thousand dollars and understandably vexed, deplored some people's "presumed rights to interpret American history."

The 1990s began with a proclamation by the columnist George Will that suggested there was no need to worry about Sally Hemings or Thomas Jefferson's problems with slavery. Mr. Will named Jefferson the "man of the millennium." The Sage of Monticello summed up "the American idea" in his character—confident, serene, tolerant, curious—the epitome of a free man. His whole life, as a politician and statesman, a scientist, educator, and architect, bore witness to his extraordinary and unique greatness. With the collapse of communism and America's emergence as the world's only superpower, was James Parton's Jefferson going global?

In the scholarly community, very different thoughts were germinating. In 1993, the 250th anniversary of Jefferson's birth, historians under the leadership of Peter S. Onuf, who had succeeded Merrill Peterson as Thomas Jefferson Memorial Foundation Professor of History at the University of Virginia, convened a six-day Jeffersonian Legacies Conference. Mr. Onuf marshaled a wide range of scholars for the task of understanding "not only Jefferson in his own world" but also "his influence in shaping ours." The tone of the meeting was, in the words of one participant, "unreverential." Few speakers tried to defend Jefferson. Instead, his role in the history of race and slavery was frequently attacked.

Rhys Isaac of La Trobe University in Australia castigated the conclave for being too kind to Jefferson, who in his view had left America a legacy of inequality for blacks, women, and native Americans. Paul Finkelman, a visiting professor at Virginia Polytechnic Institute, acidly noted Jefferson's anguish when the Missouri Compromise crisis of 1820 raised the specter of civil war and disunion. Jefferson called it "an act of . . . treason against the hopes of the world." Finkelman said treason had indeed been committed in the land of the free—by Thomas Jefferson. The speaker who got the most press attention was Robert Cooley, a descendant of Sally Hemings. He told the scholars that they should take the Hemings oral tradition seriously. The lack of documentary evidence was hardly surprising, he said; Jefferson's white children made sure any incriminating records were destroyed after his death. No one challenged Mr. Cooley's assertions.[9]

Interest in Sally Hemings remained intense in many quarters of the cultural world. In 1995, Disney/Touchstone released a motion picture, *Jefferson in Paris,* starring Nick Nolte as Jefferson. The producers were Ismail Merchant and James Ivory, a team famed for historical dramas. The film accepted the Madison Hemings story of Sally becoming Jefferson's concubine in Paris.

Many scholars were beginning to feel overwhelmed by the seemingly relentless media determination to assume the Jefferson and Sally story was true. Andrew Burstein, professor of history at the University of Tulsa, exclaimed, "I'm becoming an endangered species, a Jeffersonian scholar that accepts the traditional notion that maybe a large number of Virginia slave owners did go to bed with their slaves but maybe Thomas Jefferson was not one of them."[10]

Another holdout was Joseph Ellis. His 1997 book *American Sphinx, The Character of Thomas Jefferson* ranged over Jefferson's long life, studying him at various periods. Ellis's approach reflected the new, unreverential attitude of the scholarly community. He cast an often jaundiced eye on the evolution of Jefferson as a national icon. Ellis debunked Jefferson's supposed originality, pointing out how often he borrowed resounding phrases and lofty ideas from other people. In his political career he was often guilty of duplicity, especially in his statements about his relationship with James Thomson Callender. But Ellis drew the line when it came to Sally Hemings. "The accusations of sexual promiscuity defy most of the established patterns of Jefferson's emotional life," he wrote.[11]

Even more unreverential was another book published in 1997, *American Scripture, Making the Declaration of Independence* by Pauline Maier, professor of history at Massachusetts Institute of Technology. Maier undertook to prove Jefferson was far from the sole author of the Declaration. Other members of the five-man committee, such as Benjamin Franklin, made important contributions, and the Continental Congress as a whole spent hours adding and subtracting clauses. Moreover, Maier uncovered in her research dozens of similar declarations issued by towns and state legislatures during the period when, in John Adams's words, the sentiment for independence became a "torrent" sweeping the reluctant Congress into the decision to break with the mother country. The Declaration, in short, was "the work not of one man but of many."

Maier found additional evidence to support her conclusion that the later worship of Jefferson and the Declaration, above all Abraham Lincoln's reverential tribute to it in the Gettysburg Address, was the product of similar mass emotions. Like the fervor of 1776, they were generated by political turmoil that "prepared" American hearts to receive the document as gospel truth. In Maier's view, Jefferson becomes a sort of unconscious plagiarist by taking credit for the Declaration. Sally Hemings and Jefferson's other personal flaws go unmentioned in this demolition.[12]

That same year (1997), Annette Gordon-Reed, an African-American lawyer on the faculty of New York Law School, published *Thomas Jefferson and Sally Hemings, An American Controversy.* Gordon-Reed had grown up in a Texas town where views on blacks and slavery had not changed very much since the Civil War. She had gotten interested in Jefferson when she read Winthrop Jordan's *White Over Black.* The movie *Jefferson in Paris* had focused her attention on Jefferson and Sally Hemings. She brought to her subject a sharp intelligence and a readiness to take on Jefferson's defenders, no matter how weighty their academic prestige.

Going Fawn Brodie one better, Gordon-Reed called Samuel Wetmore's ghostwritten version of Madison Hemings's life "the Rosetta Stone" of the puzzle and argued that white scholars were much too quick to dismiss black testimony. Essentially she maintained that what was at issue was not absolute proof, which can probably never be achieved, but of "controlling public impressions of the amounts and nature of the evidence."

Douglas Adair as well as subsequent biographers of Jefferson have argued that overseer Edmund Bacon was an objective eyewitness, with no ax to grind.[13] Gordon-Reed dug out the 1862 volume in which Bacon's recollections were first published. The last chapter had been omitted in a version published in the next century. In this missing chapter, Gordon-Reed found strong statements condemning the South for the Civil War and argued that the book was trying to reclaim Jefferson as an icon of a reunited America. She noted that Bacon proudly recalled a friend saying the overseer would "go into the fire if Thomas Jefferson asked him to." In short, he, too, was motivated to "shade the truth" about Sally Hemings.[14]

Many historians were impressed by Gordon-Reed's book. Charles B. Dew of Williams College called it "the definitive work on the Thomas Jefferson–Sally Hemings issue." In the media and in the historical community, there was a growing sense that it would not take much more evidence to convince a great many people that Jefferson and Sally had a relationship. In Charlottesville, Virginia, Dr. Eugene Foster was hard at work taking blood samples for his study of Jefferson-Hemings DNA.

Notes

BOOK ONE: George Washington

THE AGONIES OF HONOR

1. *The New York Herald*, March 30, 1877, p. 2, col. 5.
2. *The Papers of George Washington*, Colonial Series (hereafter *PGWCL*), University of Virginia, Charlottesville, VA, 1983–6:12, source note. The biographer who discovered the original letter was Bernard Knollenberg, *George Washington, The Virginia Period, 1732–1775* (Durham, NC, 1964).
3. Peter R. Henriques, *Realistic Visionary, A Portrait of George Washington* (Charlottesville, VA, 2006), 68. In his biography of Washington, Fitzpatrick wrote that claiming it was a love letter to Sally "requires an imagination unresponsive to the niceties of honor and good breeding." John C. Fitzpatrick, *George Washington Himself, A Common Sense Biography* (Indianapolis, 1934), 110.
4. Wilson Miles Cary, *Sally Cary, A Long Hidden Romance of Washington's Life* (New York, 1916) (privately printed), 13–18.
5. Ibid., 20 (note).
6. Ibid., 50. Sally Fairfax writes of the family's "impression that my husband's mother was a black woman." Also see James Thomas Flexner, *George Washington*, vol. 1, *The Forge of Experience* (Boston, 1965), 27.
7. Cary, *Sally Cary*, 23–24.
8. Douglas Southall Freeman, *George Washington*, vol. 1 (New York, 1948), 33, 43, 71, 73.
9. Ibid., 103.
10. Flexner, *Washington*, vol. 1, 19–20.
11. Ibid., 21.

12. Freeman, *Washington*, vol. 1, 193–94 ff.

13. Ibid., 264–66.

14. Ibid., 261.

15. Flexner, *Washington,* vol. 1, 242.

16. Freeman, *Washington*, vol. 2, 87.

17. Rupert Hughes, *George Washington, The Human Being and the Hero*, vol. 1 (New York, 1926), 203, 226–27. Also see Flexner, *Washington*, vol. 1, 162. Washington's letter implies "he was staying away from Belvoir, that his feelings were hurt."

18. GW to Sally Fairfax, November 15, 1757, *PGWCL*, vol. 5, 36.

19. GW to Sarah Fairfax, February 13, 1758, *PGWCL*, vol. 5, 93.

20. Flexner, *Washington*, vol. 1, 184–85.

21. Henriques, *Realistic Visionary*, 74.

22. GW to Col. Stanwix, March 4, 1758, *PGWCL*, vol. 5, 102.

23. Peter R. Henriques, "Major Lawrence Washington Versus the Reverend Charles Green: A Case Study of the Squire and the Parson," *Virginia Magazine of History and Biography* 100 (1992), 233–64.

24. Flexner, *Washington*, vol. 1, 193.

25. Ibid.

26. Washington to Martha D. Custis, July 20, 1758, *PGWCL*, vol. 5, 301.

27. Freeman, *Washington*, vol. 2, Appendix II, 404–6.

28. Washington to Sally Fairfax, September 25, 1758, *PGWCL*, vol. 6, 41–43.

29. Knollenberg, *George Washington*, 167, note 21.

30. John C. Fitzpatrick, *Writings of George Washington* (hereafter *WGW*), vol. 2, 170–71. Washington asked that the suit be shipped "as soon as possible." Also see Patricia Brady, *Martha Washington, An American Life* (New York, 2005), 63.

PARTNER IN LOVE AND LIFE

1. For relative money values, see Economic History Services, http://www.eh.net/hmit. For the lawsuit, see Freeman, *Washington,* vol. 3, 225, 282.

2. Henriques, *Visionary Realist*, 88–89.

3. Brady, *Martha Washington*, 31.

4. George Washington Parke Custis, *Recollections and Private Memoirs of Washington* (Philadelphia, 1861), 20.

5. Ibid.

6. Brady, *Martha Washington*, 55.

7. GW to John Alton, April 5, 1759, *PGWCL*, vol. 6, 200.

8. GW to Robert Cary & Co, May 1, 1759, *PGWCL*, vol. 7, 315–18.

9. GW to Richard Washington, September 20, 1759, *PGWCL*, vol. 6, 359.

10. Joseph E. Fields, ed., *"Worthy Partner," The Papers of Martha Washington*, with an introduction by Ellen McCallister Clark (Westport, CT, 1994), 149.

11. Rev. Jonathan Boucher to GW, August 7, 1768, *PGWCL*, vol. 8, 122–25.

12. Ibid.

13. GW to Benedict Calvert, April 4, 1773, *PGWCL*, vol. 9, 209.

14. B. Calvert to GW, April 18, 1773, *PGWCL*, vol. 9, 215.

15. GW to Burwell Bassett, August 28, 1762, *PGWCL*, vol. 7, 147.

16. John Parke Custis to Washington, July 5, 1773, *PGWCL*, vol. 9, 264.

17. Washington to Myles Cooper, December 15, 1773, *PGCWL*, vol. 9, 406–7.

18. Rupert Hughes, *George Washington, The Rebel and the Patriot, 1762–1777* (New York, 1927), 164. Freeman, *Washington*, vol. 3, 306–7.

19. Some historians cite a letter supposedly written by Pendleton, in which he wrote that Martha talked to them "like a Spartan to her son going to battle," urging them to "stand firm" against the British. This letter has never been found. It appeared for the first time in a nineteenth-century biography. She may have had such sentiments. Certainly her husband had them and probably discussed them with her.

20. GW to MW, June 18, 1775, *Papers of George Washington, Revolutionary Series* (hereafter *PGWRV*), vol. 1, 3.

21. GW to MW, June 23, 1775, *PGWRV*, vol. 1, 27.

FROM GREAT SOMEBODY TO LADY WASHINGTON

1. Fields, ed., *"Worthy Partner,"* from MW to Elizabeth Ramsay, December 30, 1775, 164.

2. Mercy Otis Warren to Abigail Adams, April 17, 1776, *PGWRV* 3, 75n.

3. Paul K. Longmore, *The Invention of George Washington* (Berkeley, CA, 1988), 204.

4. John Parke Custis to GW, June 10, 1776, *PGWRV* 2: 484–86.

5. Flexner, vol. 2, *George Washington and The American Revolution*, 359.

6. Elswyth Thane, *Mount Vernon Family* (New York, 1968), 48–58.

7. Freeman, *Washington*, vol. 5, 281–82.

8. Fields, ed., *"Worthy Partner,"* John Parke Custis to MW, October 12, 1781, 187–88n.

9. GW to Lafayette, November 15, 1781, "I arrived . . . to see poor Mr. Custis breathe his last." Fitzpatrick, ed., *WGW*, vol. 23, 340.

10. GW to JAW, January 16, 1783, Fitzpatrick, ed., *WGW*, vol. 26, 41–45.

11. Miram Anne Bourne, *First Family: George Washington and his Intimate Relations* (New York, 1982), 101–2.

12. Freeman, *Washington*, vol. 6, 211. One senator, previously hostile to GW, wrote: "It was a great dinner, and the best of the kind I ever was at."

13. Abigail Adams to Mary Cranch, July 12, 1789. Stewart Mitchell, ed., *New Letters of Abigail Adams* (Boston, 1947), 15.

14. Stephen Decatur Jr., *Private Affairs of Washington From the Records and Accounts of Tobias Lear, Esq., His Secretary* (Boston, MA, 1933), 62. Also see Bourne, *First Family*, 130.

15. Brady, *Martha Washington*, 181.

16. *"Worthy Partner,"* MW to Fanny Bassett Washington, October 23, 1789, 219. MW to Mercy Otis Warren, December 26, 1789, 223–24.

17. Freeman, *Washington*, vol. 7, 231.

18. David Freeman Hawke, *Paine* (New York, 1974), 319–20.

19. Thomas Fleming, ed., *Affectionately Yours, George Washington, A Self Portrait in Letters of Friendship* (New York, 1967), 243–44.

20. Patricia Brady, ed., *George Washington's Beautiful Nelly* (Columbia, SC, 1991). Introduction, 7. Also see Unger, *Unexpected George Washington*, (Hoboken, NJ, 2006) 256–57.

21. MW to Lucy Knox, undated, 1797, Fields, ed., *"Worthy Partner,"* 304–5.

22. Elizabeth Willing Powel to GW, March 11, 1797, Dorothy Towhig, ed., *Papers of GW Retirement Series* (hereafter *PGWRet*), vol. 1, 28–30.

23. GW to Elizabeth Willing Powell, *PGWRet*, vol. 1, 51–53.

24. GW to Sarah Fairfax, May 16, 1798, Fitzpatrick, *WGW*, 36, 262–66.

25. MW to SF, May 17, 1798, Fields, ed., *"Worthy Partner,"* 314–15.

26. Ibid., Introduction, xxv, 1797 letter to Tobias Lear.

27. EWP to MW, January 7, 1798, Fields, ed., *"Worthy Partner,"* 311–12.

28. Freeman, *Washington*, vol. 7, 620–25. Also see Henriques, *Realistic Visionary*, 187–204. Many people, including this writer, consider the latter the best account of Washington's death.

29. Fields, ed., *"Worthy Partner,"* xxxi.

30. Ibid., xxvii.

THE OTHER GEORGE WASHINGTON SCANDALS

1. Nigel Cawthorne, *The Sex Lives of the Presidents* (New York, 1998).

2. Allen French, "The First George Washington Scandal," *Massachusetts Historical Society Proceedings*, Vol. 65 (1935), 460–74, citing Public Record Office, AD1: 485–689.

3. Worthington Chauncey Ford, *The Spurious Letters Attributed to Washington* (Brooklyn, NY, 1889), Introduction, 1–13. The letters are reprinted in full in this book.

4. Ford, *Spurious Letters*, 69–76.

5. Troy O. Bickham, "Sympathizing with Sedition? George Washington, the British Press and British Attitudes during the American War of Independence," *William and Mary Quarterly*, January 2002, Third Series, vol. 59, no. 1, 101–122.

6. Ford, *Spurious Letters*, 26–29.

7. John Thornton Posey, *General Thomas Posey, Son of the American Revolution* (East Lansing, MI, 1992), 14.

8. Ibid., 14–18.

9. John C. Fitzpatrick, "The George Washington Scandals," Bulletin No. 1 of the Washington Society of Alexandria, 1929, 4–5. This is the *Scribner's* article "with some additions."

10. Posey, *General Thomas Posey*, 272–75.

11. Linda Allen Bryant, *I Cannot Tell a Lie* (Lincoln, NE), 2004, xii.

12. Ibid., 15.

13. Ibid., 20.

14. Ibid., 41–42.

15. Mount Vernon Fact Sheet on West Ford, 2000, 1–2. Mary Thompson, the research historian at Mount Vernon, has done a very thorough study of Washington's travel data for 1784–1785. It confirms that Washington never visited Bushfield between 1783 and his brother's death in 1787.

16. Joel Williamson, *New People, Miscegnation and Mulattoes in the United States* (New York, 1980), 49ff. Peter Henriques has kindly given me a copy of a speech he gave in 2005 about West Ford and his relationship to John Augustine Washington's family. He shows convincingly that the probable father was William Augustine Washington,

John's third son. William and Venus were about the same age. William died tragically at age seventeen when a gun held by a friend accidentally discharged. West Ford later named his son William.

17. Cawthorne, *Sex Lives*, 25. Fitzpatrick, *The George Washington Scandals*, 5–6.

BOOK TWO: Benjamin Franklin

THE SINS OF THE FATHER

1. Extracts from the diary of Daniel Fisher, 1755. *Pennsylvania Magazine of History and Biography*, vol. 17, 1893, 276–77.

2. Sheila L. Skemp, Benjamin Franklin, and William Franklin, *Father and Son, Patriot and Loyalist* (New York, 1994), 18.

3. Benjamin Franklin, *Autobiography* (New York, 1948), 96. All of the preceding narration is from this source.

4. William H. Mariboe, *The Life of William Franklin*, unpublished Ph.D. dissertation, University of Pennsylvania, 1962, 19–24. A half dozen opinions are considered here.

5. *The Papers of Benjamin Franklin* (hereafter *PBF*), vol. 2, 353–54.

6. BF to Catherine Ray, March 4, 1755, *PBF*, vol. 5, 503–43.

7. Ibid.

8. BF to Catherine Ray, September 11, 1755, *PBF,* vol. 6, 184.

9. BF to Jane Mecom, June 1748, *PBF*, vol. 3, 303.

10. Sheila L. Skemp, *William Franklin, Son of a Patriot, Servant of a King* (New York, 1990), 24–25.

11. W. Strahan to Deborah F., December 13, 1757 and January 1758, *PBF*, vol. 7, 297, 369.

12. WS to DF, December 13, 1757, *PBF*, vol. 7, 295–98.

13. BF to DF, June 27, 1760, *PBF*, vol. 9, 174.

14. Claude-Anne Lopez, *The Private Franklin, The Man and his Family* (New York, 1975, 83–84). In 1770, when Polly married a young doctor named Hewson, Mrs. Stevenson asked Franklin to give her away at the wedding ceremony.

THE OLDEST REVOLUTIONARY

1. BF to Margaret Stevenson, *November* 2, 1772, *PBF,* vol. 14, 299–300.

2. The *Craven Street Gazette* is printed in full in *PBF*, vol. 17, 220–26. A good sample of it can be read in David Freeman Hawke, *Franklin* (New York, 1976), 280–89.

3. Skemp, *William Franklin*, 83.

4. BF to WF, October 7, 1773, *PBF*, vol. 20, 437.

5. BF to WF, March 22, 1775, *PBF*, vol. 21, 545–99.

6. Lopez, *The Private Franklin*, 200. Also see Skemp, *William Franklin*, 178–79. Skemp suggests BF may have used comments he wrote to William in a 1774 letter that the British appeared to lack the discretion "to govern a herd of swine." Also see Mariboe, *William Franklin*, 436–37.

7. BF to WF, May 7, 1774, *PBF*, vol. 21, 212.

8. David Freeman Hawke, *Franklin* (New York, 1976), 1–2.

9. Lopez, *The Private Franklin*, 201.

10. WF to WTF, January 22, 1776, *Benjamin Franklin Papers*, American Philosophical Society, Philadelphia, vol. 101.

11. Walter Isaacson, *Benjamin Franklin, An American Life* (New York, 2003), 322.

12. Skemp, *William Franklin*, 222–223. Also see Mariboe, *William Franklin*, 471–72; and Lopez, *The Private Franklin*, 213–14.

MON CHER *PAPA*

1. J. C. Ballagh, *Letters of Richard Henry Lee* (New York, 1911–14), vol. 2, 202.

2. Claude-Ann Lopez, *Mon Cher Papa, Franklin and the Ladies of Paris* (New Haven, CT, 1966), citing Albert Henry Smyth, *The Writings of Benjamin Franklin* (New York, 1905), vol. 7, 132.

3. Gilbert Chinard, *Abbe Lefebvre de la Roche's Recollections of Benjamin Franklin*, Proceedings of the American Philosophical Society, vol. 44 (1950), 219.

4. Lopez, *Mon Cher Papa*, 38.

5. The recent HBO TV series on John Adams has a gross distortion of this combination of bathing and chess. The scene portrayed Franklin sitting naked in the bathtub with Madame Brillon.

6. Lopez, *Mon Cher Papa*, 64–65.

7. Lopez, *The Private Franklin*, 222–25.

8. Lopez, *Mon Cher Papa*, 259.

9. Ibid., 260.

10. Ibid., 265–67.

11. Ibid., 271.

12. Ibid., 269–70.

13. Lopez, *The Private Franklin*, 263–64.

14. WF to BF, July 22, 1784, *Smyth, Writings of Benjamin Franklin*, vol. 9, 264.

15. BF to WF, August 16, 1784, ibid., 252–54.

16. Skemp, *William Franklin*, 269.

17. Lopez, *The Private Franklin*, 273.

18. Nicholas Guyatt, "Adams Ribbed," *The Nation*, June 16, 2008, 39–43.

19. Lopez, *The Private Franklin*, 274. Lopez has been an editor of the Franklin papers as well as the author of the two best books on Franklin's private life.

20. Gaillard Hunt, ed., *Margaret Bayard Smith, The First Forty Years of Washington Society* (New York, 1906), 55–59.

21. Lopez, *The Private Franklin*, 275. Sally had seven children, all told. Three were born before Franklin left America in 1776.

22. Lopez, *Mon Cher Papa*, 299.

23. Ibid., 300.

24. Carl Van Doren, *Benjamin Franklin* (New York, 1941), 242.

25. Lopez, *Mon Cher Papa*, 314.

BOOK THREE: John Adams

AN AMOROUS PURITAN FINDS A WIFE

1. JA to Abigail Adams, January 23, 1775, Margaret A. Hogan and C. James Taylor, ed., *My Dearest Friend, Letters of Abigail and John Adams* (Cambridge, MA, 2007), 64. Also see Catherine Drinker Bowen, *John Adams and the American Revolution* (Boston, 1950), 534.

2. L. H. Butterfield, ed., *Diary and Autobiography of John Adams*, vol. 1 (New York, 1964), 194.

3. Ibid., 66–67.

4. *Diary of John Adams*, vol. 1, 109.

5. JA to Abigail Smith, October 4, 1762 (electronic edition), *Adams Family Papers: An Electronic Archive*, Massachusetts Historical Society, http://www.masshist.org/digitaladams/.

6. JA to Abigail Smith, October 4, 1762, ibid.

7. Abigail Smith to JA, May 9, 1764, ibid.

8. JA to Abigail Smith, September 30, 1764, ibid.

9. Abigail Smith to JA, October 13, 1764, ibid.

10. L. H. Butterfield, ed., *Letters of Benjamin Rush*, vol. 1, 1763–1792 (Princeton, NJ, 1951), 152–53. This letter is a good summary of the True Whig ideology.

11. AA to JA, March 31, 1776, and JA to AA, April 14, 1776, Hogan and Taylor, eds., *My Dearest Friend* (Cambridge, MA, 2007), 109–111.

12. Ibid., 108 (quoted in editor's commentary).

13. JA to AA, May 22, 1776, ibid., 119–20.

14. JA to AA, July 3, 1776, ibid., 121–23.

15. AA to JA, May 7, 1776, ibid., 115–16.

16. AA to JA May 27, 1776 (electronic edition), *Adams Family Papers: An Electronic Archive*, Massachusetts Historical Society, http://www.masshist.org/digitaladams/.

17. John Thaxter to John Adams, July 13, 1777, L. H. Butterfield et al., eds., *Adams Family Correspondence* (Cambridge, MA, 1963–93), vol. 2, 282.

18. Ibid., 370–71.

PORTIA'S DUBIOUS DIPLOMAT

1. John Adams autobiography, part 2, "Travels and Negotiations," 1777–1778, sheet 9 of 37 (electronic edition). *Adams Family Papers: An Electronic Archive*, Massachusetts Historical Society, http://www.masshist.org/digitaladams/.

2. JA to AA, April 25, 1778, Hogan and Taylor, eds., *My Dearest Friend*, 206.

3. AA to JA, March 8, 1778, ibid., 205.

4. AA to JA, November 12–23, 1778, and JA to AA, December 3 and December 18, 1778, ibid., 215–20.

5. AA to JA, January 1779 (electronic edition), *Adams Family Papers: An Electronic Archive*, Massachusetts Historical Society, http://www.masshist.org/digitaladams/.

6. JQA to AA, February 20, 1779, Adams Family Correspondence, vol. 3, 175–76.

7. Edith B. Gelles, *Portia, the World of Abigail Adams* (Bloomington, IN, 1992), 58–71, "A Virtuous Affair."

8. JA to AA, February 28, 1779, *Adams Family Correspondence*, vol. 3, 181–82.

9. JA to AA, February 28, 1779, ibid., 182 (a second letter written on the same day).

10. Richard B. Morris, *The Peacemakers, The Great Powers and American Independence* (New York, 1965), 15ff.

11. Gregg L. Lint et al., eds., *Papers of John Adams* (Cambridge, MA, 1996), vol. 10, 39–40.

12. JA to AA, March 16, 1780, Hogan and Taylor, eds., *Dearest Friend*, 234.

13. JA to AA, December 18, 1780 (electronic edition), *Adams Family Papers: An Electronic Archive*, Massachusetts Historical Society, http://www.masshist.org/digitaladams/.

14. AA to JL, *Adams Family Correspondence*, vol. 4, 165–66 (no date but assumed to be June 1781 based on Lovell's responses).

15. *PBF*, vol. 36, 226, note, *Adams Family Correspondence*, vol. 4, 284–86, n. 1, 295.

16. Phyllis Lee Levin, *Abigail Adams, A Biography* (New York, 1987), 141–44.

17. AA to JA, December 9, 1781 (electronic edition), *Adams Family Papers: An Electronic Archive*, Massachusetts Historical Society, http://www.masshist.org/digitaladams/.

18. AA to Mrs. Cranch, February 20, 1785, ibid., 279.

19. AA to Miss Lucy Cranch, September 5, 1784, ibid., 251.

20. AA to Mary Cranch, September 5, 1784, ibid., 166.

21. David McCullough, *John Adams* (New York, 2001), 371.

22. Elizabeth Smith Shaw to AA, March 18, 1786, *Adams Family Correspondence*, vol. 7, p. 93.

SECOND BANANA BLUES

1. AA to JA, July 27, 1788, Caroline Amelia Smith DeWindt, ed., *Journal and Correspondence of Miss Adams* (New York, 1841), vol. 2, 90–92.

2. AA to JA, May 14 and June 6, 1789, Hogan and Taylor, eds., *My Dearest Friend*, 324, 328.

3. Paul C. Nagel, *Descent from Glory, Four Generations of the John Adams Family* (New York, 1983), 45–47.

4. Ibid.

5. January 12, 1788, *The Diaries of John Quincy Adams: A Digital Collection,* Boston: Massachusetts Historical Society, 2008, http://www.masshist.org/jqadiaries.

6. AA to Mary Cranch, September 1, 1789, *Adams Family Correspondence*, vol. 8, 402–5, n. 5.

7. Nagel, *Descent from Glory*, 52.

8. Levin, *Abigail Adams*, 275.

9. JA to AA, February 4, 1794, Hogan and Taylor, eds., *Dearest Friend*, 355.

10. Jack Shepherd, *Cannibals of the Heart, A Personal Biography of Louisa Catherine and John Quincy Adams* (New York, 1980), 58.

11. AA to WSS, March 16, 1791, DeWindt, ed., *Journal and Correspondence of Miss Adams*, vol. 2, 109.

12. JA to AA, December 6, 1795, Hogan and Taylor, eds., *Dearest Friend*, 389–90.

13. JA to AA, December 28, 1792, ibid., 336–38.

14. AA to JA, December 31, 1793, ibid., 346–47.

PARTY OF TWO

1. JA to AA, February 10, 1796, *Adams Family Papers: An Electronic Archive*, Massachusetts Historical Society, http://www.masshist.org/digitaladams/.

2. JA to AA, January 9, 1797, ibid.

3. AA to JA, January 28, 1797, ibid.

4. Levin, *Abigail Adams*, 305.

5. JA to AA, March 13, 1797, ibid. JA to AA, March 2, 1797, Hogan and Taylor, eds., *Dearest Friend*, 442. JA to AA, May 4, 1797, Hogan and Taylor, eds., *Dearest Friend*, 448.

6. AA to JA, January 1, 1797 (electronic edition). *Adams Family Papers: An Electronic Archive,* Massachusetts Historical Society, http://www.masshist.org/digitaladams/.

7. AA to JA, December 31, 1796, ibid.

8. McCullough, *John Adams*, 489.

9. Richard N. Rosenfeld, *American Aurora* (New York, 1997), 237.

10. AA to Mary Cranch, February 28, 1798, Stewart Mitchell, ed., *New Letters of Abigail Adams. 1788–1801* (Boston, 1947), 137.

11. Ibid.

12. Levin, *Abigail Adams*, 344.

13. JA to Oliver Wolcott, September 24,1798, Charles Francis Adams, ed., *The Works of John Adams, Second President of the United States with a Life of the Author*, vol. 8 (Boston, 1853), 600–3.

14. John Ferling, *John Adams: A Life* (Knoxville, TN, 1992), 362.

15. AA to MC, May 26, 1798, Mitchell, ed., *New Letters*, 179.

16. Peter Shaw, *The Character of John Adams* (Chapel Hill, NC, 1976), 258. Page Smith, *John Adams*, 982.

17. Theodore Sedgwick to AH, February 22, 1799, Harold Syrett, ed., *Papers of Alexander Hamilton* (hereafter *PAH*), vol. 22 (New York, 1974), 494.

18. Robert Troup to Rufus King, April 19, 1799, Charles R. King, ed., *Life and Correspondence of Rufus King*, vol. 2 (New York, 1894–1900), 596–97.

19. Levin, *Abigail Adams*, 370–71.

20. JA to Timothy Pickering, May 17, 1799; JA to Wolcott, May 17, 1799; JA to Charles Lee, May 17, 1799, *Works of John Adams*, vol. 3, 648–50.

21. JA to AA, October 12, 1799 and October 27, 1799, Hogan and Taylor, eds., *Dearest Friend*, 466–68. Also see Smith, *John Adams*, 1015.

22. AA to MC, December 22, 1799, Mitchell, ed., *New Letters*, 222.

23. AA to MC, January 28, 1800, ibid., 228–29.

24. AA to MC, December 11, 1799, ibid., 219–21. Also see Ron Chernow, *Alexander Hamilton* (New York, 2004), 601.

25. Smith, *John Adams*, 1027–28.

26. Levin, *Abigail Adams*, 380.

27. AA to MC, Nov. 21, 1800, Mitchell, ed., *New Letters*, 256.

28. Smith, *John Adams*, 1049.

29. AA to Sally Smith Adams, December 8, 1800, Mitchell, ed., *New Letters*, 261–62.

30. McCullough, *John Adams*, 556.

31. AA to TBA, November 13, 1800, Mitchell, ed., *New Letters*, 431.

REMEMBERING SOME OTHER LADIES

1. Levin, *Abigail Adams*, 479.

2. Nancy Rubin Stewart, *The Muse of the Revolution, The Secret Pen of Mercy Otis Warren and the Founding of the Nation* (Boston, 2008), 247–56. Also see James Grant, *John Adams, Party of One* (New York, 2005), 433ff.

3. McCullough, *John Adams*, 594. Also see Grant, *Party of One*, 433; and Levin, *Abigail Adams*, 424–26.

4. Joseph J. Ellis, *Passionate Sage*, (New York, 2001), 72.

5. Levin, *Abigail Adams*, 426–27.

6. Stuart, *Muse of the Revolution*, 262–64. Also see Ellis, *Passionate Sage*, 184.

7. Paul C. Nagel, *The Adams Women* (New York, 1987), 126.

8. Levin, *Abigail Adams*, 449–50; McCullough, *John Adams*, 602.

9. Nagel, *The Adams Women*, 145.

10. AA to TJ, September 20, 1813, Lester J. Cappon, ed., *The Adams-Jefferson Letters* (Chapel Hill, NC, 1959), 378.

11. Nagel, *The Adams Women*, 170.

12. Ibid., 168.

13. Ellis, *Passionate Sage*, 198.

14. Ibid., 198.

15. Smith, *John Adams*, 1123.

16. Ibid., 1124–25.

17. Ellis, *Passionate Sage*, 200.

18. Josiah Quincy, *Figures of the Past: From the Leaves of Old Journals* (Boston, 1883), 64–65.

19. McCullough, *John Adams*, 646. In his final moments, Adams said, "Help me! Help me!", to a granddaughter.

BOOK FOUR: Alexander Hamilton

BASTARD SON AND WARY LOVER

1. James Thomas Flexner, *The Young Hamilton* (Boston, 1978), 13. Also see Ron Chernow, *Alexander Hamilton* (New York, 2004), 11. Rachel and her mother had well-to-do relatives on St. Croix. That none of them came to her defense indicates the charge of double adultery was true.

2. Forrest McDonald, *Alexander Hamilton* (New York, 1979), 7. Also see Chernow, *Hamilton*, 16–17. He discusses the two dates at some length and also opts for 1755.

3. AH to JH, June 22, 1785, *PAH*, vol. 3, 617. Flexner, *Hamilton*, 26.

4. AH to ES, November 11, 1769, *PAH*, vol. 1, 4.

5. *PAH*, vol. 1, 6–7.

6. Ibid., 35–38.
7. AH to CL, April 11, 1777, *PAH*, vol. 1, 226.
8. Robert Hendrickson, *Hamilton* (New York, 1976), vol. 1, 54–55.
9. AH to JL, April 1779, *PAH*, vol. 1, 37–38 (begins on 34).
10. Ibid., 348.
11. AH to ES, October 5, 1788, *PAH*, vol. 2, 455.
12. AH to ES, August 1780, *PAH*, vol. 2, 398.
13. AH to MS, January 21, 1781, *PAH*, vol. 2, 539.
14. AH to PS, February 18, 1781, ibid., 563–67.

THE WOMAN IN THE MIDDLE

1. Hendrickson, *Hamilton*, 531–32.
2. AH to Angelica Church, December 6, 1787, *PAH*, vol. 4, 374–76.
3. Hendickson, *Hamilton*, 530.
4. AH to AC, November 8, 1789, *PAH*, vol. 5, 501–2.
5. EH to AC, November 8, 1789, *PAH*, vol. 5, 502.
6. Hendrickson, *Hamilton*, vol. 2, 20.
7. Ibid.
8. Forest McDonald, *Alexander Hamilton* (New York, 1979), 229.
9. Ibid., 222.
10. JM to TJ, July 10, 1791, James Morton Smith, ed., *The Republic of Letters, The Correspondence between Thomas Jefferson and James Madison* (New York, 1995), vol. 2, 695–96.
11. McDonald, *Hamilton*, 229.
12. *PAH*, vol. 21, 251–52.
13. *PAH*, vol. 21, 269. Reynolds Pamphlet, Appendix II, 1792.
14. Julian P. Boyd, ed., *Papers of Thomas Jefferson* (hereafter *PTJ*) vol. 18 (Princeton, NJ, 1971), 635.
15. TJ to GW, September 9, 1792, *PTJ*, vol. 24, 353.
16. Bernard C. Steiner, *Life and Correspondence of James McHenry* (New York, 1979), 129.
17. Katherine Schuyler Baxter, *A Godchild of Washington* (New York, 1897), 224.
18. Many people think Hamilton could not have become president because he was born outside the borders of the United States. But this proviso in the Constitution applied to immigrants who arrived in the United States only after the Constitution was written and ratified in 1787–88. Hamilton came to America in 1773, before the Revolution began. Also see Chernow, *Alexander Hamilton*, 508–509, for a discussion of various reasons why Hamilton was never a candidate. The Maria Reynolds scandal was a primary factor.
19. Callender, *History of the United States for 1796* (Philadelphia, 1797), 220–22 (Google Books, online edition).
20. *PAH*, vol. 21, 135ff. Introductory Note and letter from Oliver Wolcott Jr., July 3, 1797. Additional letters to James Monroe, Frederick A. C. Muhlenberg, Abraham B. Venable, James Thomson Callender, et al.

21. *PAH*, vol. 21, 238ff, Reynolds Pamphlet.

22. Broadus Mitchell, *Hamilton, The National Adventure* (New York, 1962), 714. Also see Julian P. Boyd's eighty-page essay in *PTJ*, vol. 18. Boyd agrees in toto with Callender and in many instances goes beyond him.

23. GW to AH, August 21, 1797, *PAH*, vol. 21, 214–15.

24. Mitchell, *Hamilton*, 417–18.

25. Hendrickson, *Hamilton*, vol. 2, 420.

26. JBC to AH, July 13, 1797, *PAH*, vol. 21, 163.

27. *PAH*, vol. 25, 436. Hosack became the Hamiltons' family doctor. Later he recalled that Hamilton was the nurse "in every important case of sickness that occurred in his family."

LOVE'S SECRET TRIUMPH

1. King, ed., *Life and Correspondence of Rufus King*, vol. 2, 330.

2. AH to EH, June 3, 1798, *PAH*, vol. 21, 482.

3. Hendrickson, *Hamilton*, vol. 2, 661–64.

4. Thomas Fleming, *Duel: Alexander Hamilton, Aaron Burr and the Future of America* (New York, 1999), 360–61.

5. Ibid., 355–57.

6. Ibid., 361.

7. James A. Hamilton, *Reminiscences of James A. Hamilton* (New York, 1869), 65.

8. Jesse Benton Fremont, *Souvenirs of My Time* (Boston, 1887), 117.

9. Baxter, *A Godchild of Washington*, 222.

BOOK FIVE: Thomas Jefferson

ROMANTIC VOYAGER

1. Henry S. Randall, *Life of Thomas Jefferson* (New York, 1958), vol. 1, 40–41.

2. Marie Kimball, *Jefferson, The Road to Glory* (New York, 1943), vol. 1, 67–68.

3. TJ to John Page, October 7, 1763, *PTJ*, vol. 1, 11–12.

4. While writing *Duel: Alexander Hamilton, Aaron Burr and the Future of America* (New York, 1999), the author consulted Robert B. Daroff, professor of neurology at Case Western Reserve School of Medicine, an internationally recognized expert on migraine and other headache disorders; Aaron Burr also suffered from migraines (*Duel*, 421, n. 17.)

5. TJ to John Page, January 19, 1764, *PTJ*, vol. 1, 13.

6. TJ to John Page, April 9, 1764, *PTJ*, vol. 1, 17.

7. Dumas Malone, *Jefferson The Virginian* (Boston, 1948), vol. 1, 84–85.

8. Ibid., 154–55.

9. TJ to John Page, February 21, 1770, *PTJ*, vol. 1, 36.

10. TJ to James Ogilvie, February 20, 1771, *PTJ*, vol. 1, 63.

11. TJ to Robert Skipwith, August 3, 1771, *PTJ*, vol. 1, 78.

12. Kimball, *Jefferson, The Road to Glory*, 174–75.

13. TJ to Thomas Adams, February 20, 1771, *PTJ*, vol. 1, 61.

14. Sarah N. Randolph, *The Domestic Life of Thomas Jefferson* (Charlottesville, VA, 1947), 24–25 (Reprint). Fawn M. Brodie, in her book *Thomas Jefferson, An Intimate History* (New York, 1974), contends John Skelton was still alive when Jefferson married Martha. Her evidence is unconvincing. The bridegroom's attempt to describe his bride as a spinster makes it even more unlikely. Kimball, Jefferson, 176, note 28, confirms this interesting change.

15. Some scholars have had doubts about the bottle of wine. It originated with Henry Randall, Jefferson's mid–nineteenth-century biographer, who talked with surviving members of the Jefferson family as part of his research.

THE TRAUMAS OF HAPPINESS

1. Negro slave wet nurses were common in the South at this time. Many wealthy families used them. Julia Cherry Spruill, *Women's Life and Work in the Southern Colonies* (New York, 1972), 56–57.

2. Frederick D. Nichols and James A. Bear Jr., *Monticello* (Charlottesville, VA, 1967), 13–14.

3. Malone, *Jefferson The Virginian*, 121–22.

4. Randolph, *Domestic Life of Thomas Jefferson*, 26–28.

5. This statement is conjectural from several points of view. Elizabeth Hemings's mother was black. The Virginia law in the Howell case was aimed at white women who had children by black men. But in later years, Jefferson wrote a letter arguing that people like the Hemingses, born of mixed marriages for two generations, should be free. Another conjecture finds doubts that Wayles was the children's father. The author feels that Jefferson's special concern for the Hemingses, over several decades, makes this claim dubious.

6. Kimball, *Jefferson, The Road to Glory*, 118–19.

7. JA to Timothy Pickering, August 6, 1822, John Francis Adams, ed., *John Adams, Life and Works*, vol. 100, 239. This old man's memory requires several grains of salt.

8. TJ to John Randolph, August 25, 1775, *PTJ*, 241.

9. Boyd, ed., *Papers of Thomas Jefferson*, vol. 8, 312, n.

10. TJ to RHL, July 29, 1776, *PTJ*, vol. 1, 477.

11. TJ to JH, October 11, 1776, *PTJ*, vol. 1, 524; RHL to TJ, November 3, 1776, *PTJ*. vol. 1, 589.

12. TJ to Timothy Matlack, April 18, 1781, *PTJ*, vol. 5, 490.

13. TJ to GW, October 28, 1781, *PTJ*, vol. 6, 185.

14. TJ to ER, September 16, 1781, *PTJ*, vol. 6, 117–18, and ER to TJ, October 9, 1781, *PTJ*, vol. 6, 128–29.

15. Some years ago, the author discussed Martha's health with the late Alvan R. Feinstein, MD, Sterling Professor of Medicine at Yale Medical School. Dr. Feinstein, while properly cautious about such a long-range diagnosis, thought diabetes was a likely source of her childbirth woes. He noted that according to a

family tradition, her babies grew larger with each birth—a common symptom of this disease.

16. TJ to JM, May 20, 1782, *PTJ*, vol. 6, 185.

17. *PTJ*, vol. 6, 196.

18. Rev. Hamilton W. Pierson, ed., *Jefferson at Monticello: The Private Life of Thomas Jefferson* (New York, 1862), 106–7 (Michigan Historical Reprint Series).

19. Randolph, *Domestic Life*, 40–41.

20. TJ to EE, October 31, 1782, *PTJ*, vol. 6, 198–99.

HEAD VERSUS HEART

1. JM to Edmund Randolph, November 12, 1782, William T. Hutchinson and William M. E. Rachal, eds., *Papers of James Madison* (hereafter *PJM*), vol. 5 (Chicago, 1967), 272–73.

2. Malone, *Jefferson The Virginian*, 407–23.

3. Howard C. Rice Jr., *Thomas Jefferson's Paris* (Princeton, NJ, 1976), 42–43, 103.

4. Kimball, *Jefferson, The Scene of Europe*, vol. 3 (New York, 1950), 9.

5. Ibid.

6. Malone, *Jefferson and the Rights of Man*, vol. 2 of *Jefferson and His Time* (Boston, 1951), 11–12.

7. TJ to Carlo Bellini, professor of modern languages at the College of William and Mary, September 30, 1785, *PTJ*, vol. 8, 568–69.

8. Douglas Adair, *Fame and the Founding Fathers* (New York, 1974), 190.

9. Rice, *Thomas Jefferson's Paris*, 67.

10. Jon Kukla, *Mr. Jefferson's Women* (New York, 2007), 156.

11. TJ to Elizabeth Trist, December 15, 1786, *PTJ*, vol. 10, 600.

12. Malone, *Jefferson and the Rights of Man*, 71.

13. Ibid., notes 14 and 15. TJ described the accident as "one of those follies from which good cannot come but ill may."

14. John P. Kaminski, ed., *Jefferson In Love, the Love Letters between Jefferson and Maria Cosway* (Lanham, MD, 2001), 44–64.

15. Ibid., 65.

16. Kukla, *Jefferson's Women*, 104–5. This is the most convincing summary of Maria Cosway's second visit to Paris, during which she largely ignored Jefferson and vice versa.

17. Kimball, *Jefferson, The Scene of Europe*, 303–4.

18. Levin, *Abigail Adams*, 298–99.

19. Kimball, *Jefferson, The Scene of Europe*, 305.

20. Annette Gordon-Reed, *The Hemingses of Monticello, the Story of an American Family* (New York, 2008), 173.

21. Malone, *Jefferson and the Rights of Man*, 246.

22. William H. Gaines Jr., *Thomas Mann Randolph, Jefferson's Son In Law* (Louisiana State University Press, 1966), 15–24.

23. MJR to TJ, April 25, 1790, *PTJ*, vol. 26, 225.

THE WAGES OF FAME

1. Malone, *Jefferson and the Rights of Man,* 436ff. The chapter, "Hamilton Vs Jefferson" (462–77) is perhaps the best summary of the dispute.

2. Noble Cunningham Jr., *In Pursuit of Reason, The Life of Thomas Jefferson* (Louisiana State University Press, 1987), 207–8.

3. Michael Durey, *With the Hammer of Truth: James Thomson Callender and America's Early National Heroes* (Charlottesville, VA, 1990), 157–8. Also see *Richmond Recorder*, September 29, 1802, for reference to Madison.

4. *Richmond Recorder*, November 17, 1802. There are several more verses.

5. Durey, *With the Hammer of Truth*, 162. Merrill D. Peterson, *Thomas Jefferson and the New Nation* (New York, 1970), 709.

6. *Richmond Recorder*, September 15 and September 22, 1802.

7. Gordon-Reed, *The Hemingses of Monticello*, 496–502.

8. Malone, *Jefferson the President, First Term*, vol. 3 of *Jefferson and His Time* (Boston, 1970), 222–23.

9. Edwin M. Betts and James A. Bear Jr., eds, *The Family Letters of Thomas Jefferson* (Columbia, MO, 1966), 240.

10. Thomas Fleming, *The Louisiana Purchase* (New York, 2003). This brief book, part of a "Turning Points in American History" series, is a good summary of Louisiana story. It includes an extensive bibliography.

11. Durey, *With the Hammer of Truth*, 165–66.

12. Isaac Newton Phelps Stokes, *Iconography of Manhattan Island, 1498–1909*, vol. 5 (New York, 1915–1928), 1422. This describes a triumphant parade in New York staged by Mayor DeWitt Clinton, hailing Jefferson and the Louisiana Purchase.

13. Malone, *Jefferson The President, First Term*, 411–15.

14. Peterson, *Thomas Jefferson and the New Nation*, 789–90.

15. Levin, *Abigail Adams*, 413–19. Also see Kukla, *Jefferson's Women*, 148–50.

16. Gaines, *Thomas Mann Randolph*, 48.

17. Ibid., 64–67.

18. Gaines, *Thomas Mann Randolph*, 78–79.

19. TJ to ER, July 10, 1805, *Family Letters*, 276.

20. TJ to ER May 21, 1805, ibid., 271.

21. Martha Jefferson Randolph to TJ, June 29, 1807, ibid., 302–3.

22. Randolph, *Domestic Life*, 294–96.

23. Ibid., 298.

24. The Thomas Jefferson Encyclopedia, http://wiki, Monticello.org, mediawiki/index. php, Nailmaking. Also see Peter S. Onuf, ed., *Jeffersonian Legacies* (Charlottesville, VA, 1993), "Those Who Labor For My Happiness, Thomas Jefferson and his Slaves," by Lucia Stanton, 153–55.

25. Malone, *The Sage of Monticello*, vol. 6 of *Jefferson and His Time*, 511.

26. Randall, *The Life of Thomas Jefferson*, vol. 3, 326–27, note.

27. Gaines, *Thomas Mann Randolph*, 142–62.

28. TJ to JM, February 17, 1826, *Republic of Letters, 1964–1967.*

29. Pauline Maier, *American Scripture: Making the Declaration of Independence* (New York, 1997), 186. This superb book describes in convincing detail the rise of the Declaration to prominence in the American psyche, and Jefferson's identification with it.

30. Alan Pell Crawford, *Twilight at Monticello, The Final Years of Thomas Jefferson* (New York, 2008), 243–44.

31. Ibid., 247–49.

32. Ibid., 249.

33. Ibid., 257–60. Also see Jack McLaughlin, *Jefferson and Monticello, The Biography of a Building* (New York, 1988), 380.

IF JEFFERSON IS WRONG, IS AMERICA WRONG?

1. *Nature,* vol. 396, no. 6706, November 5, 1998, 27–28.

2. *New York Times*, November 1, 1998. *Washington Post*, November 23, 1998.

3. Peterson, *Thomas Jefferson*, 996.

4. Peterson, *The Jefferson Image in the American Mind*, 182–83.

5. Ibid., 183–84.

6. Randolph, *Domestic Life*, Introduction, vii–viii.

7. Peterson, *The Jefferson Image*, 231–32.

8. Ibid., 233–34.

9. Cynthia H. Burton, *Jefferson Vindicated*, foreword by James A. Bear, emeritus director of the Thomas Jefferson Memorial Foundation (Keswick, VA, 2005), 115.

10. *Pike County Republican*, March 13, 1873. The full text can also be found in *Report on Thomas Jefferson and Sally Hemings*, Thomas Jefferson Memorial Foundation Research Committee, January 2000, 29–31.

11. Peterson, *The Jefferson Image*, 185, note.

12. Burton, *Jefferson Vindicated*, 116–17.

13. Recollections of Israel Gillette Jefferson, *Pike County Republican*, December 25, 1873 (original in Ohio Historical Society), Report on Thomas Jefferson and Sally Hemings, Thomas Jefferson Memorial Foundation Research Committee, January 2000, 32–34.

14. Letter of Thomas Jefferson Randolph to editor of *Pike County Republican*, undated, original, University of Virginia Library, Accession No. 8937, Report on Thomas Jefferson and Sally Hemings, 35–40.

15. For almost a century, Monticello had been owned by the Levy family. It was purchased in 1834 by Uriah Levy, a Philadelphian who rose to the rank of commodore in the U.S. Navy. Marc Leeson, *Saving Monticello, the Levy Family's Epic Quest to Save the House that Jefferson Built* (New York, 2001).

16. Peterson, *The Jefferson Image*, 358.

17. *New York Times*, October 15, 2006 (obituary of Mrs. Bennett).

18. *Nature*, vol. 396, no. 6706, November 5, 1998.

19. *New York Times* letters to the editor, November 9, 1998.

20. *New York Times* letters to the editor, November 6, 1998.

21. Report on Thomas Jefferson and Sally Hemings, 10.

22. Response to the Minority Report, prepared by Lucia C. Stanton, Shannon Senior Research Historian, April 26, 2000.

23. *Jet*, February 14, 2000.

24. Jefferson's Blood Interviews, Dr. Eugene Foster, 6.

25. *American Heritage*, February-March 2002, vol. 53, issue 1.

26. http://www.monticello.org/Matters/people/hemings-jefferson_contro.html,"Matters of Fact, The Hemings-Jefferson Controversy: A Brief Account." There is a separate biography of Sally Hemings under "Matters of Fact." For the revised statement, see: http://www.Monticello.org/planatation/hemingscontro/hemings-jefferson.

27. "A Daughter's Declaration," *John Hopkins Magazine*, September 1999, 21–27.

28. *New Yorker*, December 1, 2008, 34–38.

29. Steven Corneliussen's essay can be read online at TJscience.com.

30. Burton, *Jefferson Vindicated*, 116, citing Albermarle County Minute Book, 1830–31:123.

31. Burton, *Jefferson Vindicated*, 42.

32. Ibid., 88, citing Callender in the *Richmond Recorder*.

33. At the end of the nineteenth century, in the aftermath of the Spanish-American War, which spawned outrageous lies in newspapers, Joseph Pulitzer created a sensation when he told his reporters that henceforth they were expected to tell the truth. W. A. Swanberg, *Joseph Pulitzer* (New York, 1967), 254–55.

34. For Wetmore's service record see: Military Pension File, Enlistment Record, Microfilm T-288, roll #509, National Archives and Records Administration, Washington D.C. For the Adaline Rose lawsuit, see Docket Book #3, Pike County District Court, 1878.

35. *The Pike County Republican*, October 14, 1875, The Adaline Rose lawsuit was dismissed a year later, in March 1876, when Wetmore had long since fled the scene.

36. Burton, *Jefferson Vindicated*, 112. Noted Columbia University scholar Eric L. McKittrick thought, after reading these accounts, that Jefferson, far from being a sensualist, was more like "a tightlipped Irish pastor trying to keep the lid on a parish." McKittrick thought the real issue was not Jefferson's personal guilt but "the psychosexual dilemma of an entire society confronting slavery." "The View From Jefferson's Camp," *The New York Review of Books*, December 17, 1970.

37. Ibid., 19–20.

38. Douglas Adair, *Fame and the Founding Fathers* (New York, 1974), "The Jefferson Scandals," 160–191.

39. Bernard Mayo, ed., Preface by James A. Bear Jr., *Thomas Jefferson and His Unknown Brother* (Charlottesville, 1981), 1–6. Also see *Family Letters*, 66, 182, 343. In *Jefferson Vindicated*, Cynthia Burton also examines Randolph as a potential father, 52–60.

40. McLaughlin, *Jefferson and Monticello, the Biography of a Builder*, 150–51.

41. *Jeffersonian Legacies*, edited by Peter S. Onuf (Charlottesville, 1993), is a good example of this trend. Its essays were delivered at a 1993 conference at the University of Virginia (discussed in the appendix). Pauline Maier's book *American Scripture* (New York, 1997) raises questions about Jefferson's role in writing the Declaration.

41. The top four in the C-Span survey were Lincoln, Washington, Franklin D. Roosevelt, and Theodore Roosevelt.

BOOK SIX: James Madison

A SHY GENIUS MAKES A CONQUEST

1. Irving Brant, *James Madison, The Nationalist* (Indianapolis, IN, 1948), vol. 2, 33.

2. Ibid., 17.

3. Ibid., 284.

4. Ralph Ketcham, *James Madison, A Biography* (Newtown, CT, 1971), 110.

5. TJ to JM, August 21, 1783, Smith, ed., *The Republic of Letters*, 264.

6. JM to William Bradford, November 9, 1772, William T. Hutchinson and William M. E. Rachal et al., eds., *The Papers of James Madison* (hereafter *PJM*), vol. 1, 74–76.

7. Carl Van Doren, *The Great Rehearsal, the Story of Making and Ratifying the Constitution of the United States* (New York, 1948), 37.

8. Katherine Anthony, *Dolley Madison, Her Life and Times* (New York, 1949), 74–75.

9. Irving Brant, *James Madison, Father of the Constitution, 1787–1800* (Indianapolis, IN, 1950), 343.

10. Richard N. Cote, *Strength and Honor, The Life of Dolley Madison* (Mt. Pleasant, SC, 2005), 109.

11. Catherine Coles to DM, June 1, 1794, David C. Mattern and Holly C. Schulman, *The Selected Letters of Dolley Payne Madison* (Charlottesville, VA, 2003), 27–28.

12. Catherine Allgor, *A Perfect Union* (New York, 2006), 30–31. There is some disagreement about this advice from Mrs. Washington. Richard Cote dramatizes it as a face-to-face meeting. Whether it was advice in the mail or otherwise, Madison may well have enlisted Martha Washington. He knew her well from his many visits to Mount Vernon. She also knew Dolley, whose sister, as noted in the text, was about to marry a favorite nephew, George Steptoe Washington. Dolley lived only a few blocks away from the President's House. Cote, *Strength and Honor*, 115–16.

13. JM to DM, August 18, 1794, Mattern and Schulman, eds., *Selected Letters*, 28–29.

14. DM to Eliza Collins Lee, September 16, 1794, ibid., 31.

PARTNERS IN FAME

1. Eric McKittrick and Stanley Elkins, "The Divided Mind of James Madison," *The Age of Federalism* (New York, 1993), 133ff.

2. Cote, *Strength and Honor*, 149.

3. Allgor, *A Perfect Union*, 54–56.

4. Cote, *Strength and Honor*, 155–56.

5. Ketcham, *James Madison*, 386.

6. Sally McKean to DM, August 3, 1797, *Selected Letters*, 32.

7. Ketcham, *James Madison*, 387. General Henry "Light Horse Harry" Lee, a Federalist opponent but an old Princetonian friend, made a similar remark. Congratulating

Madison on his marriage, he hoped Dolley would "soften . . . some of your political asperities." (Anthony, *Dolley Madison*, 91).

8. Ibid., 408.

9. James Sterling Young, *The Washington Community, 1800–1828* (New York, 1966), 88–93.

10. Allgor, *A Perfect Union*, 73–74.

11. Cote, *Strength and Honor*, 207–8.

12. Anthony, *Dolley Madison*, 104–6.

13. DM to Anna Cutts, May 22, 1805, Mattern and Schulman, eds., *Selected Letters*.

14. Brant, *James Madison, Secretary of State* (Indianapolis, IN, 1953), 268.

15. DM to JM, October 30, 1805, *Selected Letters*, 68.

16. Ketcham, *James Madison*, 431.

17. Cote, *Strength and Honor*, 159ff.

18. Allgor, *A Perfect Union*, 95.

19. DM to Anna Payne Cutts, June 4, 1805, *Selected Letters*, 61.

20. DM to JM, November 1, 1805, *Selected Letters*, 70.

21. Garry Wills, *James Madison* (New York, 2002), 54–55.

22. Anthony, *Dolley Madison*, 163.

23. Brant, *James Madison, Secretary of State*, 322.

24. Allgor, *A Perfect Union*, 133. The writer had a long and valued friendship with the late Margaret Truman Daniel, President Harry S. Truman's daughter. He told her to pay no attention to anything said about him in the newspapers or on radio or television. She did so and lived a remarkably happy life.

25. Cote, *Strength and Honor*, 250.

26. Allgor, *A Perfect Union*, 137.

27. Ibid., 139–40.

28. Virginia Moore, *The Madisons, A Biography* (New York, 1979), 223.

29. Cote, *Strength and Honor*, 159ff.

30. Allgor, *A Perfect Union*, 144.

31. Ibid., 152.

32. Ketcham, *James Madison*, 477.

33. Anthony, *Dolley Madison*, 196–97.

34. Allgor, *A Perfect Union*, 167.

35. Ibid., 171. It was not officially designated The White House until 1901.

36. Ketcham, 478.

37. DM to Anna Cutts, December 22, 1811, *Selected Letters*, 154.

38. Allgor, *A Perfect Union*, 193.

HOW TO SAVE A COUNTRY

1. Brant, *James Madison, Commander in Chief* (Indianapolis, IN, 1961), 157.

2. Ketcham, *James Madison*, 553–54.

3. Allgor, *A Perfect Union*, 291.

4. JM to DM, August 7 and 9, 1809, *Selected Letters*, 121–22.

5. Allgor, *A Perfect Union*, 311.

6. Ketcham, *James Madison*, 548, 570, 575.

7. DM to Lucy Payne Washington, August 23, 1814, *Selected Letters*, 193–94. There are several versions of rescuing Washington's portrait. See Algor, *A Perfect Union*, 313–14. Dolley's letter, portraying Carroll "in a very bad humor" waiting while the servants struggled with it, seems the most reliable.

8. Brant, *Madison, Commander in Chief*, 305–6. Ketcham, *James Madison*, 379. Allgor, *A Perfect Union*, 314–18.

9. JM to DM, August 28, 1814, *Selected Letters*, 195.

10. Allgor, *A Perfect Union*, 319.

11. Ibid., 328.

12. Ketcham, *James Madison*, 586.

13. Rutland, *James Madison*, 230.

14. DM to HG, January 14, 1815, *Selected Letters*, 195.

15. Moore, *The Madisons*, 342. "Impeach this man, if he deserves the name of man," one Federalist newspaper shrilled, a few days before the good news arrived. Another paper declared, "His body is torpid and he is without feeling."

16. Cote, *Strength and Honor*, 319.

17. Ketcham, *James Madison*, 610–11.

18. DM to AC, April 3, 1818, *Selected Letters*, 228–29.

19. Ibid., DM to Sarah Coles Stevenson, February 1820, 238–39.

20. Allgor, *A Perfect Union*, 351.

21. *Selected Letters*, Introduction to "A Well Deserved Retirement," 221.

22. Drew R. McCoy, *The Last of the Fathers, James Madison and the Republican Legacy* (New York, 1989), 144–51.

23. Ibid., 223.

24. Rutland, *James Madison*, 251. Allgor, *A Perfect Union*, 377. Paul Jennings, *A Colored Man's Reminiscences of James Madison*, Electronic Edition, University of North Carolina Press.

25. MJR to DM, July 1, 1836, *Selected Letters*, 327.

26. Ibid., AJ to DM, July 9, 1836, 328.

27. Ibid., DM to ECL, 329–30.

28. Ibid., Introduction to "Washington Widow," 317ff.

29. Ibid., 320.

30. Ibid., DM to Henry W. Moncure, August 12, 1844, 374.

31. Jennings, *A Colored Man's Reminiscences*.

32. Ibid., Introduction, 324.

33. Allgor, *A Perfect Union*, 397.

34. Cote, *Strength and Honor*, 357.

APPENDIX: THE EROSION OF JEFFERSON'S IMAGE IN THE AMERICAN MIND

1. Peterson, *Jefferson Image*, 186. Also see "The Strange Career of Thomas Jefferson, Race and Slavery in American Memory, 1943–1993" by Scott A. French and Edward L. Ayers, in *Jeffersonian Legacies*, Peter S. Onuf, ed. (Charlottesville, VA, 1993), 422–23.

2. Peterson, *Jefferson Image*, 187.

3. Winthrop Jordan, *White Over Black* (Chapel Hill, NC, 1968), 466.

4. Douglass Adair, *Fame and the Founding Fathers*, Trevor Colbourne, ed. (New York, 1974), 182–83.

5. *New York Review of Books*, vol. 21, no. 89, April 18, 1974.

6. French and Ayers, in *Jeffersonian Legacies*, Peter S. Onuf, ed., 432–33.

7. Peter Nicolaisen, "Sally Hemings, Thomas Jefferson and the Question of Race: An Ongoing Debate," *Journal of American Studies*, vol. 37 (2003), 101.

8. "A Note on Evidence, The Personal History of Madison Hemings," by Dumas Malone and Steven H. Hochman, *The Journal of Southern History*, November 1975, 527.

9. Onuf, ed., *Jeffersonian Legacies*, 77–103, "The First Monticello," by Rhys Isaac,181–212; "Jefferson and Slavery," by Paul Finkelman, 450; "The Strange Career of Thomas Jefferson" (Cooley statement).

10. Mr. Burstein later changed his mind and wrote *Jefferson's Secrets, Death and Desire at Monticello* (New York, 2005).

11. Joseph A. Ellis, *American Sphinx, The Character of Thomas Jefferson* (New York, 1997), 219. Dr. Eugene Foster's DNA tests changed Ellis's mind. In the same issue of *Nature* that published Foster's results, Ellis wrote an article in collaboration with MIT geneticist Eric Lander declaring that the report "seems to seal the case" that Sally Hemings was Jefferson's concubine.

12. Pauline E. Maier, *American Scripture, Making the Declaration of Independence* (New York, 1997), Introduction, xx-xxi. For quotation, 99. The entire chapter "Mr. Jefferson and His Editors" (97–153) convincingly makes this "work of many" case. Elsewhere in her introduction, Maier states that she has no animus against Jefferson but admits she once nominated him as "the most overrated person in American history" for an *American Heritage* survey. Her reason was "the extraordinary adulation (and sometimes, execration) he has received" (xvii).

13. The biographers include the writer of this book, who published *The Man From Monticello, An Intimate Biography*, in 1969.

14. Annette Gordon-Reed, *Thomas Jefferson and Sally Hemings, an American Controversy* (Charlottesville, VA, 1997), 34–35. In 2008, Gordon-Reed published *The Hemingses of Monticello, An American Family*. The book won a National Book Award and a Pulitzer Prize. The narrative explores the lives of Sally Hemings and the other members of the Hemings family in elaborate detail. But there is little new information about Sally. From the first page, Gordon-Reed assumes that Jefferson was the father of all her children and had a four-decade-long relationship with her. Antipaternity historians have severely attacked the book. At a press conference in Richmond, Virginia, on April 13, 2009, Jefferson's 266th birthday, they insisted that the case against him remains unproved.

Index

BOOKS BY THOMAS FLEMING

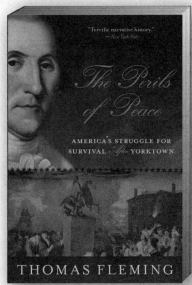

THE PERILS OF PEACE

America's Struggle for Survival
After Yorktown

ISBN 978-0-06-113911-6 (paperback)

"Fleming crafts a dynamic
account that leaves readers
as anxious as the actual
historical figures about how
things will turn out. With
astutely drawn character
sketches, he fluidly engages
such historical contingency."

— *Booklist*

WASHINGTON'S SECRET WAR

The Hidden History of
Valley Forge

ISBN 978-0-06-087293-9 (paperback)

"Fleming enhances his
position as a leading
general-audience historian
of the American Revolution
with this convincing
argument."

— *Publishers Weekly*
(starred review)